30th Winter Workshop on Nuclear Dynamics

(WWND 2014)

Journal of Physics: Conference Series Volume 535

Galveston, Texas, USA
6-12 April 2014

ISBN: 978-1-63439-344-7
ISSN: 1742-6588

Printed from e-media with permission by:

Curran Associates, Inc.
57 Morehouse Lane
Red Hook, NY 12571

Some format issues inherent in the e-media version may also appear in this print version.

Printed by Curran Associates, Inc. (2014)

For permission requests, please contact the Institute of Physics
at the address below.

Institute of Physics
Dirac House, Temple Back
Bristol BS1 6BE UK

Phone: 44 1 17 929 7481
Fax: 44 1 17 920 0979

techtracking@iop.org

Additional copies of this publication are available from:

Curran Associates, Inc.
57 Morehouse Lane
Red Hook, NY 12571 USA
Phone: 845-758-0400
Fax: 845-758-2634
Email: curran@proceedings.com
Web: www.proceedings.com

Journal of Physics: Conference Series **535** (2014) 011001 doi:10.1088/1742-6596/535/1/011001

Preface

These are the proceedings of the 30th Winter Workshop on Nuclear Dynamics, which was held in Galveston, Texas, in April 2014. As in previous years, the unique character of this conference series has allowed us to bring together nuclear scientists with very different interests to discuss recent progress and scientific achievements.

Out of the 67 contributions at WWND 2014 we have selected these 34 manuscripts. The topics capture the range of theoretical and experimental advances in our field. On the experimental side we saw very exciting results from the RHIC beam energy scan program and the p–p, p–Pb and Pb–Pb runs at the highest collision energies at the LHC. On the theory side the system size dependence of the experimental measurements led to a detailed evaluation of the initial conditions and plasma propagation using a wide variety of phenomenological approaches. These results were complemented by the most recent continuum extrapolated data from lattice in order to model the complete evolution of the relativistic heavy ion system.

These proceedings of the 30th Winter Workshop on Nuclear Dynamics again provide a snapshot of the status of the field. The articles, many of which were written by some of the most promising young scientists in the field, are documenting the excitement and achievements that are characteristic for modern day nuclear science.

Rene Bellwied (University of Houston)
Frank Geurts (Rice University)
Anthony Timmins (University of Houston)

Journal of Physics: Conference Series **535** (2014) 011002 doi:10.1088/1742-6596/535/1/011002

Peer review statement

All papers published in this volume of *Journal of Physics: Conference Series* have been peer reviewed through processes administered by the proceedings Editors. Reviews were conducted by expert referees to the professional and scientific standards expected of a proceedings journal published by IOP Publishing.

Hanbury Brown and Twiss Interferometry with respect to 2nd and 3rd-order event planes in Au+Au collisions at $\sqrt{s_{NN}} = 200$ GeV

Takafumi Niida for the PHENIX Collaboration
University of Tsukuba, 1-1-1 Tennoudai, Tsukuba, Ibaraki 305-8571, Japan
E-mail: niida@bnl.gov

Abstract. Azimuthal angle dependence of HBT interferometry have been measured with respect to 2nd and 3rd-order event planes in Au+Au collisions at $\sqrt{s_{NN}} = 200$ GeV at the PHENIX experiment. The 3rd-oder dependence of a Gaussian source radii was clearly observed as well as 2nd-order dependence. The result for 2nd-order indicates that the initial source eccentricity is diluted but still retain the initial shape at freeze-out, while the result for 3rd-order implies that the initial triangularity vanishes during the medium evolution, which is supported by a Gaussian source model and Monte-Carlo simulation.

1. Introduction

Initial density distribution in a heavy ion collision fluctuates due to finite number of participating nucleons [1], which leads to higher harmonic flow as recently measured at RHIC and the LHC [2, 3, 4]. Such spatial fluctuations may be preserved until kinetic freeze-out, depending on the strength of the initial fluctuations, the flow profile, the expansion time, and viscosity of the created matter. Detailed study of the spatial distribution at the freeze-out is needed to reveal these properties with the picture of the space-time evolution.

Hanbury Brown and Twiss (HBT) interferometry [5] is a powerful tool to study the space-time extent of a particle emitting source in the heavy ion collision. Azimuthal angle dependence of Gaussian source radii via two-pion interferometry have been recently measured with respect to 2nd and 3rd-order event planes with PHENIX in Au+Au collisions at $\sqrt{s_{NN}} = 200$ GeV [11]. We present these results and discuss triangularity of the source as well as the eccentricity at freeze-out by comparing with recent hydrodynamic models.

2. Analysis

Analysis was performed using data collected by the PHENIX experiment in 2007 running period, where the number of events were about 4.2 billion events. The event planes were determined by the Reaction Plane Detector (RXNP) [9] by measuring azimuthal anisotropy of emitted particles. The azimuthal distribution of emitted particles is defined in a form of Fourier expansion:

$$\frac{dN}{d\phi} \propto 1 + 2\sum_n v_n \cos[n(\phi - \Psi_n)], \quad (1)$$

where ϕ is azimuthal angle of emitted particles, v_n is nth-order Fourier coefficient representing the strength of anistropic flow, and Ψ_n is azimuthal angle of nth-order event plane. The resolution of the ever plane is defined as $\text{Res}\{\Psi_n\} = \langle\cos[n(\Psi_n^{obs} - \Psi_n^{true})]\rangle$, and 75% for 2nd-order and 32% for 3rd-order at maximum, which were estimated using the event plane correlation between forward and backward RXNPs. Particle identification was performed by combining time-of-flight from the collision vertex to the electromagnetic calorimeter with reconstructed momentum and flight path length in the magnetic field, where π/K separation was achieved up to ~1 GeV/c.

The experimental correlation function is defined as $C_2 = A(q)/B(q)$, where $A(q)$ is the relative momentum (q) distribution of pair particles in the same event and $B(q)$ is that in different events but with similar centrality, vertex position, and the event plane. The pair analysis was performed with the Bertsch-Pratt parameterization [6, 7] and the fitting function is based on core-halo model [8] taking into account the Coulomb interaction and the long-lived resonance decays:

$$C_2 = C_2^{core} + C_2^{halo} = [\lambda(1+G)F_c] + [1-\lambda], \quad (2)$$
$$G = \exp(-R_s^2 q_s^2 - R_o^2 q_o^2 - R_l^2 q_l^2 - 2R_{os}^2 q_o q_s), \quad (3)$$

where F_c is the Coulomb correction factor and λ determines the fraction of pairs in the core. The $R_\mu(\mu = s, o, l)$ represents Gaussian source radii called HBT radii and R_{os} is the cross term between q_o and q_s directions. It is known that the finite resolution of the event plane smears azimuthal angle dependence of HBT radii. Therefore a model independent correction [10] was applied in this analysis.

3. Results

Figure 1 shows the extracted HBT radii of charged pions as a function of azimuthal pair angle ϕ relative to 2nd and 3rd-order event planes in $0.2 < k_T < 2.0$ GeV/c for two centrality bins, where upper panels ((a)-(d)) for 2nd-order and bottom panels ((e)-(h)) for 3rd-order. In 2nd-order dependence, the oscillation of R_s^2, R_o^2, and R_{os}^2 are clearly seen, and they seems to become larger in 20-30% than in 0-10%, where R_o^2 shows the opposite sign of the oscillation to R_s^2. This trend could be understood by out-of-plane extended elliptical source as seen in the STAR experiment [12]. In 3rd-order dependence, the finite but relatively weaker oscillations are observed, where R_s^2 and R_o^2 have the same sign of the oscillation unlike the 2nd-order case.

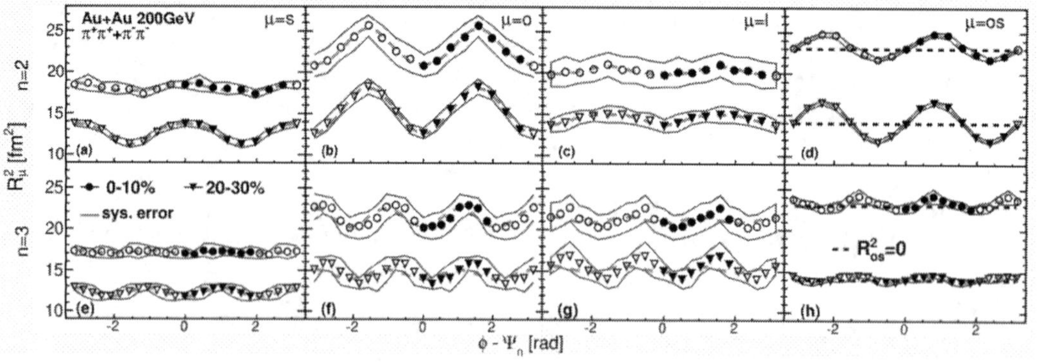

Figure 1. HBT radii of charged pions with respect to 2nd and 3rd-order event planes in $0.2 < k_T < 2.0$ GeV/c and two centrality bins. The filled symbols show the extracted HBT radii and the open symbols are reflected by symmetry around $\phi - \Psi_n = 0$. Bands of two thin lines represent the systematic uncertainties and magenta dashed lines are fit functions by Eq. (4).

In order to extract the oscillation amplitudes, the azimuthal angle dependence of HBT radii was fitted by the following formula [13]:

$$\begin{aligned} R_\mu^2(\phi - \Psi_m) &= R_{\mu,0}^2 + 2\sum_{n=m,2m} R_{\mu,n}^2 \cos[n(\phi - \Psi_m)] \quad (\mu = s, o, l), \\ R_\mu^2(\phi - \Psi_m) &= 2\sum_{n=m,2m} R_{\mu,n}^2 \sin[n(\phi - \Psi_m)] \quad (\mu = os), \end{aligned} \tag{4}$$

where $R_{\mu,n}^2$ are the Fourier coefficients. The extracted 2$^{\rm nd}$-order Fourier coefficients relative to 0$^{\rm th}$ ones are plotted as a function of initial source anisotropy ε_n calculated by Monte-Carlo Glauber simulation in Fig. 2. Final source eccentricity given by $2R_{s,2}^2/R_{s,0}^2$ [13] shows a half of ε_n, which indicates that the source eccentricity is reduced during the system evolution but still retain initial shape extended out-of-plane direction. For 3$^{\rm rd}$-order, there is no significant centrality (ε_n) dependence, but $R_{s,3}^2(R_{o,3}^2)$ is almost equal to zero or slightly negative (positive) in all centralities.

Figure 3 shows k_T dependence of 3$^{\rm rd}$-order oscillation amplitudes in two centrality bins. The $R_{s,3}^2/R_{s,0}^2$ doesn't show any clear dependency on k_T unlike 2$^{\rm nd}$-order [14], while the magnitudes of $R_{o,3}^2/R_{o,0}^2$ increase with k_T in both centralities. Calculations from the Gaussian source model including 3$^{\rm rd}$-order triangular flow and spatial deformation [15] are compared, where two extreme cases, finite triangular flow without spatial deformation ($\bar{\epsilon}_3 = 0, \bar{v}_3 = 0.25$) and non-triangular flow with spatial triangular deformation ($\bar{\epsilon}_3 = 0.25, \bar{v}_3 = 0$), are shown. The $R_{o,3}^2$ favors the deformed flow case, while $R_{s,3}$ matches only at low k_T region. Result from Monte-Carlo simulation [11] with finite triangular flow and non-spatial deformation is also compared as broken lines, which is qualitatively consistent with the Gaussian source model. We note that both calculations have a discrepancy of $R_{s,3}^2$ at higher k_T, but it may be solved by a full hydrodynamic calculation [16].

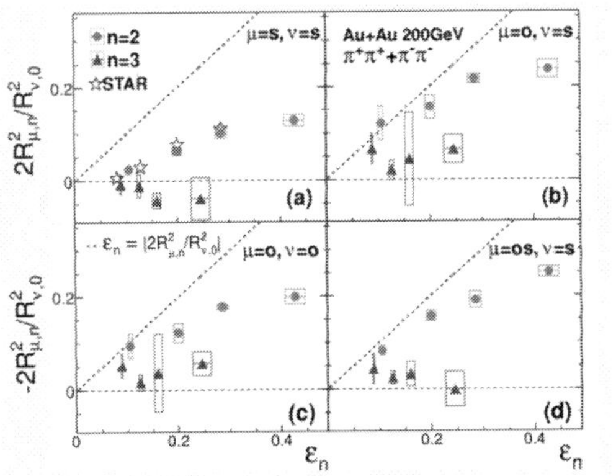

Figure 2. Oscillation amplitudes relative to their averages for 2$^{\rm nd}$ and 3$^{\rm rd}$-order event plane dependences as a function of initial spatial anisotropy.

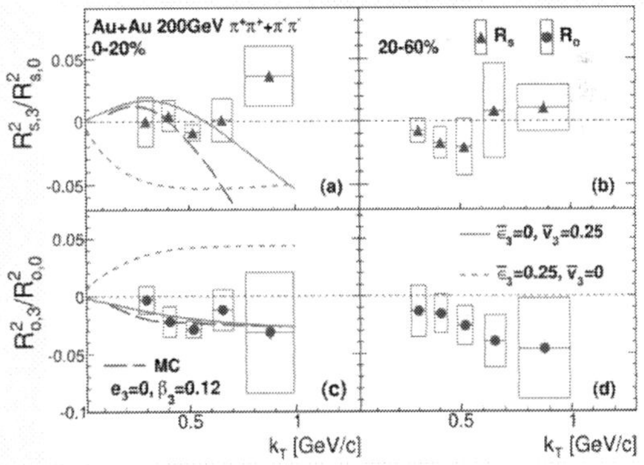

Figure 3. k_T dependence of 3$^{\rm rd}$-order oscillation amplitudes relative to their averages for R_s^2 and R_o^2, where the Gaussian source model (solid and dashed lines) [15] and a Monte-Carlo simulation (broken line) [11] are compared.

Hydrodynamic models suggests that the higher-order spatial anisotropies have different responses to the time evolution of the system [17, 18]. Especially the sign of 3$^{\rm rd}$-order anisotropy (ε_3) turns over faster than 2$^{\rm nd}$-order (ε_2) during the system evolution. Therefore the detailed comparison with the event-by-event hydrodynamics would provide us further description of the system evolution, and information on the initial conditions and the properties of the medium.

4. Summary

We have presented results on HBT measurements with respect to 2nd and 3rd-order event planes in Au+Au collisions at $\sqrt{s_{NN}} = 200$ GeV at the PHENIX experiment. The observed oscillations of HBT radii for 2nd-order event plane indicate the elliptical source shape at the end of the system evolution, while finite oscillation of R_o for 3rd-order event plane would be mainly driven by triangular flow at final state and very weak oscillation of R_s implies that the spatial triangular deformation almost vanishes during the system evolution.

References

[1] B. Alver and G. Roland, *Phys. Rev.* C **81**, 054905 (2010)
[2] A. Adare *et al.* (PHENIX Collaboration), *Phys. Rev. Lett.* **107**, 252301 (2011)
[3] K. Aamodt *et al.* (ALICE Collaboration), *Phys. Rev. Lett.* **107**, 032301 (2011)
[4] G. Aad *et al.* (ATLAS Collaboration), *Phys. Rev. C* **86**, 014907 (2012)
[5] R. Habury Brown and R. Q. Twiss, *Nature* **178**, 1046 (1956)
[6] S. Pratt, *Phys. Rev. D* **33**, 72 (1986)
[7] G. Bertsch, M. Gong, and M. Tohyama, *Phys. Rev. C* **37**, 1896 (1988)
[8] Y. Sinyukov, R. Lednicky, S. V. Akkelin, J. Pluta, and B. Erazmus, *Phys. Lett. B* **432**, 248 (1998)
[9] E. Richardson *et al.* (PHENIX Collaboration), *Nucl. Instrum. Methods Phys. Res., Sect.* A **636**, 99 (2011).
[10] U.Heinz, A.Hummel, M.A.Lisa and U.A.Wiedemann, *Phys. Rev. C* **66**, 044903 (2002)
[11] A. Adare *et al.* (PHENIX Collaboration), *Phys. Rev. Lett.* **112**, 222301 (2014)
[12] J. Adams *et al.* (STAR Collaboration), *Phys. Rev. Lett.* **93**, 012301 (2004)
[13] F. Retière and M. A. Lisa, *Phys. Rev. C* **70**, 044907 (2004)
[14] T. Niida for the PHENIX Collaboration, *J. Phys.: Conf. Ser.* **422**, 012005 (2013)
[15] C. J. Plumberg, C. Shen and U. Heinz, *Phys. Rev. C* **88**, 044914 (2013)
[16] C. J. Plumberg, talk given at Workshop on Particle Correlations and Femtoscopy, Acireale, Nov. 5-8, 2013
[17] D. Teaney and L. Yan, *Phys. Rev. C* **83**, 064904 (2011)
[18] C. Nonaka, talk given at Workshop on Initial Fluctuations and Final Correltions, Chengdu, Aug. 11-14, 2013

Journal of Physics: Conference Series **535** (2014) 012002 doi:10.1088/1742-6596/535/1/012002

Free energy and direct photon emission at finite chemical potential

S. Somorendro Singh

Department of Physics and Astrophysics, University of Delhi, Delhi, India

E-mail: sssingh@physics.du.ac.in

Abstract. We investigate the evolution of free energy and direct photon production from quark-gluon plasma (QGP) considering finite chemical potential. The evolution of QGP formation at the chemical potential is done through finite value of quark mass. The evolution rate is found to be decreasing with chemical potential. We further study the direct photon emission from the fireball of such QGP and found the result to be increasing function of chemical potential in all the channels of photon production. It also shows enhancement of photon emission in comparison to the other theoretical calculation of direct photon productions.

1. Introduction

The research on the ultra-relativistic heavy-ion collisions indicate the evolution of strongly interacting matter called quark-gluon plasma (QGP) [1]. The matter exists for a short life in the process of transformation from a confined to a deconfined state. This short period of existence as a deconfined matter in the early evolution is believed to be in the form of a super-fluid matter. Due to the complicated nature of the state, the investigation on this matter has become a core issue in the present scenarios of heavy ion reactions. Many theoretical and experimental researcher have a keen view on search of evolution of the early universe. There are many theoretical and experimental published works on QGP and the reports of these works are obtained through the path from Relativistic Heavy Ion-collider (RHIC) and Large Hadron Collider(LHC) [2, 3]. The lattice QCD calculation is one method for prediction of such matter and they also inform the presence of occurring such matter (QGP) at very high temperature [4] and at very high nuclear density. In fact, experiments at SPS/CERN, RHIC/BNL and LHC/CERN have claimed for the creation of such situation at very high temperature for the study of QGP [5]. The formation of QGP in these experiments is done through the central collision of two massive nuclei, resulting the product of many particles subsequently bringing the entire information about the collision zone. So many theorist have modeled to create the existence of QGP formation. They also studied the phase structure of QGP-hadron phase transition. Likewise, we use a simple statistical model to construct the free energy evolution of QGP adopting the mean field potential between the interacting particles. After the evolution of the QGP fireball we consider all related processes for the production of particles from these processes. These produced particles consist of a large number of various elements varying from the heavy particles to light elements called dileptons and photons [6 – 10]. The light particles produced are considered to be the most probable signals for the formation of QGP. As they interact electromagnetically they carry the entire information of the reaction zone to the detector and give the fundamental information for the formation of QGP. So photons and dileptons are assumed to be the most promising and exciting signal in the formation of QGP.

Moreover, the present experimental programs at RHIC/BNL and LHC/CERN have proved the main goal of its experimental pursuit in the research of QGP with the higher beam energies. These active programs will provide clear ideas in the experimental outcomes and in the analysis of theoretical modeling too. In addition to this, the transverse momentum distribution function of direct photon is significant in the photon induced reaction at the same energies $\sqrt{s}$ [11] for the momentum greater than 1.5 GeV/c in the central collision. It implies the importance of direct photon production dependence on the space time evolution scenarios of the finite QGP formation. So, there are many probable outcomes of direct photon and dilepton production from the finite baryonic density. Among the calculations at finite baryonic density, it is Dumitru et al.'s photon production at finite baryonic density and subsequently Strickland's calculation using the Jüttner distribution function [12 – 14]. Later on there are works on dilepton and photon production at finite baryonic potential from Hammon and Bass et al. [15, 16] etc. Even at the present experiments the consideration of finite chemical potential is taken into account at FAIR experiments as an example of CBM (compact baryon matter). On these information, we focus the photon radiation directly from the thermalized quark-gluon plasma at $T = 0.57$ GeV which at last freeze out to hadronic matter with the production of latent heat, that again reprocess heating and cooling into process of hadronization as pions at around temperature $T = 0.15$ GeV [17].

In this present work, we investigate our simple statistical model of strongly interacting matter of quark-gluon plasma at finite baryonic density incorporating the parametrized momentum factor in the quark mass and study free energy evolution at finite chemical potential. Then we study the photon production at these finite chemical potential. In most calculations we found that the photon production is done through the massless quark. Due to this massless quark we obtain infrared divergence in the production rate. So we use the finite value of quark mass for removal the infrared (IR) divergence. The finite value of quark mass is taken as the effective quark mass and its value is obtained through the momentum cut off defined as [18]:

$$m_q^2 = \frac{1}{3}\frac{8\pi}{(33 - 2n_f)}\frac{T^2}{\ln(1 + \frac{p^2}{\Lambda^2})} \tag{1}$$

where $p = (\frac{2\pi\gamma^3 T^6 \Lambda^6}{33-2n_f})^{1/12}$ is known as low momentum cut off with quark flavor n_f. γ is defined as root mean square, equal to $\sqrt{\frac{2}{\gamma_q^2} + \frac{2}{\gamma_g^2}}$ with $\gamma_g = a\gamma_q$. γ_q is equal to 1/6 with the suitable value of a searched in an ad-hoc fashion to fit the free energy evolution of the QGP with the stable droplet size [19]. So we calculate the photon radiation at finite baryonic density at the finite temperature for the quark flavor n_f. The calculation of photon production is performed at the temperature $T = 0.57$ GeV. At last we compare the result with our earlier results without chemical potential and other theoretical works.

Thus, we organize the paper as: In section 2 we study a brief idea of free energy evolution of QGP at finite chemical potential. In section 3, we present the photon radiation from QGP at finite baryonic density. In last section 4, we give the results and conclusions.

2. Free energy evolution of QGP at finite baryon density

The free energies of QGP evolution can be obtained through the density of states of the non-interacting fermions and bosons at the finite baryonic density. Here fermions are contributed by quarks and antiquarks and bosons are represented by gluons, which are considered to be active constituent particles of QGP system. So the free energy of these particles is defined as follows [20]:

$$F_i = \mp T g_i \int dp \rho_i(p) \ln[1 \pm e^{-(\sqrt{m_i^2+p^2}-\mu)/T}], \tag{2}$$

where g_i is degeneracy factor and $\rho_i(p)$ is density of states of the constituent particles of the QGP, which we define below. To calculate the density of state, we first use the thermal Hamiltonian process of thermal dependent quark mass. The thermal Hamiltonian is composed of the unperturbed Hamiltonian part and the interaction potential between the constituent particles. This interaction potential represents the mean field potential in the phase space presentation. Now the Hamiltonian is obtained as [21]:

$$H(p,T) = H_o(p,T) + \frac{1}{2p}\gamma_{q,g} g^2(p) T^2 \tag{3}$$

where first term, $H_o(p,T)$ is unperturbed Hamiltonian and second one is the effective mean field potential between the particles in which $g^2(p) = 4\pi\alpha_s$ with QCD coupling constant of α_s defined as

$$\alpha_s = \frac{4\pi}{(33 - 2n_f)\ln(1 + \frac{p^2}{\Lambda^2})}. \tag{4}$$

Now the density of state of quarks and gluons is constructed by the method of Ramanathan and Bethe et al [22] using the interacting mean field potential obtained earlier. It is obtained after analytical calculation using their model.

$$\rho(p) = \frac{v^2\gamma^3 T^6}{4\pi^2 p^3} g^4(p)[\frac{dg(p)}{dp} - \frac{g^2(p)}{2p}]. \tag{5}$$

So at last we obtain the free energies of quarks, antiquarks and gluons plugging this density of states in Eqn.2. Besides these free energy, there is pion free energy and the Weyl [23] surface energy which confine the system instead of using Bag energy. They are represented as follows

$$F_\pi = (3T/2\pi^2)v \int_0^\infty p^2 \ln[1 - e^{-\sqrt{(m_\pi^2+p^2)}/T}]dp. \tag{6}$$

and

$$F_{surface} = \gamma R^2 \int \frac{T}{p^2}\delta(p-T)dp \tag{7}$$

where R is the size of the droplets. Now, the total free energy of the QGP can be computed as sum of all the relevant energies in the system.

$$F_{total} = \sum_i F_i \tag{8}$$

where i denotes each contribution of the particles stated above.

3. Photon production from QGP at finite baryon density

The photon production from QGP at finite baryon density is very interesting theoretical problem. There are a lot of studies for the calculation of photon production at finite temperature and baryon density. In most of the studies, the production rate of photon and dilepton are measured in three classified regions. In first measurement the photon spectra is obtained at the low transverse momentum which are mainly produced due to the soft hadronic decay. In second region, the spectra is studied in the intermediate transverse momentum and subsequently, in higher transverse momentum, which is mainly contributed by the strongly interacting QGP. Here we focus on a baryon rich density and thermalized soup of QGP system after the Big Bang process. In the process the system takes a longer time compared to the time scale associated with the photon production. Moreover the production rate for the coupling parameter $\alpha_s << 1$ [24], turns out to be slow expansion near the equilibrium temperature. Depending on such issues, we estimate our work of photon production at finite baryon density considering the QCD quark antiquark annihilation process and Compton process. Due to rich baryon density the number of quark contents in the system is large enough to interact among themselves and other particles. The quark-antiquark annihilation and QCD Compton are the dominant processes for the production of photon from thermalized and baryon rich system of quarks and gluons. The production rate through the annihilation process $q\bar{q} \to \gamma g$, at thermal equilibrium temperature $T = 0.57$ GeV is calculated as:

$$\begin{aligned} E\frac{dN^a}{d^3pd^4x} &= \frac{16f_q(p)}{(2\pi)^6 4E}\sum_{f=1}^{N_f}\int dsdEf_{\bar{q}}(E_{\bar{q}}) \\ &\times [1+f_g(E_{\bar{q}})]\sqrt{s(s-4m^2)}\sigma_{q\bar{q}\to\gamma g}, \end{aligned} \tag{9}$$

where N_f is number of quark flavors with the Jüttner distribution functions. The Jüttner distributions for quarks, anti-quarks and gluons are defined as:

$$f_{q(\bar{q})} = \frac{\lambda_{q(\bar{q})}e^{\pm\mu/T}}{e^{E/T}+1} \quad and \quad f_g = \frac{\lambda_g}{e^{E/T}-1}. \tag{10}$$

We substitute the distribution functions in the above relation and obtain photon radiation rate at the finite baryon density through annihilation and it is same with other theoretical works of Ref. [13, 14, 25 − 29] :

$$\begin{aligned} E\frac{dN^a}{d^3pd^4x} &= \frac{2\alpha\alpha_s}{\pi^4}\lambda_Q^2 T^2 e^{-E/T}\sum_f e_f^2 \\ &\times [ln(\frac{4ET}{k_c^2}) - C_{euler} - 1 - \ln(n)], \end{aligned} \tag{11}$$

where $\lambda_{Q(\bar{Q})} = \lambda_{q(\bar{q})}e^{\pm\mu/T}$ and $k_c^2 = 2m_q^2$.

In the similar line we perform one loop calculation of Compton process $q(\bar{q})g \to q(\bar{q})\gamma$ in the finite baryon density as:

$$\begin{aligned} E\frac{dN^c}{d^3pd^4x} &= \frac{8N_c f_q(p)}{(2\pi)^5}\sum_{f=1}^{N_f}\int dsdE_g f_g(E_g) \\ &\times [1-f_q(E_g)](s-m^2)\sigma_{gq\to\gamma q}. \end{aligned} \tag{12}$$

Again we substitute the distribution functions of quark, anti-quark and gluon in Compton process expression. Finally we obtain photon radiation rate as [13, 14, 25 − 29]:

$$\begin{aligned} E\frac{dN^c}{d^3pd^4x} &= \frac{2\alpha\alpha_s}{\pi^4}\lambda_Q\lambda_g T^2 e^{-E/T}\sum_f e_f^2 \\ &\times [\ln(\frac{4ET}{k_c^2}) - C_{Euler} + 1/2 - \ln(n)], \end{aligned} \tag{13}$$

where $C_{Euler} = 0.577216$. Similarly, the production rates of photons due to two loops order of AWS and Bremsstrahlung processes are given as [28, 29]:

$$\begin{aligned} E\frac{dN}{d^3pd^4x} &= \frac{2N_c n_f}{3\pi^5}\alpha\alpha_s\{\frac{2}{5}\lambda_q^3 + \frac{3}{5}\lambda_g\lambda_q^2\} \\ &\times \sum_f e_f^2 TE[J_T - J_L], \end{aligned} \tag{14}$$

and

$$E\frac{dN}{d^3pd^4x} = \frac{2N_c n_f}{\pi^5}\alpha\alpha_s\{\frac{4}{7}\lambda_q^2 + \frac{3}{7}\lambda_g\lambda_q\} \times \sum_f e_f^2 T^2 e^{-E/T}[J_T - J_L]\ln(2) \tag{15}$$

where N_c is color degree of freedom. J_L and J_T values are taken as 1.13 and 1.20 for $n_f = 3$.

Further we integrate the total photon rate over the space time history of the collision to obtain the total photon spectrum for all the channels. It is expressed as [14, 29, 30]:

$$\frac{dN}{d^2p_T dy} = \int d^4x(E\frac{dN}{d^3pd^4x}) = Q\int_{\tau_0}^{\tau_f}\tau d\tau\int dy(E\frac{dN}{d^3pd^4x}) \tag{16}$$

where τ is the initial and final value of time evolution. We take rapidity $y_{nuc} = \pm 5.3$ corresponding to RHIC scale. The transverse cross-section of the considered nuclei is taken as $Q \sim 180fm^2$. p_T is the photon transverse momentum. Then the quantity on the right hand side is defined in the centre-of-mass system with the photon energy $E = p_T cosh(y' - y)$. With the values of rapidity and p_T, we get the total photon spectrum shown in figures.

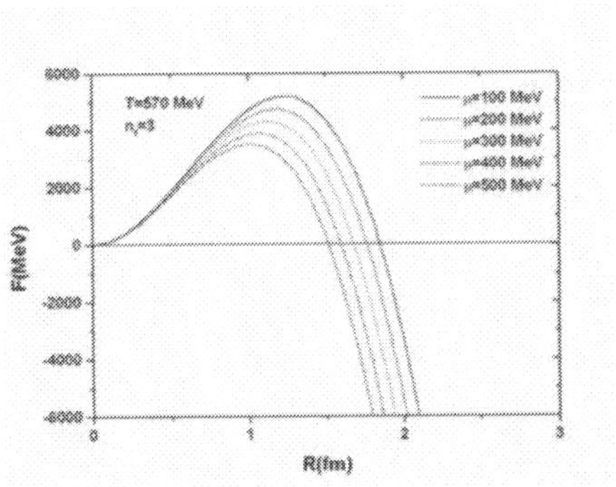

Figure 1. Free energy changes with the size of droplet for various value of chemical potential at temperature $T = 0.57$ GeV with $n_f = 3$.

4. Results and Conclusions

In the results and conclusions we represent the updated and extended calculations of free energy evolution and photon radiation through the finite chemical potential. The calculation is performed for quark flavor $n_f = 3$. The formation/ evolution of QGP droplet is shown in Fig. 1 for quark flavor $n_f = 3$. In Fig. 1, the droplet size of QGP formation is found to vary from $R = 2.0$ to 1.5 fm for the chemical potential varying from $\mu = 100 - 500$ MeV. It means that increase of chemical potential decreases the droplet size. It is just opposite pattern to the the case of finite temperature. In case of finite temperature, the size of droplet is increased with respect to the increase in temperature, which indicates good output in respect of phase structure. In QCD phase structure the phase transition with the chemical potential is just opposite to the temperature, which indicates the good output in terms of phase structure.

In Fig. 2 we show the photon emission rate at the initial temperature $T = 0.57$ GeV through the annihilation process for quark flavor $n_f = 3$. The emission rates are found to be increasing function with the chemical potential μ. The increase in the emission rate is highly effected by the temperature as well as the chemical potential of the system. It is found to be very large in the production at very high temperature and chemical potential. In Fig. 3 we again show the photon emission at temperature $T = 0.57$ GeV through the Compton mechanism. In this

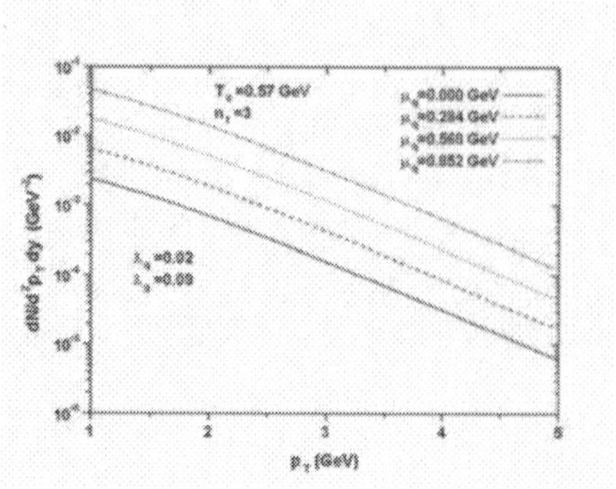

Figure 2. The photon emission rate through ann. at thermal temperature $T = 0.57$ GeV with the transverse momentum for $n_f = 3$.

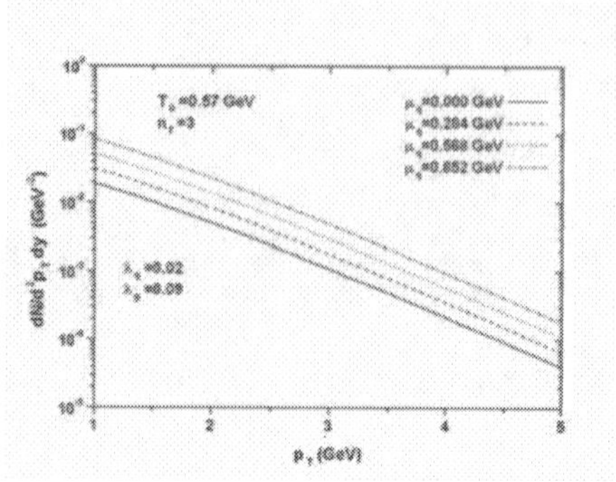

Figure 3. The photon emission rate through Compton at thermal temperature $T = 0.57$ GeV for $n_f = 3$.

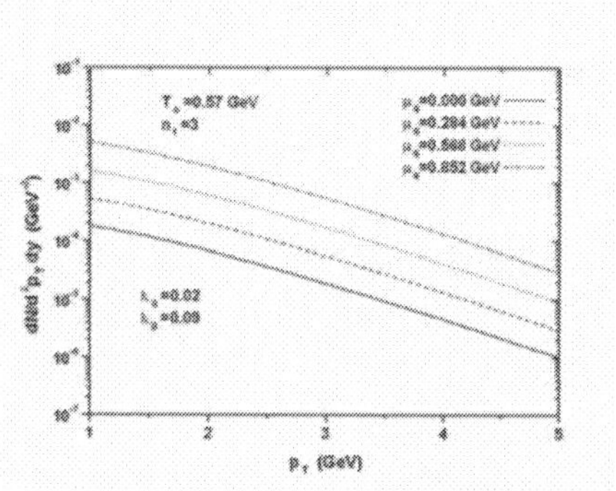

Figure 4. The photon emission rate through AWS process at thermal temperature $T = 0.57$ GeV for $n_f = 3$.

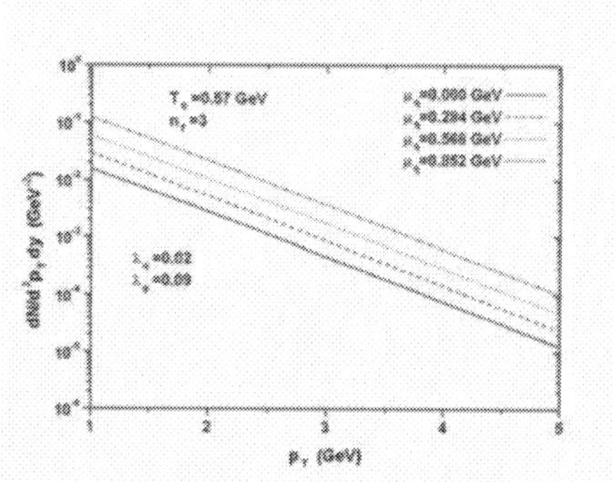

Figure 5. The photon emission rate through Bremsstrahlung process at thermal temperature $T = 0.57$ GeV for $n_f = 3$.

channel too the production rate is high in accordance with the chemical potential. The large production is found in the case of Compton process in comparison to the annihilation process for the same value of chemical potential. So the Compton shows better outcomes in terms of production rates [14, 26, 27, 28].

Now we study the production rate of AWS channel. The photon productions rate is shown in Fig. 4. It shows that production rates follow same pattern as before. Yet AWS channels produce less photons in comparison with the earlier channels. In last process, we do the Bremsstrahlung process which is shown in Fig. 5. Photon production due to this process is increased with the increased chemical potential, and again in this process the production rate is still high compared to the AWS channel and other two channels above. This implies that in all the processes/channels, our model has large advantage in the photon production rate particularly Compton and Bremsstralung processes. So the production rate of our model with the flavor $n_f = 3$ have improvement from other works in these channels. It means that our result is dominant over the production rate of Ref.[14, 27, 28]. Finally, we conclude that the evolution of the fireball through the parametrization factor decrease the size of droplet formation in this finite chemical potential. It shows that the evolution of fireball is suppressed in comparison with the evolution of fireball at the finite temperature. The suppression may be due to large presence of quark, antiquark and specially due to strange quark particles. Even though it is not encouraged in the evolution of the plasma, the condensed matter of the system enhance the interaction between the particles and therefore enhance the production rates. Overall the results show that the calculation of photon production as a function of photon transverse momentum incorporating chemical potential and the factor in the quark mass give the improved results in the photon yield products from the earlier results without the chemical potential [31]. Thus, the consideration of parametrization factor in the quark mass has important role in the evolution as well as in the photon measurements of the high energy heavy ion collisions in finite chemical potential.

4.1. Acknowledgments

We are very thankful to R. Ramanathan for the constructive suggestions and discussions in preparing the manuscript. We would like to thank University of Delhi, Delhi for giving financial assistance in terms of research strengthen grant.

References

[1] Harris J W and Müller B 1996 *Ann. Rev. Nucl. Sci.* **46** 71; Geiger K, Müller B 1992 *Nucl. Phys. B* **369** 600.
[2] Arsene I *et al.* 2005 *Nucl. Phys. A* **757** 1; Back B B *et al.* 2005 *ibid A* **757** 28.
[3] Adams J *et al.* 2005 *ibid A* **757** 102; Adcox K *et al.* 2005 *ibid A* **757** 184.
[4] Karsch F and Laermann E 2004 *QGP*, ed R C Hwa and X N Wang (World Scientific, Singapore).
[5] Proc. of Quark Matter'01 2002 *Nucl. Phys. A* **698**.
[6] Shuryak E V 1978 *Phys. Lett. B 78* **150**; 1980 *Phys. Rep.* **61** 71; Kajantie K and Miettinen H I 1981 *Z. Phys. C* **9** 341; Domikos Gand Goldman J 1981 *Phys. Rev. D* **23** 203.
[7] Mclerran L and Toimela 1985 *Phys. Rev. D* **31** 545.
[8] Gale C and Kapusta J I 1991 *Nucl. Phys. B* **357** 65.
[9] Peitzmann T and Thoma M H 2002 *Phys. Rep.* **364** 175.
[10] Cassing W and Bratkovskaya E L 1999 *Phys. Rep.* **308** 65; Aurenche P (*Preprint* hep/0201011).
[11] Wang S Y and Boyanovsky D 2001 *Phys. Rev. D* **63** 051702(R); Boyanovsky D and de Vega H J 2003 *Phys. Rev. D* **68** 065018.
[12] Dumitru A, Rischke D H, Schönfed Th, Winckelmann L, Stöker H and Greiner W 1993 *Phys. Rev. Lett.* **70** 1860; Dumitru *et al.* 1993 *Mod. Phys. Lett. A* **8** 1291.
[13] Strickland M 1994 *Phys. Lett. B* **331** 245; Biro T S, Doorn E V, Müller B, Thomas M H and Wang X N, 1993 *Phys. Rev. C* **48** 1275.
[14] Dutta D, Mohanty A K, Kumar K and Choudhury R K 2000 *Phys. Rev. C* **61** 064911.
[15] Hammon N *et al.* 1999 *Phys. Rev. C* **61** 014901.
[16] Bass S A *et al.* 2003 *Phys. Rev. Lett.* **91** 052302; Ze-Jun He *et al.* 2005 *C Phys. Lett.* **22** 1350.
[17] Srivastava D K and Sinha B 1994 *Phys. Rev. Lett.* **73** 2421.
[18] Kumar Y and Singh S S 2012 *Can. J. Phys.* **90** 955.
[19] Ramanathan R, Mathur Y K, Gupta K K and Jha A K 2004 *Phys. Rev. C* **70** 061408; Ramanathan R, Gupta K K, Jha A K and Singh S S 2007 *Pram. J. Phys.* **68** 757.
[20] Neerguard G and Madsen J 1999 *Phys. Rev. D* **60** 054011; Mustafa M G, Srivastava D K and Sinha B 1998 *E. Phys. J. C* **5** 711.
[21] Peshier A, Kämpfer B, Pavlenko O P and Soff G 1994 *Phys. Lett. B* **337** 235.
[22] Bethe H A 1937 *Rev. Mod. Phys.* **9** 69; Fermi E 1928 *Z Phys.* **48** 73; Thomas L H 1927 *Proc. Cambridge Philos. Soc.* **23** 542.
[23] Weyl H 1911 *Nachr. Akad.* Wiss Gottingen 110 .
[24] Moore G D (*Preprint* hep.ph/0403169); V.Ruuskanen 1991 *Quark-Gluon Plasma* , ed R C Hwa (World Scientific, Singapore).
[25] Kampfer B and Pavlenko O P 1994 *Z. Phys. C* **62** 491.
[26] He Z, Long J, Ma Y and Ma G 2005 *C. Phys. Lett.* **22** 1350.
[27] Long J, He Z, Ma Y and Liu B 2006 *Nucl. Phys. A* **766** 201
[28] Mustafa M G and Thoma M H 2000 *Phys. Rev. C* **62** 014902.
[29] Long J, He Z, Ma Y and Liu B 2005 *Phys. Rev. C* **72** 064907.

[30] Steffan F D and Thoma M H 2001 *Phys. Lett. B* **510** 98; Traxler C T and Thoma M H 1996 *Phys. Rev. C* **53** 1348.
[31] Singh S S and Kumar Y Accepted in *Int. J. Mod. Phys. A*

Recent results of gluon and sea quark polarization measurements in polarized proton-proton collisions at STAR

Xuan Li[1] **for the STAR Collaboration**

1) Physics Department, Temple University, 1900 North 13th street, Philadelphia, 19122, USA

E-mail: xuanli@rcf.rhic.bnl.gov

Abstract. The STAR experiment at RHIC is carrying out a comprehensive high-energy spin physics program to understand the internal structure and dynamics of the proton in polarized proton-proton collisions at $\sqrt{s}$ = 200 GeV and $\sqrt{s}$ = 500/510 GeV. STAR has the capability, with nearly full azimuthal coverage, to reconstruct leptons, hadrons and jets in the mid-rapidity region ($|\eta| < 1$). The results for inclusive jet longitudinal double spin asymmetries taken during the 2009 RHIC run indicate the first non-zero gluon contribution ($\Delta g(x, Q^2)/g(x, Q^2)$) to the proton spin for $0.05 < x < 1$ (Bjorken-x: momentum fraction of partons). Recent longitudinal single-spin asymmetry measurements of $W^{+/-}$ bosons at $\sqrt{s}$ = 500/510 GeV in polarized proton-proton collisions provide a direct probe of the polarized anti-u and anti-d quark distributions ($\Delta\bar{u}(x, Q^2)$, $\Delta\bar{d}(x, Q^2)$). These results better constrain the polarized gluon and sea quark distributions of the proton in the RHIC sensitive kinematic region. Future measurements with continuing high energy polarized proton-proton run at RHIC and detector upgrade will explore the gluonic contribution to the proton spin in extended range.

1. Introduction

The discovery of the so-called "proton spin crisis" [1] has prompted a variety of investigations to determine precisely the various constituent spin-contributions. The sum of the quark and anti-quark polarization contributions ($\Delta\Sigma$) measured in polarized deep inelastic scattering (DIS) experiments is less than one third of the total proton spin [2, 3]. The limited Bjorken-x region accessed by the fixed target polarized DIS experiments provides poor constraints on the polarized gluon distribution $\Delta g(x, Q^2)$ and flavor separated polarized anti-quark distributions $\Delta\bar{q}(x, Q^2)$. The helicity structure of the proton is being studied at the Relativistic Heavy Ion Collider (RHIC) via longitudinally polarized proton-proton collisions. RHIC is a versatile machine, capable of colliding longitudinally and tranversely polarized protons at varying center of mass energies as well as unpolarized heavy ions at different energies and atomic numbers. Since 2009, RHIC has collided longitudinally polarized proton beams at $\sqrt{s}$ = 200 GeV and $\sqrt{s}$ = 500/510 GeV with improved luminosity by a factor of over 100 relative to previous years and polarization over 50%. The Solenoidal Tracker at RHIC (STAR) has both tracking detectors and electromagnetic calorimeters with nearly full azimuthal coverage, therefore final states like jets and electrons can be well reconstructed, especially at mid-rapidity ($|\eta| < 1$).

Jets can be treated as surrogates of the hard scattering partons. STAR has measured inclusive jet and di-jet production in proton-proton collisions. A global QCD fit which includes the 2009

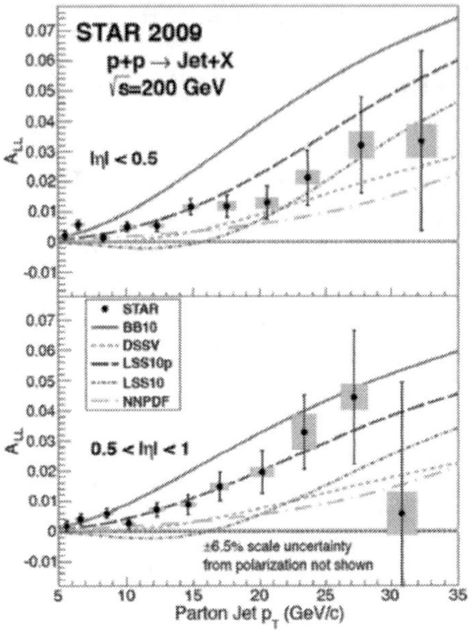

Figure 1. *STAR 2009 inclusive jet* A_{LL} *versus parton jet* p_T *in 200* GeV *p+p collisions [4].* A_{LL} *of inclusive jets with* $|\eta| < 0.5$ *is shown in the top panel, and the results with* $0.5 < |\eta| < 1.0$ *are shown in the bottom panel. Error bars are statistical and gray boxes stand for systematic errors.*

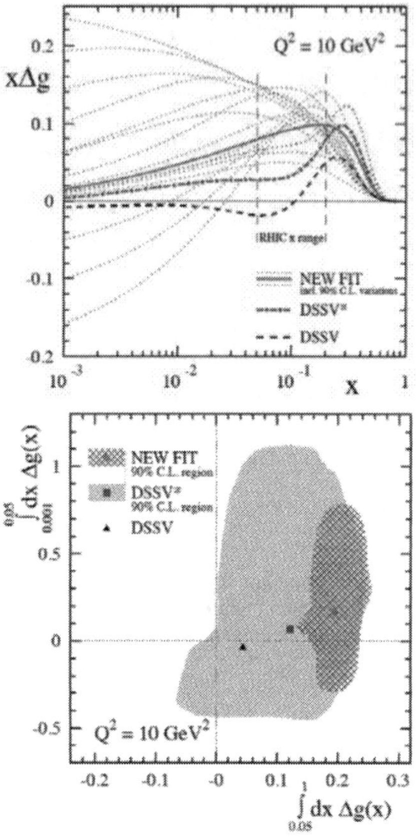

Figure 2. *Recent global analysis from the DSSV group. Top panel: the gluon helicity distribution* Δg *at* $Q^2 = 10$ GeV^2*, the red line is the newest fit including the 2009 RHIC data. Bottom panel: Truncated integral of* Δg *for* $0.001 \leq x \leq 0.05$ *and* $0.05 \leq x \leq 1$ *at* $Q^2 = 10$ GeV^2*. Figure from [5].*

STAR inclusive jet double spin asymmetry A_{LL} in 200 GeV p+p collisions, extracts a non-zero gluon spin contributions $(\Delta g(x, Q^2)/g(x, Q^2))$ to the proton in $0.05 < x < 1$ [4, 5]. W production in polarized proton-proton collisions which avoids the parton fragmentation processes is a cleaner probe than hadrons to detect the light sea quark $\bar{u}$ and $\bar{d}$ polarization contribution $(\Delta\bar{u}(x, Q^2)/\bar{u}(x, Q^2)$, $\Delta\bar{d}(x, Q^2)/\bar{d}(x, Q^2))$ to the proton. The first lepton rapidity dependent $W^{+/-}$ single spin asymmetry A_L in 500/510 GeV proton-proton collisions indicates a sizable and positive $\bar{u}$ quark polarization [6]. The latest released inclusive jet and W spin results at STAR promote our understanding of the gluon, $\bar{u}$ and $\bar{d}$ polarized distributions in the probed $0.05 < x < 1$ region. Details about the jet and W spin analyses and projections for the future measurments will be discussed in the following sections.

2. Jet production at STAR to probe the gluon helicity distribution

Next-to-leading-order (NLO) perturbative QCD (pQCD) can describe the inclusive jet and di-jet cross sections in proton-proton collisions at STAR [7]. In 200 GeV proton-proton collisions, the mid-rapidity jet production is dominated by quark+gluon processes, whereas mid-rapidity jet production in 500 GeV proton-proton collisions are dominated by the quark+gluon and gluon+gluon processes. The double spin asymmetry A_{LL} of jets/di-jets is proportional to the helicity distribution and spin asymmetry of polarized partons involved in the hard scattering. Therefore, the gluon polarization contribution to the proton can be directly probed by measuring the inclusive jet or di-jet A_{LL}.

2.1. Inclusive jet production at STAR

In 2009 STAR recorded approximately 25 pb^{-1} integrated luminosity of 200 GeV proton-proton collisions with the averaged polarization of beams around 55%. The relative uncertainty on the product of the polarization of the two beams is 6.5%. This data sample contains 20-fold more statistics than the published results in [8]. Instead of the Mid-point cone jet algorithm applied in previous results, the anti-k_T jet algorithm [9] with R=0.6 is applied in the 2009 analysis to reduce the underlying event contribution and pile up backgrounds to the reconstructed jets. Jets reconstructed from the detector response (e.g. charged tracks or neutral energy depositions) in data are defined as detector level jets. Particle level jets which are based on final state particles and parton level jets which are formed from hard scattered partons (excluding those from the underlying events and beam remnants) can be reconstructed in the simulaton. Detector level jets can be reconstructed in simulation as well with the same trigger and detector conditions as in data. In simulation, a detector level jet and a parton or particle level jet are considered to be correlated when the distance between the two in $\eta - \varphi$ space is minimal and less than 0.5. The detector jet p_T measured in data can be corrected to particle or parton jet p_T according to the associations in simulation. The measured detector jet A_{LL} values are corrected to parton jet A_{LL} in each jet p_T bin according to calculations in the simulation with polarized/unpolarized PDFs. Figure 1 shows the 2009 STAR mid-rapidity inclusive jet double spin asymmetry A_{LL} versus the parton jet p_T in 200 GeV proton-proton collisions [4]. The jet acceptance is divided into $|\eta| < 0.5$ (top panel) and $0.5 < |\eta| < 1$ (bottom panel) to access different x regions. Colored lines represent the NLO pQCD inclusive jet A_{LL} calculations with different polarized parton distributions as inputs [2, 3, 10, 11, 12]. Data points fall between LSS10p [2] and DSSV [10] model predictions.

The NNPDF polarized PDFs are based on the DIS data, and a reweighting method is developed to include new experimental results [3]. The DSSV group is the first to apply a global QCD fit on the combined data samples from DIS, Semi-Inclusive DIS (SIDIS) and RHIC pp experiments [5]. The latest DSSV update which includes the 2009 STAR inclusive jet A_{LL} [14] and 2009 PHENIX inclusive π^0 A_{LL} [15] exhibits a postive polarized gluon helicity function in the momentum fraction region $0.05 < x < 1$ as shown in the top panel of Figure 2. The integral of the gluon helicity function in this x region $\Delta G = \int_{0.05}^{1} \Delta g\, dx(Q^2 = 10\ \mathrm{GeV}^2) \approx 0.2$ is shown in the bottom panel of Figure 2 (red point within 90% confidence level (C.L.) represented by the blue band). A first non-zero gluon polarization contribution to the proton has been observed at RHIC.

RHIC is scheduled to operate another 200 GeV longitudianlly polarized proton-proton run in 2015. Future inclusive jet A_{LL} measurements will provide better constraints with improved statistics on the gluon polarization in the same x region [16]. Dotted lines in the top panel of Figure 2 stand for alternative fits that are within 90% confidence level limit, which reflects large uncertainties in derterming the gluon helicity function beyond the current x range that can be accessed by RHIC. In additional to the 200 GeV longitudinal proton-proton run, RHIC also operated 500 GeV longitudinal proton-proton collisions with around 600pb^{-1} delivered

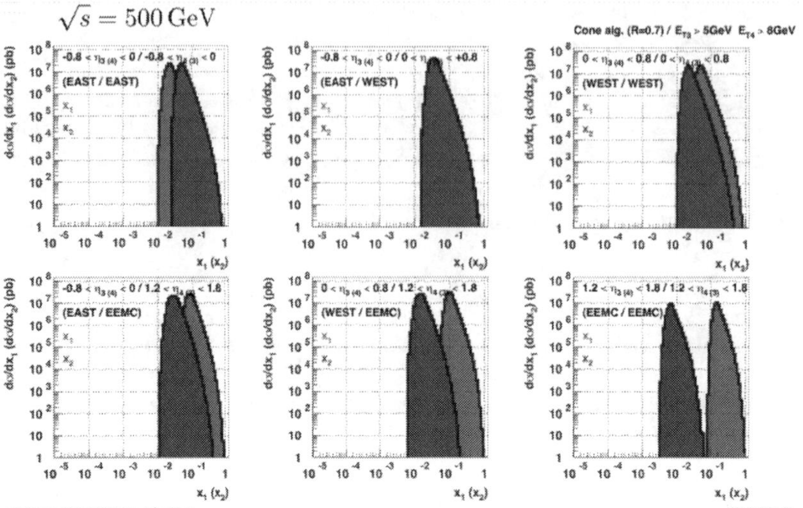

Figure 3. *x_1 (red) and x_2 (blue) of initial partons probed by di-jets reconstructed within the STAR detector acceptance ($-1 < \eta < 2$) in MC based on NLO thoery calculations. [13]*

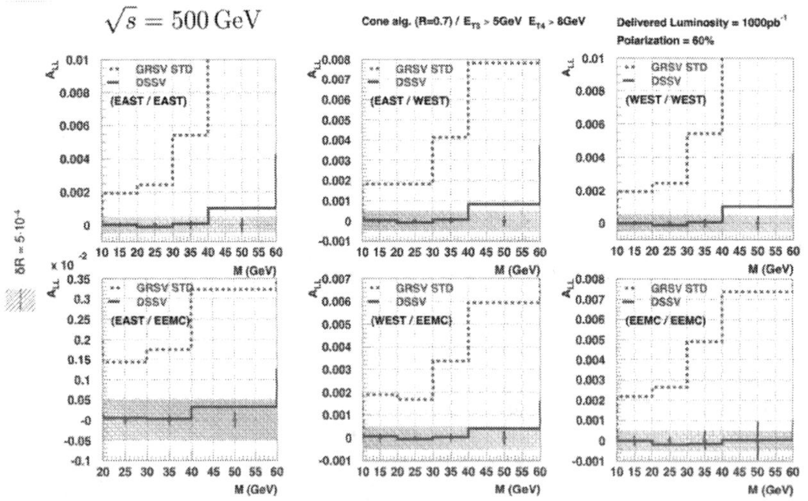

Figure 4. *Di-jet A_{LL} versus invariant mass with different configurations in the STAR detector acceptance, $-1 < \eta < 2$ from MC based on NLO thoery calculations. [13]*

luminosity in 2012 and 2013. Gluon polarization in lower x region will be probed by the ongoing 2012/2013 jet analysis in higher center of mass collision energy.

2.2. Di-jet production at STAR

Inclusive jets measured in 200 and 500 GeV proton-proton collisions can be used to probe the gluon polarization integral over a broad range in x. In the leading order QCD with collinear approximations, the momentum fractions of inital hard scattering partons x_1 and x_2

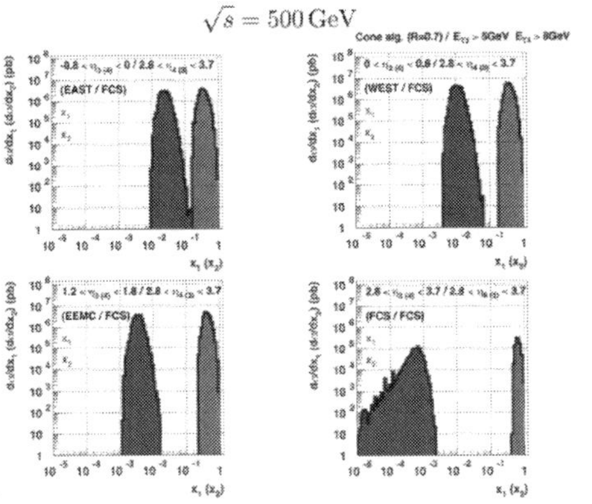

Figure 5. *x_1 (red) and x_2 (blue) of initial partons probed by forward di-jet with a proposed forward hadrnoinc calorimeter upgrade at STAR. [13]*

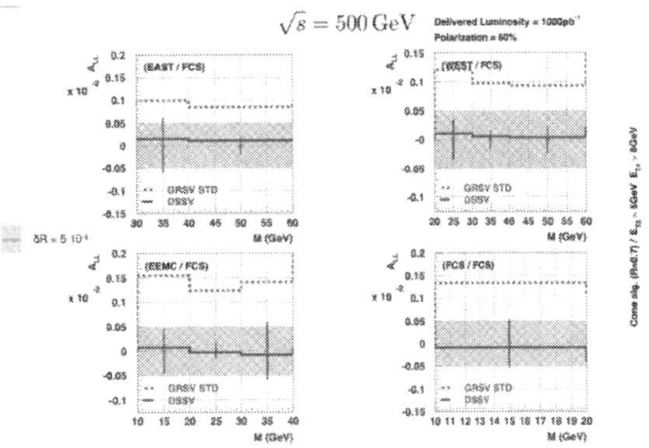

Figure 6. *Projection of forward di-jet A_{LL} asymmetries in 500 GeV proton-proton collisions of 1000 pb^{-1} delivered luminoisty with beam polarization of 60%. [13]*

are correlated with the di-jet invariant mass M and the jet pseudo-rapidities η_3, η_4 ($M^2 = x_1 x_2 s$, $ln(x_1/x_2) = \eta_3 + \eta_4$). Therefore, a certain x region can be selected by the di-jet production by varying the pseudorapidities of the jets. In the asymmetric partonic scattering, lower x gluons can be probed by forward di-jet correlations. Uncertainties of the truncated moment of $\Delta g(x, Q^2 = 10 \text{ GeV}^2)$ in the $0.001 \leq x \leq 0.05$ region (shown in the bottom panel of Figure 2) remain large compared to the fitted results in the $0.05 \leq x \leq 1$ region. Achieving measurements which can probe this low x region will help contribute to our understanding of the proton gluon spin distributions. STAR has nearly continuous, full azimuthal acceptance of electromagnetic calorimeters in $-1.0 < \eta < 4.0$: the Barrel ElectroMagnetic Calorimeter (BEMC: $-1.0 < \eta < 1.0$), the End-cap ElectroMagnetic Calorimeter (EEMC: $1.07 < \eta < 2.0$) and the Forward Meson Spectrometer (FMS: $2.3 < \eta < 4.0$). The electromagnetic calorimeters (and a possible forward hadronic calorimeter in the STAR upgrade plan [13]) together with the Time Projection Chamber (TPC: $-1.3 < \eta < 1.3$) provide a broad $\Delta\eta \times \Delta\varphi$ coverage for di-jet measurements.

Figure 3 shows the measured intial parton x_1 and x_2 by final di-jets with six different detector configurations in 500 GeV proton-proton collisions from MC based on NLO QCD framework [13]. The BEMC acceptance is divided into separate coverages: "East" ($-1.0 < \eta < 0$) and "West" ($0 < \eta < 1$). Jets are reconstructed using the Mid-point cone algorithm with cone radius R=0.7. In the di-jet pair, the leading jet p_T is required to be larger than 8 GeV/c and the sub-leading jet p_T should be larger than 5 GeV/c. The current RHIC sensitive x region extends to around 10^{-2} when both jets are reconstructed in the EEMC. The projected di-jet A_{LL} for the West-West, West-East, East-East, West-EEMC, East-EEMC and EEMC-EEMC combinations are shown in Figure 4 assuming the beam polarization is 60% and 1000 pb^{-1} delivered luminosity [13]. The dominant systematic error is the relative luminosity error $\delta R = 5 \cdot 10^{-4}$, the same value as cited in the 2009 inclusive jet A_{LL} results. The statistical and systematic errors of the projected A_{LL} as shown in Figure 4 are much smaller than the separation between DSSV [5] and GRSV-STD [17] theory calculations, and both predict non-zero gluon helicity distribution. A forward hadronic calorimeter (FCS) is proposed in the STAR forward upgrade plan [16]. Installation of this forward instrument will allow the di-jet measurements to go to the forward pseudorapidity and reach lower x values. Projected initial parton x_1, x_2 values probed by forward di-jets detected in

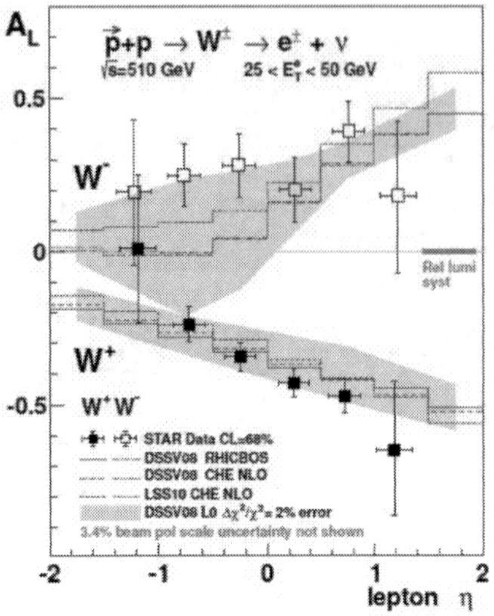

Figure 7. *2011+2012 $W^{+/-}$ A_L as a function of W decay lepton η ($|\eta| < 1.3$) [6]. Color lines/band stand for theory predictions, see details in the text. The grey band presents the systematic error from the relative luminosity.*

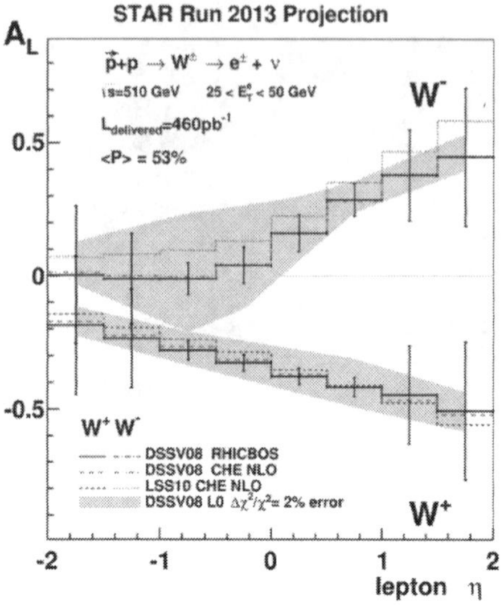

Figure 8. *Projections of $W^{+/-}$ A_L in the 2013 510* GeV *p+p collisions at STAR with 460 pb^{-1} integrated luminosity and average polarization* 53%. *Central values are evaluated by the CHE model with DSSV global QCD fit.*

the FCS and associated detectors (West, East, EEMC, FCS) and their A_{LL} asymmetries in NLO QCD MC are shown in Figure 5 and Figure 6. Fiducial volume cuts are applied as well for the forward di-jet simulation. The left-top panel is for the FCS and east Barrel di-jet production; the right-top is for the FCS and west Barrel di-jets; the left-bottom panel is for the FCS and EEMC di-jets and the right-bottom panel is for the FCS and FCS di-jets. The FCS-FCS di-jets probe the lowest x region which is below 10^{-3}. Reducing the systematic errors will be a primary goal to realize such measurements for studying gluon helicity distribution at low x .

3. W production at STAR to probe the sea quark helicity distribution

The $\bar{u}$ and $\bar{d}$ momentum fraction distributions have been found asymmetric through Drell-Yan process by the E866 experiment [18]. This phenomenon can not be fully described by perturbative QCD and indicates a non-perturbative mechanism may play a role in this field. In the parity violating weak processes, intial u (d) quark and $\bar{d}$ ($\bar{u}$) quark couple to W^+ (W^-) in proton-proton collisions. The longitudinally polarized 500 or 510 GeV proton-proton collisions at RHIC open a path to probe the light sea quark helicity functions via W productions ($p{+}p \to W^{+/-}{+}X \to l^{+/-}{+}X$). This process gets rid of the final state fragmentation functions that are encountered in semi-inclusive DIS processes. The W boson measured in high energy polarized proton-proton collisions is a unique probe for the polarized sea quark distrubtion at the W mass scale [2, 19].

STAR has successfully reconstructed W^+/W^- via their e^+/e^- decay channels in the mid-

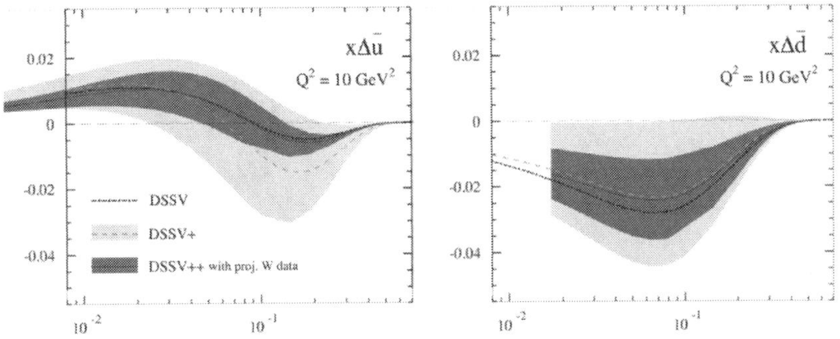

Figure 9. *x dependent sea quark polarized PDF ($\Delta\bar{u}$ is shown in the left, and $\Delta\bar{d}$ is shown in the right) extracted from the DSSV global fit. Uncertainties change from the yellow band to the red bands after including the projection of combined 2009 to 2013 RHIC W A_L asymmetries. Figure from [16].*

rapidity ($-1 < \eta < 1$) from 2009 500 GeV proton-proton collisions [20]. Measured mid-rapidity $W^{+/-}$ single spin asymmetries in 2009 data are consistent with predictions of the DSSV08 global QCD fit analysis which only include DIS and SIDIS results. But statistics were limited to separate different theoretical model descriptions. In RHIC runs of 2011 (500 GeV) and 2012 (510 GeV) longitudinally polarized proton-proton collisions, STAR collected around 94 pb^{-1} integrated luminosity of data, with average proton beam polarizations of 49% and 56%. The good statistics of this data sample allowed for the measurements of lepton pseudorapidity dependent W single spin asymmetry A_L. To obtain this goal, several updates were applied to the 2011 and 2012 combined data set. The detector acceptance to measure the W has been extended from $-1 < \eta < 1$ to $-1.3 < \eta < 1.3$, with the help from the shower maximum detector (SMD) of the EEMC and inner sectors of the TPC. The profile likehood method was used to combine W A_L results which were measured independently in run 2011 and 2012 [6]. $W^{+/-}$ events are reconstructed from $W^{+/-}$ decays to $e^{+/-}$ with transverse energy $25 < E_T^e < 50$ (GeV). A significant signal to background ratio is seen in the reconstructed $W^{+/-}$ invariant mass distributions.

The STAR W decay lepton pseudorapidity dependent single spin asymmetry A_L from the 2011+2012 data is shown in Figure 7. Results of the $W^{+/-}$ A_L are extracted from the likelihood function with 68% confidence level. The grey band stands for the relative luminosity systematic error. Colored lines stand for the NLO QCD calculations on the $W^{+/-}$ A_L in RHICBOS and CHE models with DSSV08/LSS10 [2, 19] helicty functions as inputs. The green band reflects the DSSV08 $\Delta\chi^2/\chi^2 = 2\%$ uncertainties. This result probes the flavor separated light quark and anti-quark helicity distributions in the $0.05 < x < 0.2$ region. The W^+ A_L which is consistent with the theory predictions indicates a negative $\bar{d}$ polarization. The W^- A_L which is sensitive to the $\bar{u}$ quark polarization is larger than the theory predictions for the $\eta_e < 0$ range. This suggests a positive $\bar{u}$ helicity distribution. Therefore, a postive $\bar{u} - \bar{d}$ helcity dependent function is favored by this result.

The mid-rapidity $W^{+/-}$ A_L asymmetries probe the polarization contributions carried by both valence quark and sea quark. Extending the pseudorapidity of the reconstructed $W^{+/-}$ to forward/backward regions will enhance the sensitivity for sea quark polarizations. A forward tracking system, the Forward GEM Tracker (FGT, $1 < \eta < 2$) was fully installed at STAR before the 2013 run. This update enhances the charge seperation capabilities of STAR for forward/backward electron/positrons. In 2013, RHIC delivered around 460 pb^{-1} integrated luminosity 510 GeV proton-proton collisions which is the largest data sample so far. The tracking software with the FGT included is under development, and the FGT W analysis at STAR with the 2013 data is ongoing. The projection of lepton pseudorapidity dependent $W^{+/-}$ A_L measured at STAR from the 2013 data is shown in Figure 8. Central values of the projected W A_L are based on the DSSV pQCD new fit which fits the STAR 2011+2012 W A_L asymmetries. From 2009 to 2013, RHIC has delivered around 600 pb^{-1} integrated luminosity 500/510 GeV longitudinally polarized proton-proton collisions. The extracted light sea quark polarized PDF $\Delta\bar{u}$ and $\Delta\bar{d}$ from the DSSV analysis including the combined 2009 to 2013 W A_L projection from STAR and PHENIX are shown in Figure 9 [16]. The yellow bands represent the uncertainties constrained by the DIS data. Uncertainties of the helicity dependent PDFs reduce significantly after including the combined RHIC $W^{+/-}$ A_L projection as shown in the red bands of Figure 9. The RHIC W program with the combined 2009 to 2013 data will improve our understanding of the light sea quark polarization contributions to the proton spin structure.

4. Summary and Outlook

Latest measurements of the STAR longitudinal progam improve our understanding of the proton spin structure. A non-zero gluon spin contribution in the range of $0.05 < x < 1$ is indicated from the 2009 inclusive jet double spin asymmetry A_{LL} results. The total gluon spin contribution might be at the same order of the quark polarization contribution to the proton; this needs future measurements to prove. The W decay lepton pseudorapidity dependent $W^{+/-}$ single spin asymmetries provide constraints on the $\bar{u}$ and $\bar{d}$ polarized PDFs at the W mass scale. A postive $\bar{u}$ polarized PDF is suggested based on the STAR data and theoretical model comparisons in $0.05 < x < 0.2$ kinematic region.

Uncertainties of both polarized gluon and sea quark PDFs at low x remain large compared to the currently probed region. Longer 500 GeV longitudinally polarized proton-proton operations at RHIC and possible forward upgrade will provide opportunities to measure di-jet and W productions at STAR in the forward pseudorapidity with high precison data. Such observables are under study and future analysis will impact the dermination of the gluon, $\bar{u}$ and $\bar{d}$ helicity dependent PDFs and provide new inputs for the proton spin consistuation studies in the extended x regions.

5. References

[1] Ashman J *et al.* 1988 *Phys. Lett. B* **206** 364
[2] Leader E, Sidorov A V and Stamenov D B 2010 *Phys. Rev. D* **82** 114018
[3] Ball R D *et al.* 2013 *Nucl. Phys. B* **874** 36
[4] Adamczyk L *et al.* 2014 (*Preprint* hep-ph/1405.5134)
[5] de Florian D, Sassot R, Stratmann M and Vogelsang W 2014 (*Preprint* hep-ph/1404.4293)
[6] Adamczyk L *et al.* 2014 (*Preprint* hep-ph/1404.6880)
[7] Sakuma T 2010 *Inclusive Jet and Dijet Production in Polarized Proton-Proton Collisions at 200 GeV at RHIC* Ph.D. thesis MIT
[8] Adamczyk L *et al.* 2012 *Phys. Rev. D* **86** 032006
[9] Cacciari M, Salam G P and Soyez G 2008 *JHEP* **0804** 063
[10] de Florian D, Sassot R, Stratmann M and Vogelsang W 2008 *Phys. Rev. Lett.* **101** 072001
[11] Blumlein J and Bottcher H 2010 *Nucl. Phys. B* **841** 205
[12] Mukherjee A and Vogelsang W 2012 *Phys. Rev. D* **86** 094009
[13] pp/pA Letter of Intent [STAR Collaboration] 2014 A polarized p+p and p+a program for the next years
[14] Djawotho P and for the STAR Collaboration 2013 (*Preprint* hep-ph/1303.0543)
[15] Adare A *et al.* 2009 *Phys. Rev. Lett.* **103** 012003
[16] Aschenauer E C *et al.* 2012 (*Preprint* nucl-ex/1304.0079)
[17] Glck M, Reya E, Stratmann M and Vogelsang W 2001 *Phys. Rev. D* **63** 094005
[18] Awes T C *et al.* 1998 *Phys. Rev. D* **58** 092004
[19] de Florian D, Sassot R, Stratmann M and Vogelsang W 2009 *Phys. Rev. D* **80** 034030
[20] Aggarwal M *et al.* 2011 *Phys. Rev. Lett.* **106** 062002

Journal of Physics: Conference Series **535** (2014) 012004 doi:10.1088/1742-6596/535/1/012004

Recent STAR Results from U+U and Au+Au Collisions

Hui Wang for STAR Collaboration

Physics Department, Building 510A, Brookhaven National Lab, Upton, NY, 11973

E-mail: wanghui6@bnl.gov

Abstract.

Unlike Au or Pb nuclei which are more spherical, the uranium nuclei have a relatively large deformation on average. The prolate shape of uranium nuclei provides the feasibility to study how the initial geometry of the nuclei affects the azimuthal distributions of produced particles. It also provides a unique opportunity to understand the initial condition for particle production at mid-rapidity in heavy ion collisions as well as path length dependence of jet quenching and quarkonium in-medium effects.

In this proceedings, the two- and four- particle cumulant, $v_2\{2\}$ and $v_2\{4\}$, from U+U collisions at $\sqrt{s_{NN}}$ = 193 GeV and Au+Au collisions at $\sqrt{s_{NN}}$ = 200 GeV for inclusive charged hadrons at midrapidity will be presented. The STAR Zero Degree Calorimeter is used to subdivide the 0-1% centrality bin into even finer centralities. Differences were observed between the multiplicity dependence of $v_2\{2\}$ for most central Au+Au and U+U collisions. The observed $v_2\{2\}$ slope results were compared to Monte Carlo Glauber model predictions and it was seen that this model cannot explain the presented results on the multiplicity dependence of $v_2\{2\}$ in central collisions.

1. Introduction

Heavy nuclei are collided at the Relativistic Heavy Ion Collider (RHIC) in order to create matter that is hot and dense enough to create a Quark Gluon Plasma (QGP). One evidence of the formation of QGP is the strong v_2, which is defined as the second coefficient of the following Fourier expansion:

$$\frac{dN}{d\phi} \propto 1 + \sum_n 2v_n \cos n(\phi - \psi),$$

where ψ is a reference axis often identified with either the participant plane or reaction plane and ϕ is the azimuthal angle of the particles. In non-central heavy-ion collisions, the overlap area has an almond shape with a major axis and a minor axis. Model calculations indicate that different pressure gradients along the two axes will produce a large v_2.

The unique geometry shape of uranium nucleus provides opportunities to further test our understanding of the particle production mechanism and v_2. The larger size of uranium nucleus provides higher energy density in U+U collisions. On the other hand, the uranium nucleus is not spherical and have a prolate shape, which makes the collision geometry vary from body-on-body to tip-on-tip configurations, even in nearly fully overlapping collision. In the case of tip-on-tip collisions, the produced particle multiplicity should be higher due to the larger number of binary

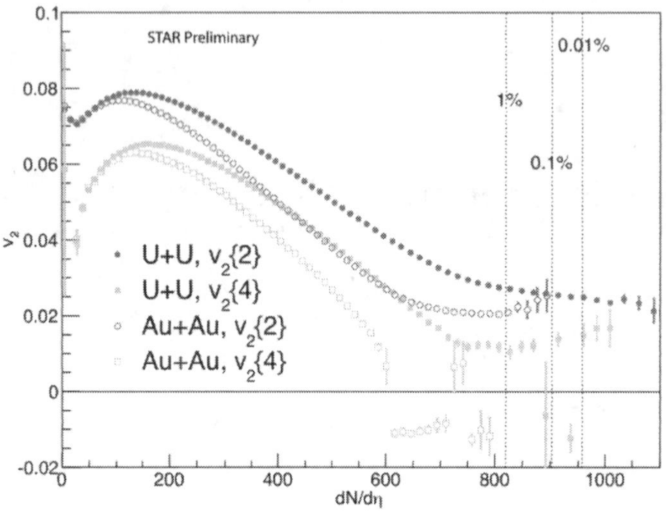

Figure 1. (Color online) The two- and four- particle cumulant $v_2\{2\}$ and $v_2\{4\}$ from 200 GeV Au+Au and 193 GeV U+U collisions as a function of $dN/d\eta$ at midrapidity for all charged particles. Dash lines represent multiplicity based centrality cuts for U+U collisions. Negative $v_2\{4\}^4$ are shown as negative $v_2\{4\}$.

collisions, and v_2 is smaller since the overlap region is symmetric, while in the case of body-on-body collisions, one should observe a smaller multiplicity associated with larger v_2. Thus if our understanding of particle production is correct, a strong correlation between multiplicity and v_2 should exist in U+U collisions compared with more symmetric systems like Au+Au.

2. Data and Analysis Method

The data sets for this analysis were collected by the STAR detector from Au+Au at $\sqrt{s_{NN}}$ = 200 and U+U collisions at $\sqrt{s_{NN}}$ = 193 GeV in year 2011 and 2012, respectively. All charged particles within pseudo-rapidity $|\eta| < 1$ and transverse momentum $0.2 < p_T < 2.0$ GeV/c were analyzed by the Time Projection Chamber (TPC). The Zero Degree Calorimeters (ZDCs) were used to select fully overlapping events based on spectator neutrons. Only events with primary vertices within 30 cm of the TPC center in the beam direction were analyzed. The reconstruction efficiency was obtained from GEANT based Monte Carlo simulation and the results presented here are corrected for reconstruction efficiency.

In this analysis, v_2 is calculated via the two- and four- particle cumulant methods ($v_2\{2\}$ and $v_2\{4\}$). The Q-cumulant method [4] allows us to calculate $v_2\{4\}$ without using generating functions or calculating an event plane thus it is simpler to perform. The two-particle cumulant is calculated from $v_2\{2\}^2 = \langle\cos[2(\varphi_1-\varphi_2)]\rangle - \langle\cos(\varphi_1)\rangle\langle\cos(\varphi_2)\rangle - \langle\sin(\varphi_1)\rangle\langle\sin(\varphi_2)\rangle$. In order to remove correlations from HBT, Coulomb effects and track-merging, we reject pairs with $|\Delta\eta| < 0.1$. This is not practical for the four-particle cumulant but few particle correlations in $v_2\{4\}$ are largely suppressed by combinatorics.

3. Results

In this paper, the results are presented for both minimum-bias and central events. Figure 1 shows the two- and four- particle cumulant $v_2\{2\}$ and $v_2\{4\}$ from 200 GeV Au+Au and 193 GeV U+U collisions as a function of $dN/d\eta$ at midrapidity for all charged particles. The $v_2\{4\}$

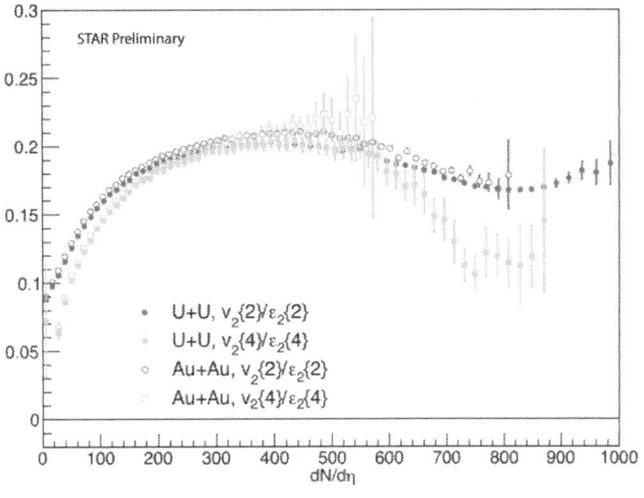

Figure 2. (Color online) v_2 scaled by participant eccentricity from 200 GeV Au+Au and 193 GeV U+U collisions as a function of $dN/d\eta$.

results are calculated from $v_2\{4\}^4$ and by taking the fourth root. For illustration purposes, we plot the imaginary $v_2\{4\}$ results as negative values. The Au+Au $v_2\{4\}$ results show a centrality dependence and turn negative (imaginary) at central collisions. This is allowed due to the fact that $v_2\{4\}$ results contains flow fluctuations and the flow signal from central collisions is small due to symmetrical collision system. Unlike $v_2\{4\}$ results in Au+Au, the $v_2\{4\}$ results in U+U are always positive at central collisions. The non-zero $v_2\{4\}$ in central U+U collision could result from the intrinsic prolate shape of the uranium nucleus: although the impact parameter is small in central U+U collisions, the prolate shape of uranium nucleus on average still provides an almond shaped overlap region which also produces this non-zero $v_2\{4\}$ result.

As discussed previously, high-multiplicity collisions would have predominantly tip-tip collisions, which also produces relatively smaller v_2. It has been predicted that a unique knee structure will be present in high multiplicity collisions if one plots v_2 vs. charged multiplicity [3]. However, the $v_2\{2\}$ results from U+U data show no indication of such a knee structure. The non-zero $v_2\{4\}$ and lack of knee structure in $v_2\{2\}$ together could indicate that although the prolate geometry shape of uranium nuclei do show up in the data, there might be additional fluctuation that washes out this knee structure in $v_2\{2\}$ results [5].

Another interesting topic is the v_2 divided by the initial anisotropy ε in coordinate space. It has been found previously that v_2/ε increases with $dN/d\eta$ and saturates at central collisions. This saturation is consistent with the picture that the system collectivity approaches the hydrodynamical limit. Figure 2 shows the v_2/ϵ from 200 GeV Au+Au and 193 GeV U+U collisions. Both U+U and Au+Au follow the same trend for v_2/ε, however, a turn-over is observed at central collisions where the initial geometry of the overlap region plays a dominant role. The turn-over may indicate an overestimation of ε in central collisions while other origins like hydro fluctuation are also possible.

In order to take advantage of the prolate shape of uranium nuclei, one needs to separate fully overlapped body-on-body and tip-on-tip collisions. Unfortunately, RHIC is not capable of providing polarized uranium beams so it relies on experimental observables to disentangle the body-on-body vs. tip-on-tip collisions. One strategy is to use spectator neutrons measured by the ZDC to select fully overlapping collision, then one could use charged particle multiplicity

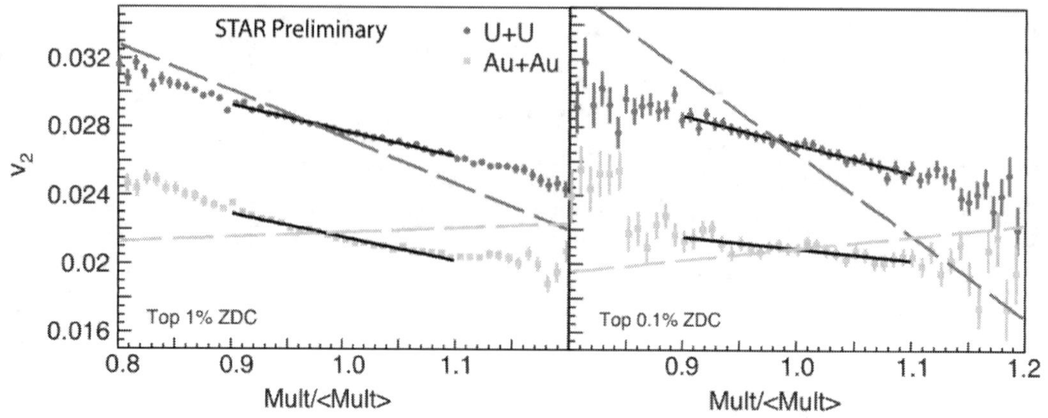

Figure 3. (Color online) The v_2 of all charged particles as a function of the normalized multiplicity. The left panel shows the results for top 1% ZDC central events, while the right panel shows the results for top 0.1% ZDC central event. A linear fit between normalized multiplicity 0.9 to 1.1 is applied to extract the slope parameter. The dash lines represent Glauber-simulation slopes calculated via eccentricity.

to select body-on-body or tip-on-tip enhanced samples [1]. In case of body-on-body collision, the multiplicity will be lower due to smaller number of binary collisions, while the v_2 will be larger due to the almond overlapping region. For tip-on-tip collisions, there will be a higher multiplicity associated with small v_2. Thus a strong correlation between multiplicity and v_2 could serve as an evidence of selection of geometry orientations.

Figure 3 shows the v_2 of all charged particles as a function of the normalized multiplicity. The left panel shows the results for top 1% ZDC central events. Since the deformation of gold nuclei is small, in case of fully overlapping collisions, the multiplicity differences are mainly from fluctuations and there should be no correlations between multiplicity and v_2. However, a strong negative slope is observed in both Au+Au and U+U collisions, which indicates the impact parameter still dominates the observed correlation. A slightly off-center collision produces smaller multiplicity and stronger v_2 due to almond shaped overlapping region. In order to see the effects from initial geometry (body-on-body vs. tip-on-tip), one needs to reduce the effects from impact parameters and select further fully overlap collisions. The right panel of Fig. 3 shows the same results for top 0.1% ZDC central events. The slope magnitude in Au+Au collisions is smaller, indicating the effects from non-central collisions are reduced, and the multiplicity differences are mainly driven by fluctuations in Au+Au, while the slope for U+U becomes steeper. This comparison between Au+Au and U+U serves as evidence of selection of geometry orientation enhanced sample based on multiplicity. On the other hand, the Glauber model predicts a steeper slope for U+U collisions and a positive slope for Au+Au due to its oblate shape [6].

To further test the correlation between multiplicity and v_2, a linear fit is applied and the slope parameter is extracted. Figure 4 shows the slope parameter for 200 GeV Au+Au and 193 GeV U+U events as a function of ZDC centrality. With tighter ZDC cuts, the slope magnitude decreases in Au+Au, indicating that the effects from the impact parameter are reduced, while the U+U slope parameter increases (more negative). This stronger correlation between v_2 and multiplicity in U+U is due to the fact that with more central collisions selected by the ZDCs, the diffusion effects from impact parameter are reduced (at large impact parameter, central body-on-body and off-center tip-on-tip collisions produce the same multiplicity but different v_2). The

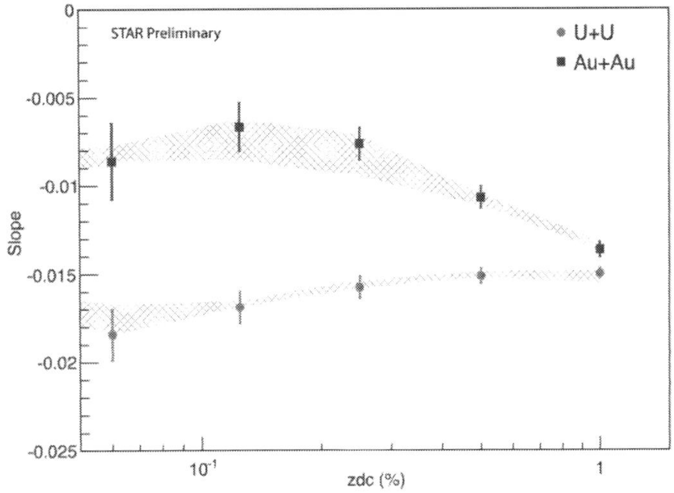

Figure 4. (Color online) The slope parameter for 200 GeV Au+Au and 193 GeV U+U events as a function of ZDC centrality. Grey band represent systematical uncertainties.

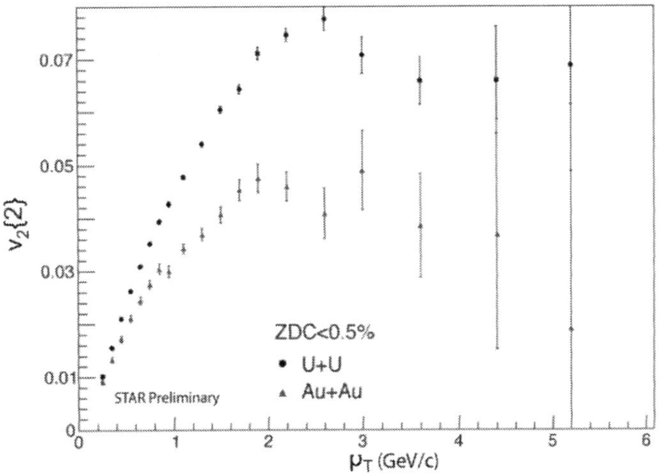

Figure 5. (Color online) p_T dependence of the two-particle cumulant $v_2\{2\}$.

comparison between Au+Au and U+U indicates that $dN/d\eta$ can help to select tip-on-tip vs. body-on-body enhanced samples.

Once selection of initial geometry (body-on-body vs. tip-on-tip) is possible, the central U+U collisions could serve as an ideal test ground for many observables like the path-length dependence of jet quenching or local parity violation [2, 3]. As the first step to study the path length dependence of jet quenching, Fig. 5 shows the p_T dependence of the two-particle cumulant $v_2\{2\}$. Both U+U and Au+Au show a maximum at p_t around 2.5 GeV/c and turn over at higher p_T. To take advantage of uranium nucleus's prolate shape, it is necessary in the future to divide the U+U data into body-on-body vs. tip-on-tip enhanced samples.

4. Conclusion

In summary, the two- and four- particle cumulant, $v_2\{2\}$ and $v_2\{4\}$, for inclusive charged hadrons have been presented from U+U collisions at $\sqrt{s_{NN}} = 193$ GeV and Au+Au collisions at $\sqrt{s_{NN}} = 200$ GeV . The non-zero $v_2\{4\}$ results in U+U collisions suggest that we observe the prolate shape of uranium nuclei. However, no knee structure is visible in $v_2\{2\}$ results from central U+U collisions. We also observed an interesting turn over of v_2/ϵ_2 in central collisions for both Au+Au and U+U collisions. It has been demonstrated that by combining cuts on ZDC and charged particle multiplicity, one can select body-body or tip-tip enhanced samples of central U+U collisions. The capability of selecting initial geometry in U+U collisions provides new opportunities to study path-length dependent jet quenching and local parity violation in the future.

References

[1] A. Kuhlman and U. Heinz, Phys. Rev. C **72**, 037901 (2005)
[2] U. Heinz and A. Kuhlman Phys. Rev. Lett. **94**, 132301 (2005)
[3] Sergei A. Voloshin, Phys. Rev . Lett. **105**, 172301 (2010)
[4] Ante Bilandzic, Raimond Snellings, and Sergei Voloshin Phys. Rev. C **83**, 044913 (2011)
[5] Maciej Rybczynski, Wojciech Broniowski and Grzegorz Stefanek, Phys.Rev. C **87**, 044908 (2013)
[6] Hiroshi Masuia, Bedangadas Mohanty and Nu Xu, Physics Letters **B 679** (2009) 440444

Journal of Physics: Conference Series **535** (2014) 012005 doi:10.1088/1742-6596/535/1/012005

Flow in the RHIC Beam Energy Scan from STAR

Xu Sun for the STAR Collaboration

Department of Physics, Harbin Institute of Technology, Harbin 150001, People's Republic of China

Nuclear Science Division, Lawrence Berkeley National Laboratory, Berkeley, California 94720, USA

E-mail: XuSun@lbl.gov

Abstract.

The first phase of the beam energy scan (BES) program at RHIC was successfully taken in the years 2010 and 2011. Data were collected at $\sqrt{s_{\rm NN}}$ =7.7, 11.5, 19.6, 27, 39 and 62.4 GeV, covering a wide range of baryon chemical potentials from $\mu_{\rm B}$ 420 to 40 MeV. The main goals are the search for the QCD critical point and to find signatures for a phase transition between the hadron gas and the QGP. Directed (v_1), elliptic (v_2) and triangular (v_3) flow are important observables to study the early evolution of the created matter in relativistic heavy-ion collision experiments. In particular, it is expected that flow is sensitive to the phase transition from a quark gluon plasma to a hadron gas.

We will present measurements of identified hadrons for v_1 ($\pi^{\pm}, p, \bar{p}$), v_2 ($\pi^{\pm}$, $K^{\pm}$, K_s^0, p, $\bar{p}$, ϕ, Λ, $\bar{\Lambda}$, Ξ^-, $\bar{\Xi}^+$,Ω^- and $\bar{\Omega}^+$) for all BES energies, and v_3 ($\pi^{\pm}, K^{\pm}, p, \bar{p}$) for $\sqrt{s_{\rm NN}}$ = 39 GeV. We will discuss the beam-energy dependent difference of v_2 between particles and anti-particles. Furthermore, we will address mass ordering at low p_T and number-of-constituent-quark scaling at intermediate p_T of v_2 and v_3 for identified hadrons.

1. Introduction

One of the main goals of heavy ion collision experiments at RHIC is to study the properties of the Quark-Gluon Plasma (QGP). To understand the formation of QGP and the structure of the phase diagram, the Beam Energy Scan (BES) phase one was started in the years 2010/2011 at RHIC [1], and recorded Au + Au collisions at $\sqrt{s_{\rm NN}}$ = 7.7, 11.5, 19.6, 27, 39 and 62.4 GeV. This paper reports the azimuthal anisotropy of identified particles of BES experiments by using the STAR detector at RHIC.

The azimuthal anisotropy is one of the most important observables for relativistic nuclear collisions. In the non-central Au+Au collision, the overlap region is an almond shape with the major axis perpendicular to the reaction plane which is defined by the impact parameter and the beam direction. Due to initial state fluctuations, the participant plane [2] in each event is not necessarily the same as the reaction plane. As the system evolves, the pressure gradient pushes the anisotropy from coordinate space to momentum space. The produced particle distribution [3, 4] can be written as:

$$E\frac{d^3N}{dp^3} = \frac{1}{2\pi}\frac{d^2N}{p_T dp_T dy}(1 + \sum_{n=1}^{\infty} 2v_n \cos[n(\phi - \Psi_n)]) \quad (1)$$

where ϕ is the azimuthal angle of a particle, and Ψ_n is the n-th order event plane angle reconstructed by the final produced particles, which is an estimation of the participant plane. The first harmonic coefficient of the Fourier expansion v_1 is called directed flow, the second harmonic coefficient v_2 is called elliptic flow and the third harmonic coefficient v_3 is called triangular flow. In this paper, we will discuss the recent experimental results for v_1, v_2 and v_3 of identified particles from the BES.

This paper is organized as follows. In Sec. 2, the recently published results from BES I of v_1, v_2 [5, 6], and the new results for v_3 are discussed. Then the summary is presented in Sec. 3.

2. Recent Flow Results

Collective flow is one of the most widely studied observables in relativistic heavy ion experiments. Different harmonic coefficient reflects different dynamic properties of the fireball. For instance, v_1 reflects the compressibility of the created dense matter, v_2 is dominated by the initial geometry and v_3 is generated by the initial state fluctuations. A systematic study of flow coefficients as a function of beam energy, particle ID, and centrality, may reveal important properties of the QGP and the phase diagram.

2.1. Directed Flow

The STAR collaboration recently published directed flow results for protons, anti-protons, and pions for BES I in Au+Au collisions. In Ref. [5], the directed flow of protons, anti-protons and pions for 0%-10%, 10%-40% and 40%-80% central Au+Au collisions are shown, and the slopes (dv_1/dy) at 10%-40% central Au+Au collisions are calculated. Since a three-fluid hydrodynamic model with a first order phase transition from a QGP phase to a hadronic phase predicted a minimum in dv_1/dy of net baryons, which is considered a "softest point collapse" [7], the STAR collaboration also analyzed the beam-energy dependent slopes of protons, anti-protons and net protons which are shown in Figure 1.

Figure 1 shows that for intermediate-centrality collisions, the slope of anti-protons increases with increasing beam energy and remain negative, but protons and net protons show a non-monotonic behavior. Proton dv_1/dy decreases from 7.7 to 11.5 GeV and changes sign from positive to negative, increases from 19.6 to 200 GeV but remains negative and shows a minimum between 11.5 and 19.6 GeV. Net protons dv_1/dy shows a similar trend but crosses zero between 27 and 39 GeV, and also shows a minimum between 11.5 and 19.6 GeV. For both protons and net protons, UrQMD model cannot reproduce this non-monotonic behavior. This observed minimum for protons and net protons is a possible signature for the softest point of the EoS and the first order phase transition.

2.2. Elliptic Flow

One of the most important observations from BES I is the difference in v_2 between particles and anti-particles [6]. Figure 2 shows the p_T-independent difference in v_2 between particles and anti-particles of different particle species (see legend in Figure 2) as a function of $\sqrt{s_{\rm NN}}$ for 0%-80% central Au+Au collisions (please check Ref. [6] for details). Larger $v_2(X)$ values are found at lower beam energies for all particle species except pions which is the opposite. The magnitude of the difference between particles and anti-particles ($\Delta v_2 = v_2(X) - v_2(\bar{X})$) increases with decreasing beam energy. There are several theoretical calculations which try to explain this difference such as hybrid model [8] and the Nambu-Jona-Lasinio (NJL) model [9], but they can only qualativly describe the experimental data so far.

Since the v_2 of particles and anti-particles are different, we cannot expect a single number-of-constituent-quark (NCQ) scaling for both particles and anti-particles, therefore, we test the NCQ scaling for particles and anti-particles separately. The upper panels of Figure 3 shows the $v_2(m_T - m_0)$ scaled with number of constituent quarks on both axes for 11.5 and 62.4 GeV

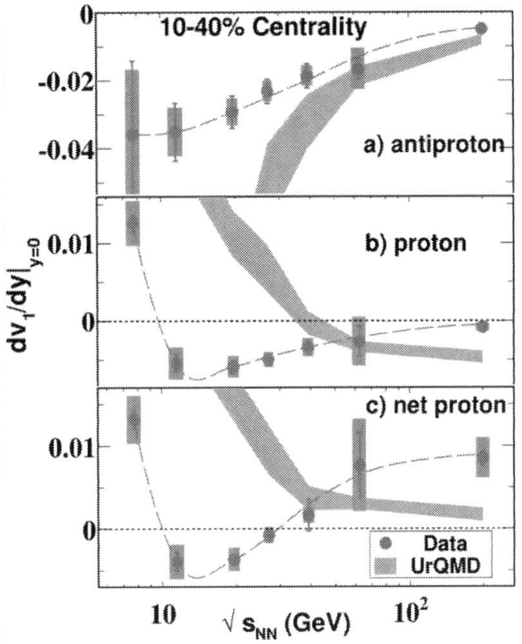

Figure 1. Directed flow slope (dv_1/dy) near midrapidity versus beam energy for 10%-40% central Au+Au collisions. The dashed curves are smooth fits to guide the eye. Figure 1 is taken from Ref. [5]

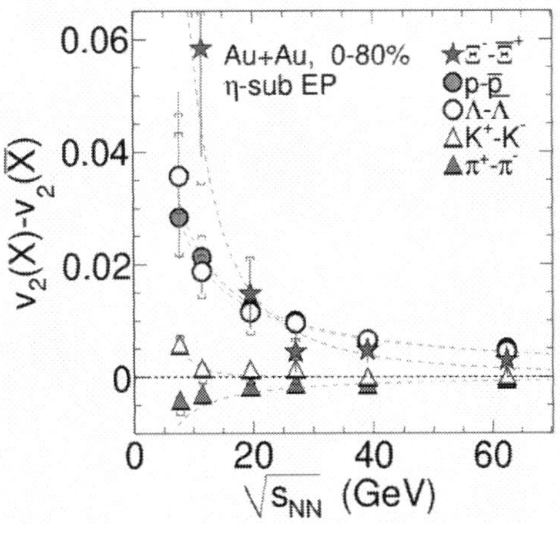

Figure 2. The difference in v_2 between particles (X) and their anti-particles ($\bar{X}$) as a function of beam energy for 0%-80% central Au+Au collisions. Figure 2 is taken from Ref. [6].

(for more energies, please see Ref. [10]). The ratios of data to fit are shown in the lower panels of Figure 3. Most of the ratio values are within ±10% of unity, which shows that the NCQ scaling holds for particles and anti-particles separately. The ϕ meson is an exception, since the

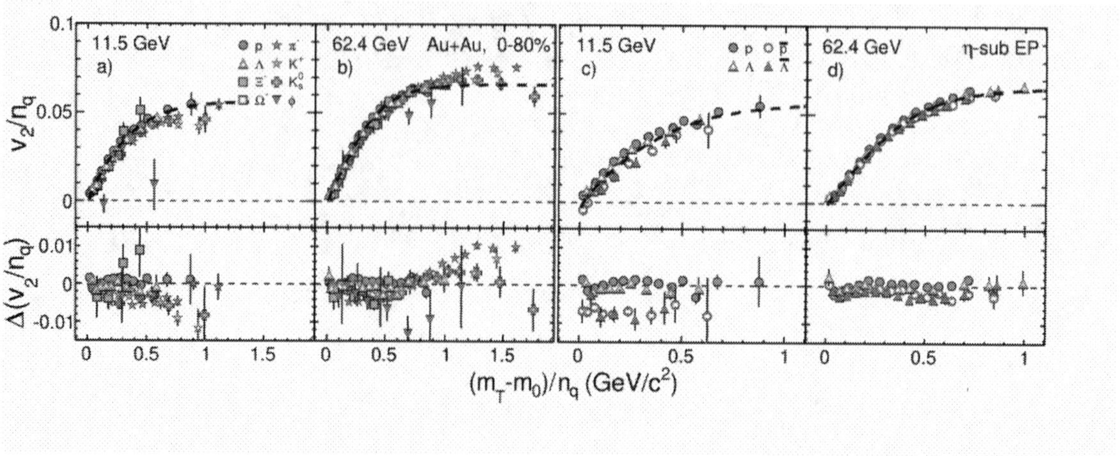

Figure 3. The NCQ scaling of particles (frame a and b) and corresponding anti-particles (frame c and d) for 11.5 GeV and 62.4 GeV. Figure 3 is taken from Ref. [6]

v_2 value of ϕ mesons are lower by 2.5 σ at 11.5 GeV (p_T = 1.9 GeV/c). The smaller v_2 value of $\phi(s\bar{s})$ meson, which has a smaller hadronic interaction cross section[11], may indicate that the hadronic interactions become more important than partonic interactions at lower energies ($\sqrt{s_{\mathrm{NN}}} \leq 11.5$ GeV).

2.3. Triangular Flow

A surprising result was observed for inclusive charged hadron v_3 results recently published by the the STAR collaboration [12]. "A good agreement is found not only between RHIC experiments but also between RHIC and LHC experiments. This is surprising because of the somewhat different $\Delta\eta$ ranges." [12] In order to investigate detailed properties of v_3, the v_3 of identified particles needs to be studied.

Figure 4 shows $v_3(p_T)$ of $\pi^{\pm}, K^{\pm}, p$ and $\bar{p}$ for 0%-80% central Au+Au collisions at $\sqrt{s_{\mathrm{NN}}}$ = 200 and 39 GeV. Since we already observed the difference of v_2 values between particles and anti-particles, it is natural to separate particles and anti-particles for v_3. For both 200 and 39 GeV, we see a mass ordering at low p_T range which means the particle with a lower mass has a higher v_2. The region of mass ordering is $p_T < 1.6$ GeV/c for 200 GeV and $p_T < 1.2$ GeV/c for 39 GeV. The difference of mass ordering regions indicates different magnitudes of radial flow at different beam energies, which means the radial flow at 200 GeV is larger than that at 39 GeV. We also observe a baryon-meson splitting at the intermediate p_T region of 200 GeV. The baryons (p and $\bar{p}$) and mesons ($\pi^{\pm}$ and $K^{\pm}$) follow different trends. There are some hints that baryon-meson splitting also exist at 39 GeV but the uncertainties are too large. Therefore more particle species are needed to investigate the baryon-meson splitting (Λ and $\bar{\Lambda}$).

To understand the partonic behavior of v_3, we check the NCQ scaling, which is observed for v_2 for particles and anti-particles separately as mentioned above, for v_3. As shown in Figure 5, we show $v_3(m_T - m_0)$ scaled with number of constituent quarks on both axes for 200 and 39 GeV. No NCQ scaling is observed with the same scaling factor n_q used for v_2; we can see clearly the baryon band deviates from the meson band for both 200 and 39 GeV. This indicates that partonic interactions might have different influences on v_3 compared to v_2.

On the other hand, theoretical calculations suggest that we should use a new scaling factor which is $n_q^{n/2}$, where n is the n-th harmonic [13]. And an AMPT calculation also indicates that

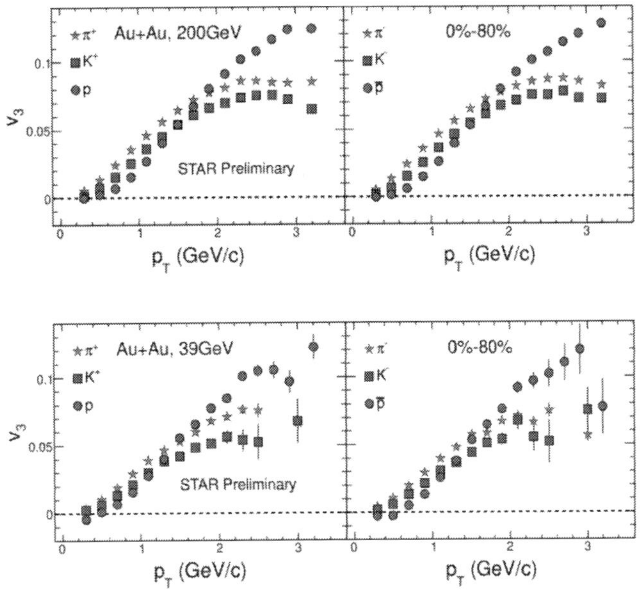

Figure 4. Triangular flow of $\pi^{\pm}, K^{\pm}, p$ and $\bar{p}$ for 0%-80% central Au+Au collisions at $\sqrt{s_{\mathrm{NN}}}$ = 200 and 39 GeV. Statistical uncertainties only.

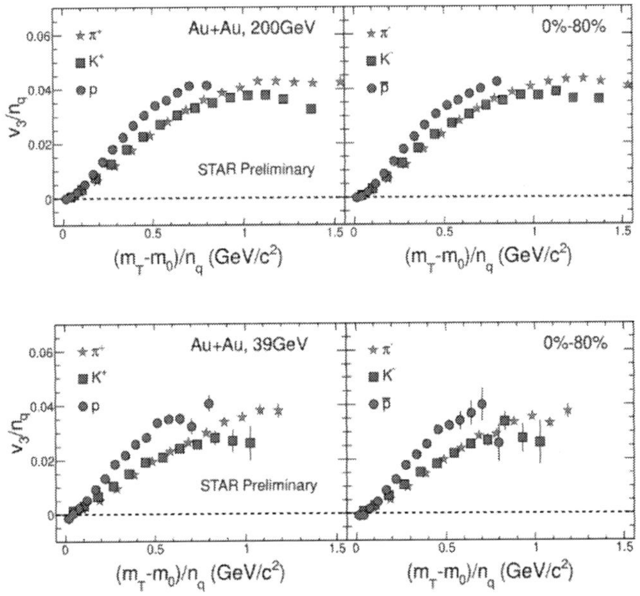

Figure 5. $v_3(m_T - m_0)$ scaled with number of constituent quark (n_q) of $\pi^{\pm}, K^{\pm}, p$ and $\bar{p}$ for 0%-80% central Au+Au collisions at $\sqrt{s_{\mathrm{NN}}}$ = 200 and 39 GeV. Statistical uncertainties only.

NCQ scaling works for v_3 within 20% by using a $n_q^{3/2}$ scaling factor [14]. We did the same thing to the experimental data. Figure 6 shows $v_3(m_T - m_0)$ scaled with $n_q^{3/2}$ on the y-axis and n_q on the x-axis for 200 and 39 GeV. We do observe the NCQ scaling at 200 GeV by using this $n_q^{3/2}$ factor; the baryon band and meson band come together. For 39 GeV we see some hint for NCQ scaling. With the future BES II program we will collect enough data to cover the important $m_T - m_0$ region above 1 GeV/c^2 where is the region we expect NCQ scaling and analyze more particle species.

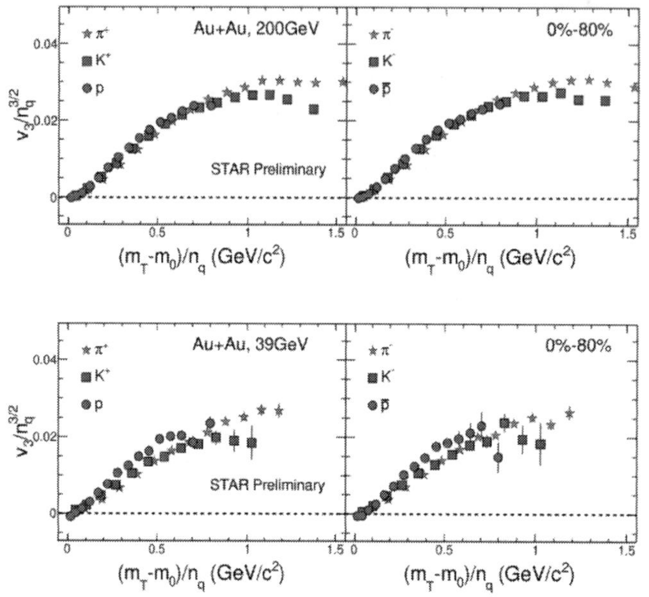

Figure 6. $v_3(m_T - m_0)$ scaled with the new scaling factor ($n_q^{3/2}$) for $\pi^\pm, K^\pm, p$ and $\bar{p}$ for 0%-80% central Au+Au collisions at $\sqrt{s_{NN}}$ = 200 and 39 GeV. Statistical uncertainties only.

3. Summary

We present the recent flow results from BES I at RHIC from the STAR collaboration.

For directed flow (v_1), we show the slope (dv_1/dy) as a function of beam energy, and a minimum between 11.5 and 19.6 GeV for protons and net protons is observed, which is a possible signature for the softest point of the EoS.

For elliptic flow (v_2), we show the difference of v_2 values for particles and anti-particles. As the v_2 value of particles and anti-particles is different, we expect a break-down of the NCQ scaling between particles and anti-particles. But the NCQ scaling holds for particles and anti-particles separately.

For triangular flow (v_3), the mass ordering at the low p_T region is observed, where the larger mass ordering region at 200 GeV indicates larger radial flow. We also observe that the NCQ scaling breaks by using the same scaling factor as what we used for v_2 (n_q). But the NCQ scaling seems to hold at 200 GeV by using a $n_q^{3/2}$ factor. We also see some hints that NCQ scaling holds at lower energy by using this $n_q^{3/2}$ factor but we need more particle species and more beam energies.

References

[1] M. M. Aggarwal *et al.* [STAR Collaboration], arXiv:1007,2613 [nucl-ex].
[2] D. Kharzeev and M. Nardi, Phys. Lett. B **507**, 121 (2001).
[3] S. Voloshin and Y. Zhang, Z. Phys. C **70**, 655 (1996).
[4] A. M. Poskanzer and S. A. Voloshin, Phys. Rev. C **58**, 1671 (1998).
[5] L. Adamczyk *et al.* [STAR Collaboration], Phys. Rev. Lett. **112**, 162301 (2014).
[6] L. Adamczyk *et al.* [STAR Collaboration], Phys. Rev. Lett. **110**, 142301 (2013).
[7] H. Stocker, Nucl. Phys. A **750**, 121 (2005).
[8] J. Steinheimer, V. Koch and M. Bleicher, Phys. Rev. C **86**, 044903 (2012).
[9] J. Xu, T. Song, C. Ko and F. Li, Phys. ReV. Lett. **112**,012301 (2014).
[10] L. Adamczyk *et al.* [STAR Collaboration], Phys. Rev. C **88**, 014902 (2013).
[11] B. Mohanty and N. Xu, J. Phys. G **36**, 046022 (2009).
[12] L. Adamczyk *et al.* [STAR Collaboration], Phys. Rev. C **88**, 014904 (2013).
[13] R. Lacey, J. Phys. G: Nucl. Part. Phys. **38**, 124048 (2011).
[14] L. Han *et al.* Phys. Rev. C **84**, 064907 (2011).

Journal of Physics: Conference Series **535** (2014) 012006 doi:10.1088/1742-6596/535/1/012006

Dielectron production in 200 GeV p+p and Au+Au collisions at STAR

Yi Guo for the STAR collaboration

E-mail: yiguo@rcf.rhic.bnl.gov

Abstract. Leptons do not interact strongly with the hot dense medium created in relativistic heavy ion collisions. They can escape the interaction region undistorted and thus carry direct information about the space-time evolution of the expanding system. In the low mass region (LMR, $0.3 < M_{ee} < 1.1$ GeV/c^2), dielectron mass spectra can provide the in-medium vector meson properties, while in the intermediate mass region (IMR, $1.1 < M_{ee} < 3$ GeV/c^2), the slope of dielectron transverse mass spectra is expected to have connection with the QGP temperature.

In this paper, we present the centrality and p_T dependence of the dielectron mass spectra measured in STAR experiment at RHIC. The data sets used in the analysis include large statistics samples collected during years 2010 for 200 GeV Au+Au collisions and 2012 for 200 GeV p+p collisions. In order to extract underlying physics, we will compare our results with model calculations.

1. Introduction

Dileptons are clean and penetrating probes for the hot and dense nuclear matter created by the high energy nuclear collisions because they do not suffer from strong interactions. They can be produced during all stages of a heavy ion collision, and their sources are expected to have different contributions to dileption invariant mass spectra. Therefore, a systematic measurement of the dilepton pair distribution can reveal the properties of medium created by high energy nuclear collisions.

Dilepton measurements have been pursued for decades in heavy ion collisions [1]-[7]. The CERES measurement of e^+e^- mass spectra showed a clear enhancement in the mass region below ~0.7 GeV/c^2 compared to the known hadronic sources [4]. High precision data from NA60 suggested that this enhancement is consistent with in-medium broadening of the ρ spectral function instead of a dropping of its pole mass hypothesis [5], [8]-[11]. In addition, slope parameters of dimuon transverse mass spectra showed a sudden drop above the ϕ mass after removing the correlated charm contributions. This is argued to be an indication of thermal dilepton from partonic source by the NA60 collaboration [6]. At RHIC energy, the result from PHENIX showed a significant enhancement in mass region 0.3~0.76 GeV/c^2 [7]. However, the huge enhancement could not be reproduced by those model calculations which successfully explained SPS data [8]-[11].

In this paper, we will present the newest STAR results on dielectron production from 200 GeV p+p and Au+Au collisions, and compare the Au+Au results with model calculations.

2. Analysis

Data used in the analysis are obtained from 200 GeV p+p and Au+Au collisions, which were collected by the Solenoidal Tracker At RHIC (STAR) detector [12] in year 2012 and year 2010, respectively. The main subsystems used in the analysis are Time Projection Chamber(TPC) [13] and the Time of Flight (TOF) [14].

In addition of track reconstruction and momentum information, TPC also provides identification capabilities for charged particles by their ionization energy loss (dE/dx) in the TPC gas. With the fully installed TOF system, the PID capabilities are greatly improved, especially in low p_T region. With the combination of dE/dx from TPC and velocity (β) from TOF, the electron purity is ~98% in p+p collisions, ~95% in Au+Au minimum bias collisions and ~93% in Au+Au central collisions.

Two methods were used in this analysis to reconstruct the background. In LMR, due to the correlated background, e.g. cross pair and jet contribution, we used the like-sign method background with an acceptance correction. The acceptance correction is to account for the slight difference in the acceptance between like-sign and unlike-sign pairs [15]. While in IMR, the mix-event technique was used to achieve better statistics. The difference between these two methods were taken as systematic uncertainty. Figure 1 left panel (a) shows the raw dielectron mass distribution, reconstructed background and signal after background subtraction in 200 GeV Au+Au minimum bias collisions. The signal-to-background ratios from p+p minimum bias collisions, Au+Au minimum bias and central collisions are shown in Fig. 1 left panel (b). In this analysis, we also subtracted the contribution from the photon conversion by the method used in Ref. [16].

The dielectron continuum results in this paper were obtained in STAR acceptance ($p_T^e > 0.2$ GeV/c, $|\eta^e| < 1.0$, $|y^{ee}| < 1.0$) and corrected for the efficiency.

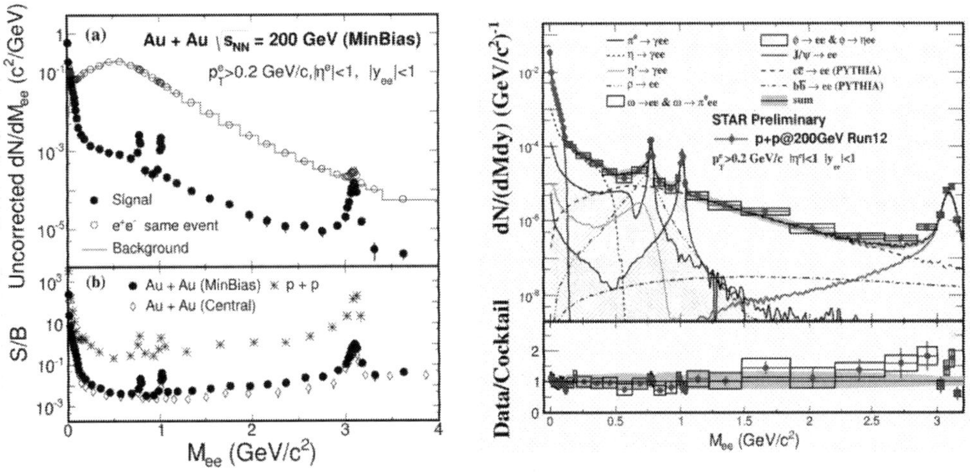

Figure 1. Left panel (a): e^+e^- raw invariant mass distribution (open circles), the reconstructed backgrounds (red histogram) and the signal (solid dots) in 200 GeV Au+Au minimum bias collisions [18]. Left panel (b): The signal to background ratio in p+p and Au+Au collision at $\sqrt{s_{NN}} = 200$ GeV [18]. Right panel: invariant mass spectra from $\sqrt{s} = 200$ GeV p+p collisions taken from year 2012. The black open box represents systematic error from data while the grey band depicts systematic uncertainty of cocktail.

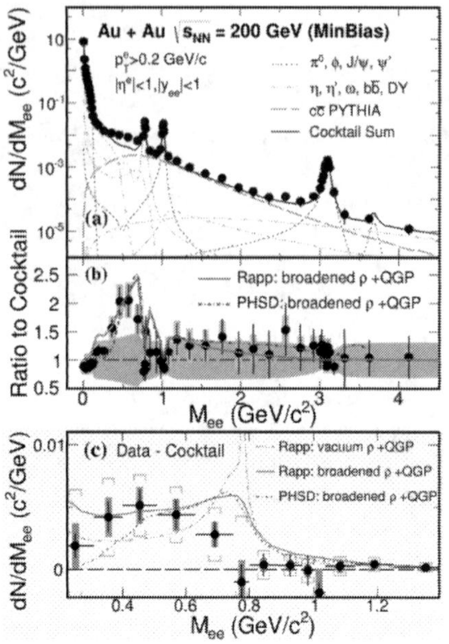

Figure 2. (a) Invariant mass spectra, (b) ratio of data to cocktail and (c) excess spectra in LMR from $\sqrt{s_{NN}}$ = 200 GeV Au+Au minimum bias collisions [18] . Two model calculations are also included. In panel (b), the grey box represents the systematic uncertainty from data while the light green band shows the systematic uncertainty from cocktail. In panel (c), green brackets depict the total systematic uncertainties including those from cocktail.

3. Results

Figure 1 right panel shows the dielectron invariant mass spectra from 200 GeV p+p collisions taken in year 2012. The cocktail is taken from the STAR published result [15], and the charm cross section is updated to $797 \pm 210(\text{stat.})^{+208}_{-295}(\text{sys.})\mu b$ with respect to the newest published result from STAR [17]. The cocktail simulation can reproduce the new preliminary result very well. With a full TOF coverage and more data taken, year 2012 result has greatly improved statistics which is ~7 times more than STAR published result [15]. The large statistics new results at p+p 200 GeV provide a better baseline for Au+Au collisions.

The $\sqrt{s_{NN}}$ = 200 GeV Au+Au results taken in year 2010 has been accepted by Phys.Rev.Lett [18]. In this paper, we will briefly review the key content of the submitted paper.

In LMR, an enhancement of 1.77±0.11(stat.)±0.24(sys.)±0.41(cocktail) is observed with respect to the cocktail without ρ, in the mass region 0.3~0.76 GeV/c^2 in minimum bias collision (Fig. 2 (b)). In addition, two model calculations [19, 20] are included to compare with our data (Fig. 2 (b), (c)): Model I by Rapp *et al.* is an effective many-body calculation [8, 19]; Model II by Linnyk *et al.* is a microscopic transport model, Parton-Hadron String Dynamics (PHSD) [11, 20]. Both models involve in-medium broadened ρ spectral function hypothesis and can successfully reproduce the NA60 results. The models, however, failed to reproduce the enhancement in central collisions reported by the PHENIX experiment [7, 20]. In the mass region below 1 GeV/c^2, both models describe our data reasonably well within uncertainties. Our measurements disfavor a pure vacuum ρ mass distribution for the excess dielectron ($\chi^2/NDF = 25/8$ in 0.3-1GeV/c^2, Fig. 2 (c)).

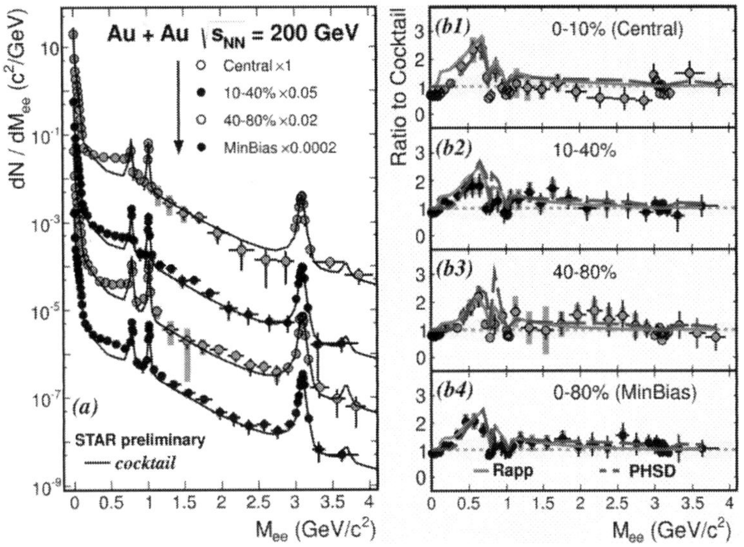

Figure 3. Left panel shows dielectron invariant mass spectra in different centralities. The solid curves represent the hadronic cocktail. The charm contribution is calculated by PYTHIA and scaled by N_{bin}. Right panel shows the ratio of data to cocktail in different centralities. The yellow band represents the systematic uncertainty of cocktail.

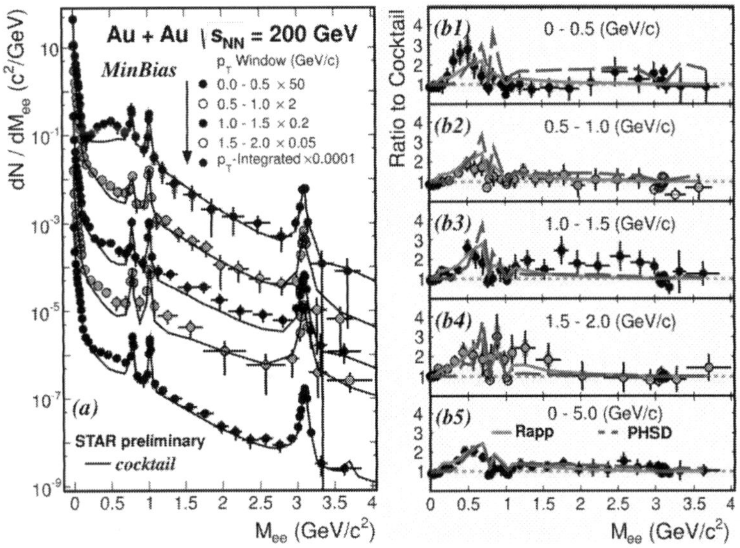

Figure 4. Left panel shows dielectron invariant spectra in different p_T^{ee} ranges. The solid curves represent the hadronic cocktail. Right panel shows the ratio of data to cocktail in different p_T^{ee} ranges. The yellow band represents the systematic uncertainty of cocktail.

Figures 3 and 4 show the dielectron spectra measured in various centrality bins and p_T

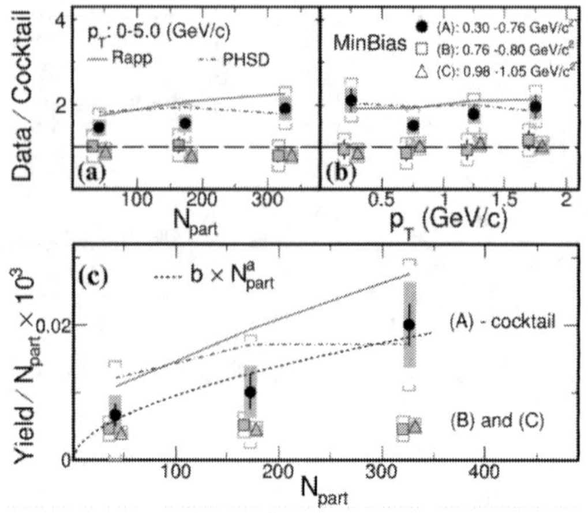

Figure 5. Panel (a) and (b) shows the integrated dielectron yields in mass regions of 0.3-0.76(ρ-like), 0.76-0.80(ω-like) and 0.98-1.05(ϕ-like) GeV/c^2 as a function of centrality and p_T^{ee}. Panel (c) shows the yields scaled by $N_{\rm part}$ for the ρ-like with cocktail subtraction, and the ω-like and ϕ-like without cocktail subtraction [18]. The dashed curve is a power-law fit to the yield/$N_{\rm part}$ for the ρ-like region subtracted by cocktail. Systematic uncertainties from data are shown as grey boxes, while the green brackets represent the total systematic uncertainties including the cocktail contribution. The ω-like and ϕ-like data points are slightly displaced horizontally for clarity.

ranges, respectively. The ratios between data and cocktail are shown in the right panels. Model calculations are also included as a comparison. Both models are able to describe the LMR excess in all p_T and centrality bins within uncertainty.

We also report the ratios of data to cocktail within STAR acceptance in three different mass regions: 0.3-0.76 (ρ-like), 0.76-0.8(ω-like) and 0.98-1.05(ϕ-like) GeV/c^2 as a function of centrality Fig. 5 (a) and dielectron p_T Fig. 5 (b). The hadronic cocktail can reproduce the dielectron yield in the ω-like and ϕ-like regions. In the ρ-like region, a significant excess is observed and the ratio of data to cocktail shows a weak dependence on $N_{\rm part}$ and dielectron p_T. Figure 5 (c) shows the yields in the ρ-like region subtracted by cocktail, and the ω-like and ϕ-like regions without cocktail subtraction. Dielectron yields in the ω-like and ϕ-like regions show a $N_{\rm part}$ scaling. The dashed curve is a power fit ($\propto N_{\rm part}^a$) to the excess yield/$N_{\rm part}$ in the ρ-like region, and the fit result shows $a = 0.54 \pm 0.18$ (stat.+uncorrelated sys.), indicating the dielectron excess yields in the ρ-like region are sensitive to the QCD medium dynamics, as expected by the theoretical calculations [19, 21].

In Fig. 6, we overlay the dielectron mass spectra from minimum bias and most central (0-10%) collisions. The spectra are scaled by the number of participant nucleons ($N_{\rm part}$). The ratio in the bottom panel starts from unity in the π^0 and η mass region and begins to increase in mass region 0.5-1 GeV/c^2 towards the $N_{\rm bin}$. This is due to the fact that correlated charm contribution starts to dominate in this mass region and the charm quark production at RHIC energy is expected to follow the $N_{\rm bin}$ scaling. The hadronic medium also has a significant contribution in this mass region and is expected to increase faster than $N_{\rm part}$ [19, 21]. In the IMR region, the ratio shows a moderate deviation from the $N_{\rm bin}$ scaling (1.8σ deviation for the data point at 1.8-2.8 GeV/c^2). The difference in mass region 1-3 GeV/c^2 indicates a potential de-correlating effect on charm pairs while traversing the QCD medium or other contribution from medium (e.g thermal radiation).

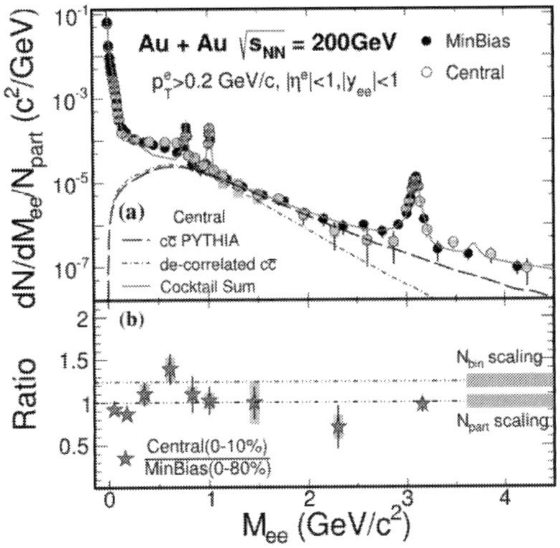

Figure 6. (a) Dielectron invariant mass spectra from minimum bias (0-80%) and central (0-10%) collisions [18]. The spectra are scaled by the number of participant nucleons (N_{part}). The solid line represents the hadronic cocktail for central collisions. The dashed line depicts the correlated charm from PYTHIA, while the dot-dashed line assumes a fully randomized azimuthal correlation between charm pairs and the p_T suppression factor on single electron spectrum from RHIC is also included [22]. (b) The ratio of N_{part} scaled dielectron yields between the central and minimum bias collisions. Systematic uncertainties are shown as the grey bands.

References

[1] R.J. Porter *et al.*, Phys. Rev. Lett. **79**, 1229 (1997).
[2] G. Adamova *et al.*, Phys. Rev. Lett. **98**, 052302 (2007); Phys. Rev. C **84**, 014902 (2011).
[3] A.L.S Angelis *et al.*, Eur. Phys. J. C **13**, 433 (2000).
[4] G. Agakichiev *et al.*, Phys. Rev. Lett. **91**, 042301 (2003); Eur. Phys. J. C **41**, 475 (2005).
[5] R. Arnaldi *et al.*, Phys. Rev. Lett. **96**, 162302 (2006); Phys. Rev. Lett. **100**, 022302 (2008);
[6] R. Arnaldi *et al.*, Eur. Phys. J. C **59**, 607 (2009).
[7] A. Adare *et al.*, Phys. Rev. C **81**, 034911 (2010).
[8] H. van Hees and R. Rapp, Phys. Rev. Lett. **97**, 102301 (2006); H. van Hees and R. Rapp, Nucl. Phys. A **806**, 339 (2008).
[9] J. Ruppert, C. Gale, T. Renk, P. Lichard and J. Kapusta, Phys. Rev. Lett. **100**, 162301 (2008); T. Renk and J. Ruppert, Phys. Rev. C **77**, 024907 (2008).
[10] K. Dusling, D. Teaney and I. Zahed, Phys. Rev. C **75**, 024908 (2007).
[11] O. Linnyk *et al.*, Phys. Rev. C **84**, 054917 (2011).
[12] Special Issue on RHIC and Its Detectors, edited by M. Harrison, T. Ludlam, and S. Ozaki, Nucl. Instr. Meth. A **499**, No. 2-3 (2003).
[13] M. Anderson, *et al.*, Nucl. Instr. Meth. A **499**, 659 (2003).
[14] W.J. Llope (for the STAR Collaboration), Nucl. Instr. Meth. A **661**, S110 (2012).
[15] L. Adamczyk *et al.*, Phys. Rev. C **86**, 024906 (2012).
[16] A. Adare *et al.*, Phys. Rev. C **81**, 034911 (2010).
[17] L. Adamczyk *et al.*, Phys. Rev. D **86**, 072013 (2012).
[18] L. Adamczyk *et al.*, arXiv:1312.7397.
[19] R. Rapp, PoS CPOD2013, 008 (2013); R. Rapp, private communications.
[20] O. Linnyk *et al.*, Phys. Rev. C **85**, 024910 (2012); O. Linnyk, private communications.
[21] U. Heinz and K.S. Lee, Phys. Lett. B **259**, 162 (1991).
[22] A. Adare *et al.*, Phys. Rev. Lett. **98**, 172301 (2007).

Journal of Physics: Conference Series **535** (2014) 012007 doi:10.1088/1742-6596/535/1/012007

Recent results on event-by-event fluctuations from the RHIC Beam Energy Scan program in the STAR experiment

Nihar Ranjan Sahoo (for the STAR Collaboration)

Texas A&M University, College Station, Texas 77843, USA

E-mail: nihar@rcf.rhic.bnl.gov

Abstract. Event-by-event fluctuations of global observables in relativistic heavy-ion collisions are studied as probes for the QCD phase transition and as tools to search for critical phenomena near the phase boundary. Dynamical fluctuations in mean transverse momentum, identified particle ratios and conserved quantities (such as net-charge, net-baryon) are expected to provide signatures of a de-confined state of matter. Non-monotonic behavior in the higher-moments of conserved quantities as a function of beam energy and collision centrality are proposed as signatures of the QCD critical point. To study the QCD phase transition and locate the critical point, the STAR experiment at RHIC has collected a large amount of data for Au+Au collisions from $\sqrt{s_{NN}} = 7.7 - 200$ GeV in the RHIC Beam Energy Scan (BES) program. We present the recent beam energy scan results on dynamical fluctuations of particle ratios and two-particle transverse momentum correlations at mid-rapidity. Higher-moments of the net-charge and net-proton multiplicity distributions as a function of beam energy will be presented. We give a summary of what has been learnt so far and future prospectives for the BES-II program.

1. Introduction

Importance of the Event-by-Event (E-by-E) fluctuations in heavy-ion collisions has been studied and discussed in a series of papers [1, 2, 3, 4]. The E-by-E fluctuation observables could provide information about the QCD phase transition and a possible signature of QCD critical point. RHIC, at Brookhaven National Laboratory, started the Beam Energy Scan (BES) program since 2010 in order to explore the QCD phase diagram. The STAR experiment, at RHIC, has collected data at $\sqrt{s_{\rm NN}} = 39$ GeV down to 7.7 GeV for this BES program.

2. RHIC Beam Energy Scan Program

The main goals of the BES program are to (1) locate the existence of the QCD critical point, (2) find evidence of the first order phase transition in the QCD Phase diagram, and (3) understand the properties of the QGP as a function of $\mu_{\rm B}$. The variation of RHIC colliding energy from $\sqrt{s_{\rm NN}} = 7.7$ to 39 GeV, along with 62.4 and 200 GeV, covers the baryonic chemical potential from 410 to 20 MeV [5]. The RHIC beam energy scan program started in the year 2010 and the first phase of the beam energy scan has been finished in year 2014 with $\sqrt{s_{\rm NN}} = 14.5$ GeV data taking. Details about the BES program are listed in Table 1.

The STAR experiment is able to measure various E-by-E observables to study the signatures of QGP-to-Hadron-Gas phase transition. Full 2π azimuthal coverage, enriched particle identification capabilities and large uniform acceptance of STAR provides a suitable environment for the E-by-E characterizations of heavy-ion-collisions. In order to achieve the goals of the BES program, the STAR collaboration have analyzed various E-by-E fluctuation measures. For the search of QCD critical point, higher-moments of the conserved charge, like net-charge, net-protons (proxy for net-baryon) and net-kaon (proxy for net-strangeness) distributions, particle ratio fluctuation, and transverse momentum fluctuations have been analyzed. Dynamical charge fluctuations for the signature of QGP have been studied as well. An elaborate discussion can be found in the subsequent sections.

$\sqrt{s_{NN}}$(GeV)	μ_B	Year	Events ($\times 10^6$)	Beam times (weeks)
39	115	2010	130	2.0
27	155	2011	70	1.0
19.6	205	2011	36	1.5
14.5	206	2014	20	3.0
11.5	315	2010	12	2.0
7.7	420	2010	4	4.0

Table 1. Beam Energy Scan program details. The μ_B values are taken from Ref. [5]

3. Search for QCD Critical Point

Lattice QCD calculations reveal that at vanishing μ_B, the transition from QGP to hadron gas is a simple crossover [6], whereas at large μ_B, the phase transition is of first order [6, 7, 8, 9, 10, 11, 12, 13]. Therefore, one expects the existence of a critical point at the end of the first order phase transition. Search for the QCD critical point has been one of the major thrusts of the RHIC BES program.

At the critical point, thermodynamic susceptibilities and the correlation length (ξ) of the system are expected to diverge for large samples in equilibrium. The finite volume effect puts a constraint on the divergence of the thermodynamic susceptibilities and ξ of the system. The phenomenon of critical slowing down in the vicinity of the critical point drives the system away from thermodynamic equilibrium, so ξ reaches a maximum value of around $1.5 - 3$ fm [14, 15]. Besides these challenges, in heavy-ion-collision experiments, the signature of critical point could be observed if the critical point is close enough to the freeze-out curve [16]. Various observables such as, higher-moments of the conserved charge, transverse momentum fluctuations, particle ratio fluctuations, have been proposed in order to search for the QCD critical point. Detailed discussions about these analysis and their observables can be found in the following sections.

3.1. Higher-moments of net-charge and net-proton distributions

Moments of the conserved charge distributions (such as the mean (M), standard deviation (σ), skewness (S) and kurtosis (κ)) have been proposed as important observables for the signature of the QCD critical point in the current wisdom. The higher-moments of conserved charge distributions are related to the respective higher order thermodynamical susceptibilities and also ξ of the system [15, 17]. In order to cancel the undetermined volume term in the higher-moments, ratio (or product) like σ^2/M, $S\sigma$, and $\kappa\sigma^2$ are used. The signature of non-monotonicity of these observables is expected if there is a nearby critical point in QCD phase transition. Recently, the STAR experiment reported net-charge [18] and net-proton [19] results from the BES program.

In an analysis of higher-moments, various sophisticated techniques, like the finite bin width effect [20], finite efficiency correction [21] for the higher-moments, have been incorporated. In order to get precise statistical uncertainty, different statistical error estimation methods (like the Delta Theorem, Bootstrap method, etc.) have been throughly studied and utilized. On the other hand, new centrality definitions (like uncorrected charge multiplicity within $0.5 < |\eta| < 1.0$ for net-charge analysis and that of other than identified protons and antiprotons within pseudo-rapidity $|\eta| < 1.0$ for net-proton) have also been studied and implemented.

The positive (N_+) and negative (N_-) charged particle multiplicities are counted within $|\eta| < 0.5$ and $0.2 < p_T < 2.0$ GeV/c (after removing protons and anti-protons with $p_T < 400$ MeV/c) to calculate net-charge ($N_+ - N_-$) in each event. The protons (N_p) and anti-protons ($N_{\bar{p}}$) are counted at mid-rapidity ($|y| < 0.5$) in the range $0.4 < p_T < 0.8$ GeV/c. In addition to these, various track selection cuts have been used. Details about these cuts can be found elsewhere in Ref [19, 18]. Figure 1 shows the net-charge multiplicity distribution for BES energies along with the Skellam distribution (the difference of two Poisson distributions assuming N_+ and N_- are independent Poisson distributions) [22]. The Skellam distribution qualitatively follows the net-charge distributions for all energies. Similarly, the net-proton distributions can be found in Ref. [19].

Reconstruction efficiency corrected values for the products of higher-moments of net-charge and net-proton number distributions have been shown in Fig. 2. In order to understand these results, different baseline measurements like Poisson and Negative Binomial Distribution (NBD) baseline for the net-charge higher-moments analysis are studied. Similarly, Poisson, Independent particle production expectation and UrQMD model simulation have been performed for the net-proton analysis. These baselines lack any QCD critical point phenomena, therefore it is expected that any deviation from these baselines could be due to the possible presence of new physics in the data. In general, both NBD and Poisson baseline calculations overestimate the net-charge data. For central collisions, within the statistical and systematic errors, the values $\kappa\sigma^2$ of the net-charge distribution at all energies are consistent with each other except for $\sqrt{s_{\rm NN}} = 7.7$ GeV. The significance of deviation of data with respect to both the baselines remains within 2σ in case of $S\sigma$ and $\kappa\sigma^2$. This implies that the products of moments of net-charge distributions do not show non-monotonic behavior as a function of beam energy within statistical errors. The large statistical uncertainty plays an important role for the net-charge higher-moments analysis and also puts a constraint for any definite conclusion.

On the other hand, the higher-moments of net-proton distribution results show the significance of deviations w.r.t Poisson baseline and Hadron Resonance Gas model, for 19.6 and 27 GeV, with 3.2 and 3.4 for $\kappa\sigma^2$, and 4.5 and 5.6 for $S\sigma$ at 0-5% centrality, respectively. In order to understand the effect of baryon number conservation effect on higher moments of net-proton distributions, UrQMD model simulation for 0-5% centrality have also been compared. It shows that $\kappa\sigma^2$ and $S\sigma$/(Skellam), for UrQMD model simulations, monotonic decrease with decreasing beam energy. Independent particle production approach (by considering independent production of protons and anti-protons in a given ensemble of events) have also been studied by constructing net-proton cumulants using the expression $C_n(N_p - N_{\bar{p}}) = C_n(N_p) + (-1)^n C_n(N_{\bar{p}})$, where $C_n(N_p)$ and $C_n(N_{\bar{p}})$ (n = 1, 2, 3 and 4) are the independently calculated cumulants of protons and anti-protons, respectively. The results based on independent particle production follow the data at all energies implying a weak correlation strength between protons and anti-protons in this analysis acceptance.

Large statistics and more energy points are needed to pin-point the exact position of the deviation from baseline as a function of beam energy. On the other hand, various final state effects like diffusion of conserved charge fluctuation, the effect of resonance decay within a given rapidity window, the effect of volume fluctuations, finite time and volume effects may play significant role in this observable. Large collection of events is needed, and systematic studies both on the experimental and theoretical sides need to be explored.

Besides being a signature of the critical point, the experimentally measured higher-moments of net-charge and net-proton distributions provide the information to extract freeze-out conditions

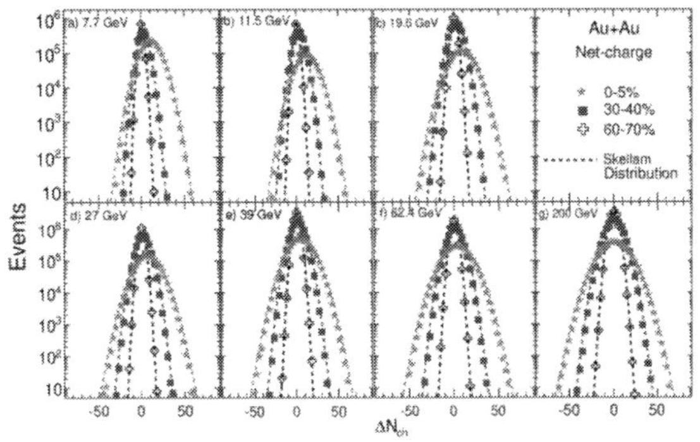

Figure 1. The net-charge distributions, for Au+Au collisions at 7.7 to 200 GeV within $|\eta| < 0.5$ and $0.2 < p_T < 2.0$ GeV/c, drawn with the Skellam distribution for three different centralities $(0-5\%, 30-40\%, 60-70\%)$.

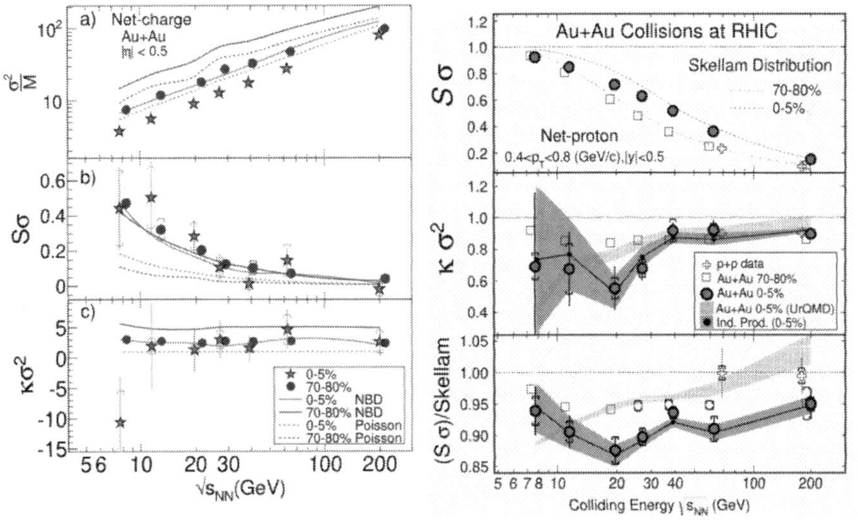

Figure 2. (left panel) Beam-energy dependence of (a) σ^2/M, (b) $S\sigma$, and (c) $\kappa\sigma^2$, after all corrections, for most central (0-5%) and peripheral (70-80%) bins [18]. The error bars are statistical and the caps represent systematic errors. Results from the Poisson and the NBD baselines are superimposed. The values of $\kappa\sigma^2$ for Poisson baseline are always unity. (Right panel) Collision energy and centrality dependence of the net proton $S\sigma$ and $\kappa\sigma^2$ from Au+Au and p+p collisions at RHIC [19]. Skellam distributions (dash lines in top panel), UrQMD (red shaded band) and indepedent particle production assumption expectation (shaded solid band) for corresponding collision centralities are shown.

Journal of Physics: Conference Series **535** (2014) 012007 doi:10.1088/1742-6596/535/1/012007

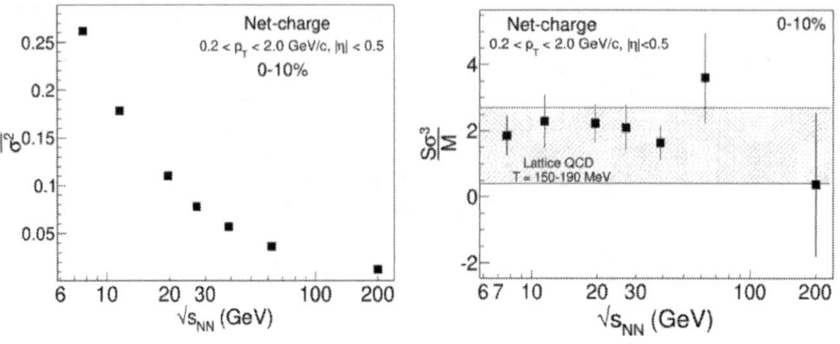

Figure 3. The beam energy dependance of efficiency corrected $\frac{M}{\sigma^2}$ (left) $\frac{S\sigma^3}{M}$ (right) are shown at $0-10\%$ centrality for Au+Au collisions [23]. The shaded region in right side plot is from Lattice QCD estimation [24].

in heavy-ion-collisions using lattice QCD calculation [24]. In Fig 3, $\frac{M}{\sigma^2}$ (*Baryometer*) decreases as a function collision energy, which implies with the increase in colliding energy baryon chemical potential decreases at freeze-out surface. The values of $\frac{S\sigma^3}{M}$ are close to 2.0 for $\sqrt{s_{\rm NN}} = 7.7$ to 62.4 GeV within statistical uncertainty. Due to the large statistical uncertainty, μ_{Bf} and T_f in heavy-ion-collisions can not be precisely constrained with this method.

3.2. Transverse momentum fluctuations

Transverse momentum (p_t) fluctuations could be envisioned as a source of temperature fluctuations of the bulk properties of the system [3]. Two-particle p_t correlation scaled by the ensemble average of $< p_t >$, $\sqrt{< \Delta p_{t,i}, \Delta p_{t,j} >}/ << p_t >>$, is also related to the specific heat, C_V, of the system. Hence, p_t fluctuations as a function of $\sqrt{s_{\rm NN}}$ may provide information about the order of QCD phase transition. The beam energy dependence of this observable has been studied for the BES program at mid-rapidity $|\eta| < 0.5$ [29].

In Fig. 4, the left panel shows the STAR measurement of this fluctuation observable significantly decreases below $\sqrt{s_{\rm NN}} = 19.6$ GeV for Au+Au collisions at $0-5\%$ centrality, whereas it remains flat for higher energies (upto ALICE energy). The effects from jet, resonance decay and finite acceptance effect may contribute to this observable. Further study is required to draw conclusions about the phase transition on it.

3.3. Particle ratio fluctuations

The non-monotonic behavior of the K/π yield ratio at $\sqrt{s_{\rm NN}}$~7.6 GeV, in SPS energy range for central Pb+Pb collisions, appears to be a unique characteristic of heavy-ion-collisions [25]. This observation is speculated to be a signature of the phase transition from hadronic to QGP state. To understand this behavior as a function of collision energy, dynamical fluctuations of K/π ratio and other particle's yield ratios (like K/p, p/π) have been proposed [26, 27]. STAR collaboration performed similar analyses in the BES program [29].

In Fig 4, the right panel shows the measure of dynamical fluctuation K/π yield ratio at BES energies at mid-rapidity. This results shows monotonic behavior as a function of $\sqrt{s_{\rm NN}}$ at $0-5\%$ centrality for Au+Au collisions. Similarly, no non-monotonic behavior is observed in case of K/p fluctuations. It is important to understand the sensitivity of these observables towards critical fluctuations. A further understanding of the various mechanisms of particle production,

Journal of Physics: Conference Series **535** (2014) 012007 doi:10.1088/1742-6596/535/1/012007

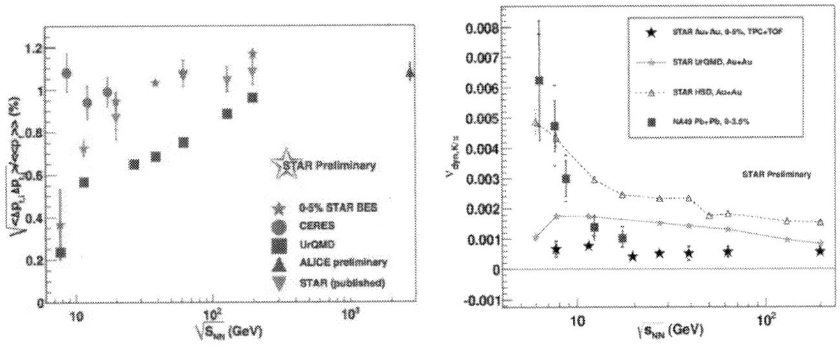

Figure 4. (Left) $\sqrt{<\Delta p_{t,i}, \Delta p_{t,j}>}/<<p_t>>$ as a function of $\sqrt{s_{\rm NN}}$ at $0-5\%$ centrality. (Right) Dynamical fluctuation of kaon to pion ratio, $\nu_{dyn,k/\pi}$, as a function of $\sqrt{s_{\rm NN}}$ at $0-5\%$ centrality [29].

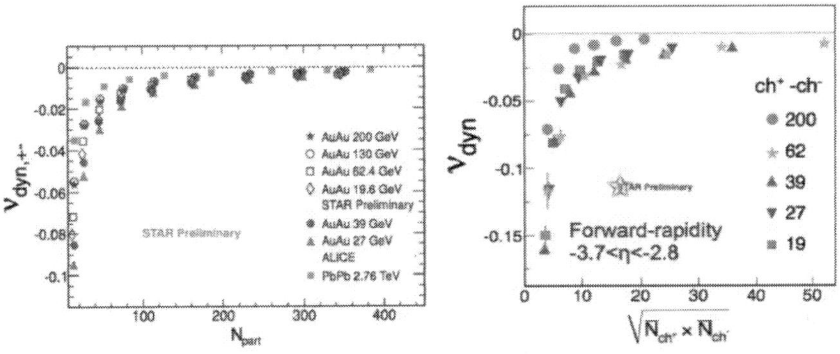

Figure 5. Dynamical Charge fluctuation as a function of centralities at mid-rapidity (left panel) and Forward rapidity (right panel) [29].

charge dependence of different particles species, resonance decays and re-scattering effects on these observables is necessary.

4. Dynamical charge fluctuations

The significant net-charge fluctuations may occur when the QGP phase (state comprises fractional charge carriers) transits to the Hadron Gas phase (state where the charges are integral) [28]. Such charge fluctuations can be detected by dynamical charge fluctuations if it could survive the process of evolution of the system in heavy-ion-collisions. In order to probe such a signature, ν_{dyn} for positively and negatively charged particles at mid-rapidity ($|\eta| < 0.5$) and forward-rapidity ($-3.7 < \eta < -2.8$) have been measured [29], as shown in Fig 5. It is observed that, both at mid and forward-rapidity, ν_{dyn} decreases from central to peripheral collisions. This observable increases with increasing collision energies. This observation may imply the correlation between positively and negatively charged particles increases with decreasing beam energy or multiplicity. A further study as a function of different rapidity-windows may help to understand the diffusion of charge fluctuation in heavy-ion-collisions [30]. Detailed study is ongoing to understand more about this observable.

5. Future perspective for the BES-II program and Upgrades

One of the main goals of the BES-I program is to search for the QCD critical point from event-by-event fluctuation. In order to draw a definite conclusion for the location of the QCD critical point, observables like $\kappa\sigma^2$ and $S\sigma$ of net-charge and net-proton distributions need a large collection of data. Besides, limited acceptance in pseudo-rapidity and the centrality determination in STAR experiment constrain further study of these event-by-event observables.

The results from BES-I program corroborate for high statistics and additional beam energies along with detector upgrades for final conclusion for the QCD critical point. Phase-II of BES program has been proposed for the years 2018-2019 from $\sqrt{s_{\rm NN}}$ = 7.7-20 GeV. For better particle momentum resolution, dE/dx resolution and improved pseudo-rapidity acceptance ($\eta| < 1.7$), the upgrade of inner sectors of the current STAR TPC (iTPC) is proposed. An Event Plane and centrality Detector (EPD) is proposed for the dedicated measurement of event-plane and centrality determination at forward region $2 < |\eta| < 4$. To achieve large statistics at low energy, a significant increase of the current luminosity is planned through the electron cooling.

6. Summary

The STAR experiment has successfully collected and analyzed large collection of data from $\sqrt{s_{\rm NN}}$ = 7.7 to 39 GeV for its first phase of the BES program. The $\kappa\sigma^2$ and $S\sigma$ values of net-proton distribution show significant deviation from Poisson expectation and Hadron Resonance Gas model prediction at $\sqrt{s_{\rm NN}}$ = 19.6 and 27 GeV, with large statistical uncertainty at low beam energies. On the other hand, those of net-charge distribution show no non-monotonic behavior within large statistical uncertainty. Besides, dynamical fluctuation of $K/p, K/\pi$ and p/π show monotonicity as a function of beam energy. The two-particle p_t correlation scaled with average $< p_t >$ fluctuation significantly decreases below $\sqrt{s_{\rm NN}}$ = 19.6 GeV for Au+Au collisions at $0-5\%$ centrality, whereas it remains flat for higher energies. Limited statistics and detector acceptance in BES-I program constrain the ability to draw final conclusion for the exact location and/or existence of the QCD critical point. The proposed BES-II program with STAR detector upgrades, like iTPC and EPD, is necessary for further understanding and drawing final conclusion on the QCD critical point and phase diagram.

Acknowledgments

This conference proceedings is supported by the US Department of Energy under the grant DE-FG02-07ER41485.

References

[1] Henning Heiselberg, Physics Reports **351**, 161 (2001).
[2] G. Baym, B. Blattel, L.L. Frankfurt, H. Heiselberg, M. Strikman, Phys. Rev. C 52 (1995) 1604.
[3] L. Stodolsky, Phys. Rev. Lett. **75** (1995).
[4] M. A. Stephanov, K. Rajagopal and E. V. Shuryak, Phys. Rev. Lett. **81** (1998) 4816; M. A. Stephanov, Int. J. Mod. Phys. **A**, 20, 4387 (2005) [arXiv:hep-ph/0402115].
[5] J. Cleymans, H. Oeschler, K. Redlich and S. Wheaton, Phys. Rev. **C 73** (2006) 034905.
[6] Y. Aoki et al., Nature **443**, 675 (2006).
[7] S. Ejiri, Phys. Rev. **D 78**, 074507 (2008).
[8] E. S. Bowman and J. I. Kapusta, Phys. Rev. **D 79**, 015202 (2009).
[9] M. A. Stephanov, Prog. Theor. Phys. Suppl. **153**, 139 (2004); Int. J. Mod. Phys. **A 20**, 4387 (2005).
[10] Z. Fodor et al., JHEP **0404**, 50 (2004).
[11] R. V. Gavai, S. Gupta, Phys. Rev. **D 78**, 114503 (2008).
[12] M. Cheng et al., arXiv:0811.1006 [hep-lat].
[13] B. Mohanty, J. Alam, S. Sarkar, T.K. Nayak and B.K. Nandi, Phys. Rev. **C 68**, 021901 (2003).
[14] B. Berdnikov and K. Rajagopal, Phys. Rev. **D 61**, 105017 (2000).
[15] C. Athanasiou, K. Rajagopal,and M. Stephanov, Phys. Rev. **D 82**, 074008 (2010).
[16] R. V. Gavai, S. Gupta, Phys. Lett. **B 696** 459, (2011).

[17] M. A. Stephanov, Phys. Rev. Lett. **102** 032301 (2009).
[18] L. Adamczyk *et al.* (STAR Collaboration), arXiv:1402.1558, [nucl-ex].
[19] L. Adamczyk *et al.* (STAR Collaboration), Phys. Rev. Lett. **112** 032302 (2014).
[20] X. Luo, J. Xu, B. Mohanty and N. Xu, Jour. Phys. G: Nucl. Part. Phys. **40**, 105104 (2013); N. R. Sahoo, S. De and T. K. Nayak, Phys. Rev. **C 87**, 044906 (2013).
[21] A. Bzdak and V. Koch, Phys. Rev. C 86, 044904 (2012).
[22] N. R. Sahoo (for the STAR Collaboration), Acta Phys. Polon. Supp. 6, 437 (2013); arXiv:1212.3892 [nucl-ex]; D. McDonald (for the STAR Collaboration), arXiv:1210.7023 [nucl-ex].
[23] N. R. Sahoo, Ph. D. thesis, Homi Bhabha National Institute, 2013.
[24] A. Bazavov et al., Phys. Rev. Lett. **109** 192302 (2012); S. Borsanyi et al., Phys. Rev. Lett. **111** 062005 (2013).
[25] S. V. Afanasiev et al. (The NA49 Collaboration), Phys. Rev. **C 66** 054902 (2002).
[26] B. I. Abelev *et al.* (STAR Collaboration), Phys. Rev. Lett. **103** 092301 (2009).
[27] C. Alt *et al.* (NA49 Collaboration), Phys. Rev. **C 79**, 044910 (2009).
[28] S. Jeon and V. Koch, Phys. Rev. Lett. **85**, 2076 (2000); Masayuki Asakawa, Ulrich Heinz, and Berndt Muller, Phys. Rev. Lett. **85**, 2072 (2000).
[29] P. Tribedy [STAR Collaboration], Nucl. Phys. A 904-905, 463c (2013); [arXiv:1211.0171 [nucl-ex]].
[30] B. Abelev et al. (ALICE Collaboration), Phys. Rev. Lett. **110**, 152301 (2013).

Journal of Physics: Conference Series **535** (2014) 012008 doi:10.1088/1742-6596/535/1/012008

Centrality Dependence of Low Momentum Direct Photon Production in $\sqrt{s_{NN}} = 200GeV$ Au+Au Collisions in PHENIX

R Petti

Department of Physics and Astronomy, Stony Brook University, Stony Brook, NY 11794-3800, USA

E-mail: richard.petti@gmail.com

Abstract. Interest in the production of direct photons at low momentum in heavy ion collisions has increased recently with the rather puzzling recent PHENIX results on the observation of a large excess in the yield of direct photons in Au+Au collisions compared to the N_{coll} scaled p+p yield and the observation of a large direct photon v_2 at p_T below 3GeV/c. The two measurements are difficult to reconcile within the current understanding of the fireball evolution and the assumption that low momentum direct photons are dominantly emitted early in the collision. This has sparked much theoretical interest, with new calculations refining old works as well as calculations of newly proposed production mechanisms. Experimental constraints on newly proposed production mechanisms are crucial in understanding the evolution of the fireball. One would also desire experimental evidence whether the direct photons below p_T of 3GeV/c are dominantly emitted early or late in the collision. In this proceedings, we show new PHENIX measurements of the detailed centrality dependence of direct photon production in the range of $0.4 < p_T < 5GeV/c$ in Au+Au collisions at $\sqrt{s_{NN}} = 200GeV$ based on the high statistics of the combined 2007 and 2010 datasets. Photons are cleanly identified through their external conversion in the backplane of the Hadron Blind Detector to dielectron pairs. We observe the shape of the excess direct photon invariant yield in the range of $0.6 < p_T < 2GeV/c$ is well described by an exponential fit with an inverse slope of $240MeV/c^2$, consistent with previous measurements. Additionally, the shape is seen to be the same for all centralities within uncertainties and the integrated yield follows a power law as a function of N_{part}. The new results will offer powerful constraints to current calculations which must predict the same centrality dependence.

1. Introduction

Direct photons are an essential probe of heavy ion collisions. Because of their negligible interaction cross-section with the partonic medium, photons can escape the medium virutally unmodified, carrying with them information about the conditions of the system at the time of the photon's production. This allows experimental access to the entire space-time evolution of the system. The difficulty lies in the inability to experimentally separate various contributions to the direct photon yield since we measure a time-integrated snapshot of photons from all sources. Direct photons are defined as any photon not originating from a hadron decay. Theoretical calculations must account for the measured spectra. These theoretical calculations need to include the various production rates for each mechanism of direct photon production integrated over the space-time evolution of the system.

We focus on low momentum ($p_T < 5GeV/c$) direct photons in this proceedings. These soft photons should be sensitive to the temperature [1] and the underlying dynamics of the collision through the yield and the observed azimuthal anisotropy of emission relative to the reaction plane quantified by v_2 [2][3](v_2 is the second order coefficienct in the Fourier decompostion of the azimuthal angle distribution [4]). These soft photons also allow us to experimentally probe possible effects from large background magnetic fields [5][6] or very early time production (possibly in the Glasma [7]). For these reasons, PHENIX has a suite of measurements dedicated to soft photon production in Au+Au collsions [8][9][10][11]. We report on the latest measurement in this proceedings. The new measurement takes advantage of the excellent electron identification capabilities of the PHENIX Ring Imaging CHerenkov (RICH) detector to measure real photons via their external conversion in material to di-electron pairs. This method viturally eliminates the miss-identified hadron background that plagues previous analyses which measure the photon energy deposition directly in the electromagnetic calorimeters [4]. This allows us to push the measurement to lower momentum than was previously possible. To further reduce systematic uncertainties, we measure the direct photon fraction $R_\gamma = \gamma^{incl}/\gamma^{hadron}$, or the ratio of the yield of inclusive photons to photons from hadron decays, as a double ratio, which explicits cancels major sources of systematic uncertainty.

2. Experimental Setup

The direct photon measurement discussed here is measured using the PHENIX detector, which is located at the Relativistic Heavy Ion Collider (RHIC) at Brookhaven National Lab (BNL) in Upton, NY. The PHENIX central arms make up a two arm spectrometer, with each arm covering 90^0 and separated by 90^0 in azimuth and covering $-0.35 < \eta < 0.35$ in psuedo-rapidity. The details of the detector and the various subsystems are explained elsewhere [12]. We only briefly mention the main detectors used in this particular measurement.

The Beam-Beam Counter (BBC) is the main detector system used to measure the global properties of the collision, such as centrality and z-vertex, and resides in the north and south muon arms located in forward and backward directions along the beamline. The BBC detector consists of arrays of quartz radiators attacted to photo-multiplier tubes arranged around the beampipe and sit about 1.5 meters down the beamline in both the north and south direction ($3.1 < |\eta| < 3.9$).

The main photon measurment is made using the PHENIX tracking system, the RICH detector, and the electromangetic calorimeters. The Hadron Blind Detector (HBD) [13] was also installed in 2007, but is used only as a source of photon conversions in this analysis. The bulk of the photons that convert in the HBD occur in the backplane, where readout electronics are located ($X/X_0 \approx 3\%$), which resides at a radius of about 60cm from the interaction point.

The PHENIX tracking system in each arm consists of drift chambers (DCH) and various layers of pad chambers (PC), which allow us to track particles through the magnetic field, give a measurement of momentum, and provide pointing to the other detector systems. The RICH detector is the main electron ID detector. Only electrons Cherenkov radiate in the Ar/CO_2 gas mixture inside the detector up to about 5GeV/c, where pions also start to radiate. We reconstruct the rings from the Cherenkov radiation of the electrons that are reflected onto arrays of photo-multiplier tubes residing outside of the experimental aperture. The electromagnetic calorimeters (EMCAL) are the outter most detector in the central arms of PHENIX. Here we measure the energy and direction of photons, as well as electrons which assists in the electron identification through the energy to momentum ratio.

3. Photon Identification

A novel approach in PHENIX is used to identify photons in this analysis. Hadron mis-identifcation is a major source of systematic uncertainty in analyses that measure photons in the EMCAL directly. We avoid this issue in the current analysis by measureing photons that externally convert in the backplane of the HBD detector (radius $\approx 60cm$). The novel approach utilizes the feature that far off-vertex particles will be reconstructed with a slightly wrong momentum, due to the fact that PHENIX does all of its tracking outside of the magnetic field, and so assumes that all particles originate from the event vertex in the track reconstruction. The effect is illustrated in Fig. 1. The figure shows a photon (white line) being emitted from the event vertex of the collision. The photon is shown to convert in the HBD and the e^+ and e^- (indicated by the red lines) leave hits in the detector systems described in the previous section. If we take the hits left by the tracks and incorrectly trace them through the field to the event vertex (shown by the yellow lines), we see in the exaggerated cartoon that the pair obtains an artificial opening angle. This translates to an apparent mass for the pair. The magnitude of the mass is proportional to the radius of conversion.

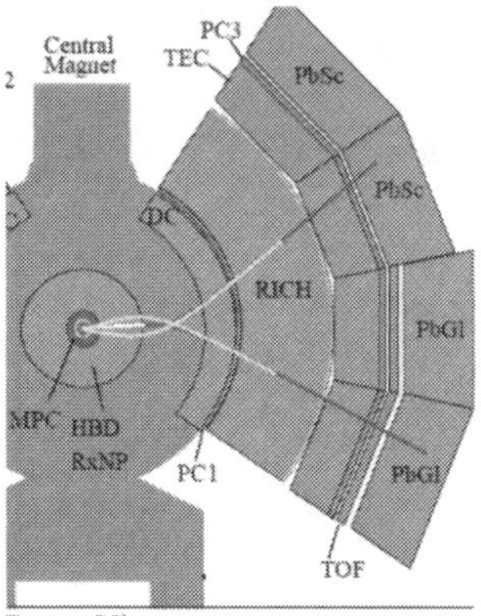

Figure 1. A cartoon illustrating the effect of the false assumption of the event vertex origin on a di-electron pair from a photon conversion in the backplane of the HBD, resulting in an artificial opening angle. The true track trajectory is shown in red, the (incorrect) traceback to the event vertex is shown by the yellow lines.

Detailed GEANT Monte Carlo simulations have been done to study the miss-reconstruction effect on conversion pairs. Conversions that occur at a radius of 60cm obtain a pair mass of about $12MeV/c$ in the field setting used in 2007 and 2010. We use the simulations to correct for the miss-reconstruction effect. To achieve this, we reparameterize the momenta of each track under an alternate track model assumption, that each track originated from 60cm rather than from the event vertex. This gives us two masses for each pair. A view of the 2D mass distribution in the data is shown in Fig. 2. The vertical axis shows the pair invariant mass calculated under the alternate track model of the HBD backplane origin (M_{hbd}). The horizontal axis shows the pair invariant mass calculated under the standard reconstruction assuming the event vertex origin (M_{vtx}). A correlation can be seen between the two for HBD backplane conversions. The large peak sitting at $M_{hbd} \approx 2MeV/c^2$ and $M_{vtx} \approx 12MeV/c^2$ is due to the backplane conversions. The smaller peak observed to the left is due to Dalitz pairs from π^0s. These peaks sit on a low background due to combinatorial background. The contamination of our photon sample is estimated to be 1%. This method allows for a very clean separation of di-electron pairs from photon conversions in the backplane of the HBD from other sources of di-electron pairs, and additionally eliminates miss-identified hadron contamination. This powerful photon identification technique is essential to the measurement.

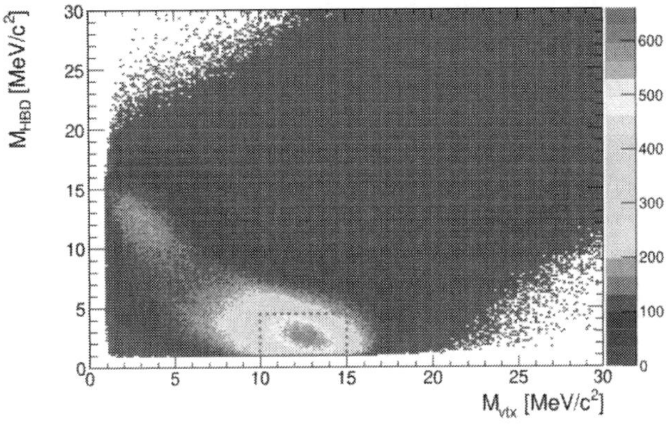

Figure 2. The distribution of di-electrons pairs in the M_{vtx},M_{hbd} space. This is also the space which defines the cut that is used to isolate the photons (shown by the red box).

4. Analysis

We start the analysis by quantifying the direct photon signal present through the direct photon fraction. The signal is quantified by the ratio $R_\gamma = \gamma^{incl}/\gamma^{hadron}$, or the ratio of the inclusive photon yield to the yield of photons coming from hadron decays. To reduce systematic uncertainties, we express R_γ as a double ratio, shown in Eqn. 1. The numerator of the double ratio consists of a ratio that is measured from the data, the ratio of the observed inclusive photon yield (N_γ^{incl}) to the observed yield of inclusive photons that are tagged as coming from a π^0 decay ($N_\gamma^{\pi^0}$). A subset of the inclusive photons are tagged as coming from a π^0 decay by pairing the converted photons with photons measured directly in the EMCAL to reconstruct the pions. Everything is performed and measured as a function of the converted photon p_T. This is important because when we form the ratio of the inclusive to pion tagged photon yield, the e^+e^- pair reconstruction efficiency and acceptance explicitly cancel, since these corrections are common factors to both the inclusive and pion tagged yields. Additionally, the probability of photon conversion also drops out. This significantly reduces the overall systematic uncertainty on the measurement and allows us to push down to very low p_T. The remaining corrections shown in Eqn. 1 concern the acceptance (f) and reconstruction efficiency (ε_γ) of getting the second leg of the pion decay into the EMCAL. The <> around the correction indicates that we average over all EMCAL photon p_T, since we estimate the corrections as a function of the converted photon p_T. The largest source of systematic uncertainty (4%) concerns the $\langle\varepsilon_\gamma f\rangle$ correction and the influence of the minimum p_T cut on the EMCAL photon of 600MeV (400MeV) for the 2007 (2010) analysis and the energy scale match in the simulation and the data. Details of the cuts and systematic uncertainties can be found in [11].

$$R_\gamma = \frac{\gamma^{incl}}{\gamma^{hadron}} = \frac{\langle\varepsilon_\gamma f\rangle \left(\frac{N_\gamma^{incl}}{N_\gamma^{\pi^0}}\right)_{data}}{\left(\frac{\gamma^{hadron}}{\gamma^{\pi^0}}\right)_{sim}} \qquad (1)$$

The π^0 reconstruction is shown for two example converted photon p_T bins in Fig. 3. As can be seen, the combinatorial background is successfully described and subtracted using a mixed event technique in which converted photons from one event are paired with EMCAL photons from completely different events within the same event class (characterized by z-vertex and centrality).

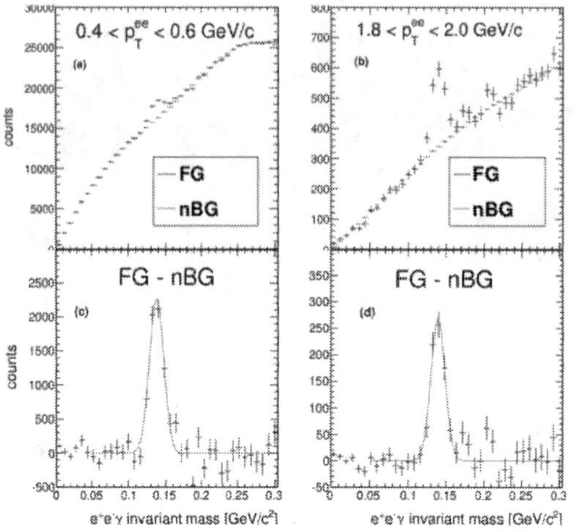

Figure 3. The π^0 reconstruction illustrated in two p_T bins. Panels (a) and (b) show the converted photon-emcal photon pairs before combinatorial background subtraction, with the raw data (FG) in blue and the combinatorial background (nBG) in red. Panels (c) and (d) show the extracted π^0 peak after background subtraction. The peak is fit with a Gaussian to estimate the peak position and width.

The $\langle \varepsilon_\gamma f \rangle$ correction is calculated with two different implemenations. The 2007 analysis calculates the correction with a full GEANT Monte Carlo simulation of π^0 decays where one photon is forced to convert at 60cm. Occupancy effects on the reconstruction are taken into account by embedding simulated EMCAL hits into real data and reperforming the reconstruction. The 2010 analysis uses a "Fast Monte Carlo" approach, where effects from a full GEANT simulation are parameterized for faster event generation. The fast Monte Carlo was cross checked with the full Monte Carlo, the two were found to be consistent within uncertainties. The combined 2007+2010 result uses the fast Monte Carlo simulation for the correction on each dataset.

As described, we only measure the pion decay photon contribution to the inclusive yield directly. We estimate the remaining (sub-dominant) contribution from other hadrons in a hadron decay photon cocktail which includes π^0, η, η', and ω decays. We follow the same procedure as in [14]. The spectral shape of the pions is parameterized by fitting existing π^0 data with a modified Hagedorn function. The shape of the remaining hadron species is then determined by m_T scaling, where the p_T in the Hagedorn function is replaced with m_T. The relative yields of each meson species is benchmarked to the pions and is taken from existing measurements. From the cocktail, we calculate the ratio of the yield of all decay photons to those just from π^0 decays and insert into the denominator of Eqn. 1 to calculate R_γ.

As stated we combine the 2007 and 2010 results for a single R_γ measurement (after confirming that the individual results are consistent). From the high statistics R_γ measurement, we calculate the invariant yield of direct photons by scaling the hadron decay photon yield calculated in the same cocktail as described above appropriately by R_γ as shown in Eqn. 2. The results and a further analysis is shown in the next section.

$$\gamma^{direct} = (R_\gamma - 1)\gamma^{hadron} \tag{2}$$

5. Results

The combined 2007+2010 R_γ measurement is shown in Fig. 4 for four different centrality bins. The previous measurement from the virtual photon anaysis [9] is shown for comparison. The two results are consistent. As can be seen, R_γ is greater than one for each centrality, indicating a signal of direct photons in each centrality bin. Fig. 5 shows the invariant yield determined from the measured R_γ in the same centrality bins, shown by the points. Also included in the plots in each bin is the paramterized contribution of prompt processes, shown by the line. This parameterization is done by fitting a modified power law $\left(a\left(1+\frac{p_T^2}{b}\right)^c\right)$ to the direct photon yield measured in p+p collisions (which presumably only has prompt contributions) and then scaling by N_{part} for each corresponding centrality bin. An excess of direct photons above the scaled p+p yield is observed in all centralities.

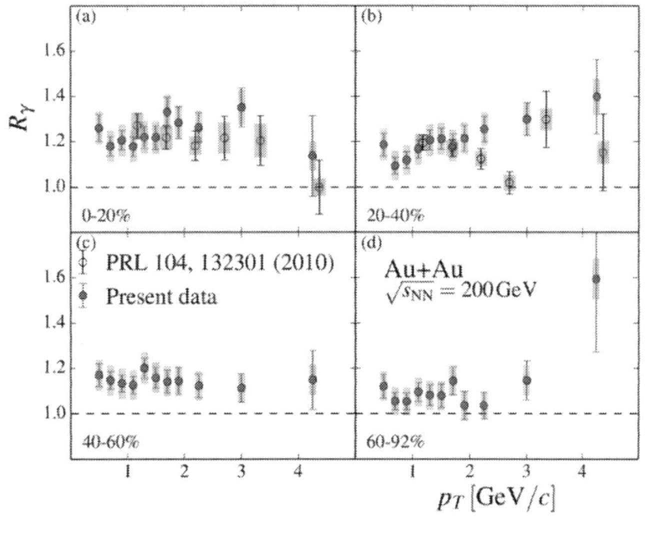

Figure 4. The measured R_γ shown for four centrality bins as indicated in the plots. This is the combined 2007+2010 analysis results.

We are not concerned with prompt contributions in this analysis; we want to focus on thermal radiation. Thus we subtract the scaled p+p yield shown by the line in Fig. 5 and display the observed excess yield in Fig. 6. We note that the entire spectrum cannot be described by a single exponential fit, although a large portion of it can (from 0.6-2.0 GeV/c). We perform the fit in this region and display the results in each panel of the figure. Note that within uncertainties, the extracted inverse slope from the fits are independent of centrality. We do a further analysis on the excess yield by looking at integrated yield as a function of N_{part}, as shown in Fig. 7. Since the first p_T point dominates a yield in these steeply falling spectra, we do the integration with different low p_T cutoffs, shown by the different colors in the figure. We observe that the shape of the excess yield as a function of N_{part} is the same at all p_T and can be described by a power law $(A \cdot N^\alpha)$ with a power $\alpha = 1.48 \pm 0.08(stat) \pm 0.04(syst)$.

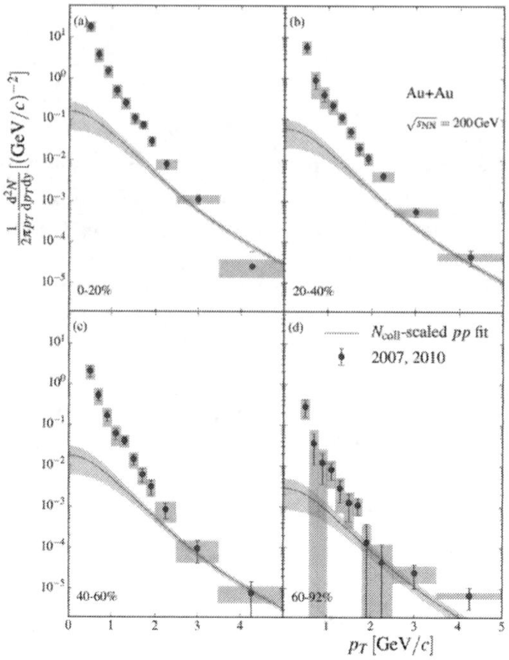

Figure 5. The direct photon invariant yield.

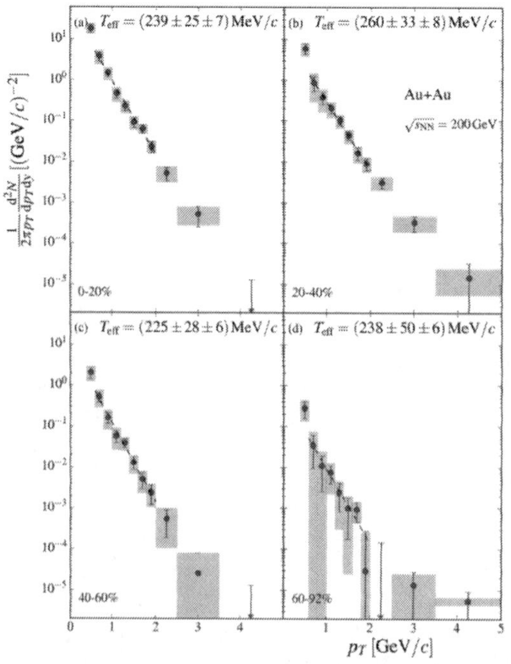

Figure 6. The isolated excess direct photon yield after scaled p+p contributions are subtracted.

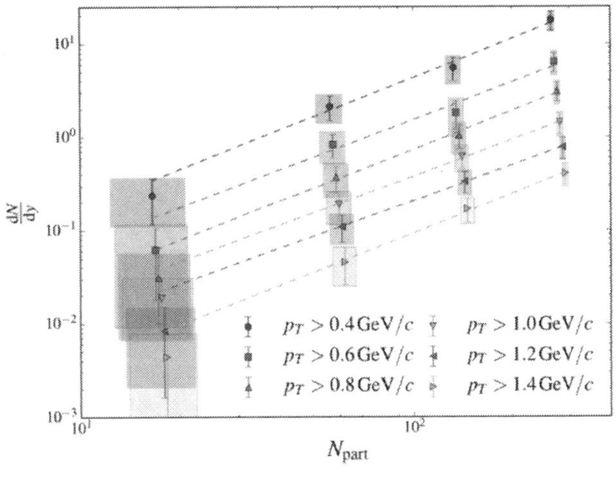

Figure 7. The integrated excess direct photon yield as a function of N_{part}. The different colors represent integrating with different low p_T thresholds.

6. Conclusions

We have discussed in this proceedings a new measurement from PHENIX of soft direct photon production in $\sqrt{s_{NN}} = 200GeV$ Au+Au collisions at RHIC. We employ a novel photon identification procedure by measuring photons via their external conversion to di-electron pairs, taking full advantage of the excellent electron id capabilities of the PHENIX RICH detector. This virtually eliminates contamination from miss-identified hadrons. We also measure the direct photon fraction R_γ using a double ratio, which further reduces systematic uncertainties, allowing us to push the measurement all the way down to 400MeV/c in p_T, a first in high energy heavy ion collisions. This allows us to explore the momentum region that is expected to be dominated by thermal radiation. We measure the direct photon invariant yield and isolate the excess production above the scaled p+p yield. We observe that the shape of the excess is independent of centrality and has a characteristic power law dependence as a function of N_{part}, with a power of $\alpha = 1.48 \pm 0.08(stat) \pm 0.04(syst)$. This measurement offers powerfull contraints to theory calculations that claim to describe the evolution of the fireball produced in heavy ion collisions coupled with the direct photon production.

References

[1] Shen C, Heinz U W, Paquet J F and Gale C 2013 (*Preprint* 1308.2440)
[2] Chatterjee, Srivastava, *Phys. Rev.* **C79**, 021901 (2009)
[3] Linnyk O, Konchakovski V, Cassing W and Bratkovskaya E 2013 (*Preprint* 1304.7030)
[4] Adare A *et al.* (PHENIX Collaboration) 2012 *Phys.Rev.Lett.* **109** 122302 (*Preprint* 1105.4126)
[5] Tuchin K 2013 *Phys.Rev.* **C87** 024912 (*Preprint* 1206.0485)
[6] Basar, Gokce, Kharzeev and Skokov 2012 *Phys.Rev.Lett.* **109** 202303 (*Preprint* 1206.1334)
[7] Chiu M, Hemmick T K, Khachatryan V, Leonidov A, Liao J *et al.* 2013 *Nucl.Phys.* **A900** 16–37 (*Preprint* 1202.3679)
[8] Adler S *et al.* (PHENIX Collaboration) 2005 *Phys.Rev.Lett.* **94** 232301 (*Preprint* nucl-ex/0503003)
[9] Adare A *et al.* (PHENIX Collaboration) 2010 *Phys.Rev.Lett.* **104** 132301 (*Preprint* 0804.4168)
[10] Afanasiev S *et al.* (PHENIX Collaboration) 2012 *Phys.Rev.Lett.* **109** 152302 (*Preprint* 1205.5759)
[11] Adare A *et al.* (PHENIX Collaboration) 2014 (*Preprint* 1405.3940)
[12] Adcox K *et al.* (PHENIX Collaboration) 2003 *Nucl. Instrum. Methods* **A499** 469
[13] Anderson W *et al.* (PHENIX Collaboration) 2011 *Nucl. Instrum. Methods* **A646** 35
[14] Adare A *et al.* (PHENIX Collaboration) 2010 *Phys.Rev.* **C81** 034911 (*Preprint* 0912.0244)

Journal of Physics: Conference Series **535** (2014) 012009 doi:10.1088/1742-6596/535/1/012009

Beam Energy Dependence of Clan Multiplicity at RHIC

Aihong Tang (for the STAR Collaboration)
Physics Department, Brookhaven National Laboratory, PO BOX 5000, Upton, NY 11973-5000
E-mail: aihong@bnl.gov

Abstract. In this paper, STAR's measurement of clan multiplicity is presented for AuAu collisions at $\sqrt{s_{NN}}$= 7.7, 11.5, 19.6, 27, 39, 62.4 and 200 GeV, for a variety of centrality classes. The mean number of particles per clan is found to decrease with decreasing centrality. Within the same centrality class, the mean number of particles per clan exhibits a reduction between 19.6 GeV and 62 GeV, with the minimum around 27 GeV. The structure is visible for most centralities, and most prominent for central collisions.

1. Introduction

Particle distributions in high energy collisions are frequently found to be of Negative Binomial Distribution (NBD) form for both total multiplicities and multiplicities in finite pseudorapidity intervals. One satisfactory explanation of this observation is that particle are produced via cascade process, and such interpretation leads to the development of clan concept of clustering in particle production [1].

Let's recall the negative binomial probability density function,

$$f(n) = \binom{k+n-1}{n}(1-p)^n p^k, \tag{1}$$

where $p \in [0,1]$, and the real number $k > 0$. Its mean and variance are given by $\mu = \frac{(1-p)k}{p}$ and $\sigma^2 = \frac{(1-p)k}{p^2}$, respectively. In order to interpret the wide occurrence of NBD, a parameterization alternative to the standard NBD parameterization has been proposed [1]. It consists of the average number of groups of particles of common ancestor, also called the average number of clans,

$$\bar{N} = k\ln(1+\frac{\mu}{k}), \tag{2}$$

and the average number of particles per clan,

$$\bar{n}_c = \frac{\mu}{\bar{N}}. \tag{3}$$

With this set of parameterization, particle production can be regarded as an independent production of clans followed by the decay of clans into particles according to a logarithmic multiplicity distribution. This can be seen by investigating the generating functions of relevant distributions. The generating function for the logarithmic distribution is,

$$G_{\ln}(x) = \frac{\ln(1 - zx)}{\ln(1 - z)}, \tag{4}$$

where z is set to be $z = \frac{\mu}{\mu+k}$. Because clans are independently produced, they can be described by Poisson distribution, for which the generating function is

$$G_{\text{Poisson}}(x) = e^{\bar{N}(x-1)}. \tag{5}$$

The final distribution is a compound distribution of logarithm and Poisson, for which the generating function is given by

$$\begin{aligned} G_{\text{compound}} &= G_{\text{Poisson}}(G_{\ln}(x)) \\ &= e^{\bar{N}(G_{\ln(x)}-1)} \\ &= \left(\frac{1-z}{1-zx}\right)^k, \end{aligned} \tag{6}$$

by which we have recovered the generating function of NBD.

Note that with the introduction of the clan concept, the NBD, if arranged properly, yields a similarity to the grand-canonical partition function. Such analogy led to the development of clan thermodynamics [2, 3]. Here, adopting the concept of clan does not necessarily imply the endorsement of clan thermodynamics. The clan parameters have been used to identify abnormalities due to phase transition [4, 5, 6, 7, 8, 9]. For a similar purpose, this study is to examine the averaged charged multiplicity per clan as a function of collision energy.

2. Data Sets and Cuts

The data set consist of minimum bias events of Au+Au collisions at $\sqrt{s_{\text{NN}}}$= 7.7, 11.5, 19.6, 27, 39, 62.4 and 200 GeV, taken by the STAR experiment. Total events used in this analysis are, from the lowest energy to the highest one, 5, 7, 40, 71, 56, 73, and 25 million, respectively. Events from energies lower than 62.4 GeV were taken as part of the Beam Energy Scan program [10] at RHIC, and events from 62.4 and 200 GeV were taken in year 2010. All events are required to have at least one Time of Flight [11] hit matching to a track, a vertex within 30 centimeters to the center of STAR's Time Projection Chamber [12] in z direction, and within 2 centimeters to the beam in transverse direction. To avoid complications arising from having more than one collisions in the time window of detector readout, which usually happens when the luminosity is high, events from 62.4 GeV and 200 GeV are selected from those taken with low luminosity beam (Beam Beam Counter coincidence rate less than 25k). Tracks are required to have a minimum of 15 hits, and the ratio of number of hits to number of maximum possible hits to be greater than 0.52. The distance of the closest approach (DCA) of a track to the primary vertex is required to be less than 1 centimeter. To avoid auto correlation, particles used in the study of multiplicity distribution and those used in the centrality definition are taken from different pseudorapidity ranges, which is $|\eta| < 0.5$ for the former and $|\eta| > 0.5$ for the latter. In this paper the multiplicity in the latter is also called "reference-multiplicity-2" to distinguish itself from the usual reference multiplicity defined with particles in $|\eta| < 0.5$. In the picture of particles being produced by clans, one is interested in produced particles only. Thus protons are removed in this study, because they are contaminated by transported ones which can introduce trivial energy dependence, in particular at low energies where transported protons become dominant. The removal of protons is achieved by requiring tracks' energy loss per unit length (dE/dx) to be larger than a threshold set to 2 standard deviations below the theoretically expected value for protons at the same rigidity.

3. Analysis Procedure and Results

For events with a given reference-multiplicity-2, the corresponding multiplicity distribution in $|\eta| < 0.5$ can be well described by NBD. This can be seen in Figure 1, in which the multiplicity distribution in $|\eta| < 0.5$ is plotted for a selected set of reference-multiplicity-2 numbers. The curves are fits to the NBD and they describe the data very well.

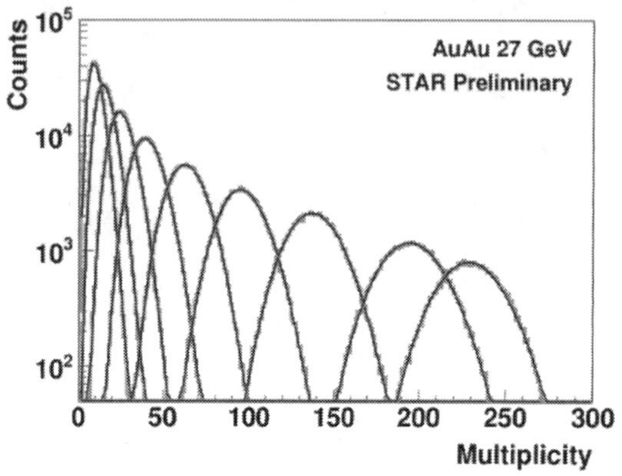

Figure 1. From left to right, the multiplicity distribution in $|\eta| < 0.5$ for reference-multiplicity-2 = 11, 18, 30, 48, 76, 116, 169, 242 and 289, respectively. Curves are fits to NBD. All histograms are for Au+Au collisions at $\sqrt{s_{NN}}$= 27 GeV.

The fit is applied to multiplicity distributions with the finest bin of reference-multiplicity-2, and the corresponding σ^2 and μ can be obtained. With both σ^2 and μ known, $\bar{N}$ and $\bar{n}_c$ can be calculated with the transformation mentioned above. In the top panel of Figure 2, σ^2, μ and $\bar{N}$ are plotted as a function of reference-multiplicity-2, and in the bottom panel, the average multiplicity for charged particles per clan ($\bar{n}_c$) is shown. Note that $\bar{N}$ and $\bar{n}_c$ can be calculated with either directly calculated σ^2 and μ, shown by black symbols, or that from fitting the NBD, shown by non-black symbols. The two sets of results agree with each other well and in the follow-up analysis only the set based on fitting NBD is used, because fitting is less sensitive to contaminations from rare events with abnormally large or small multiplicity. In general σ^2 is found to be greater than μ which is a feature of NBD. With decreasing centrality (increasing reference-multiplicity-2), $\bar{N}$ keep increasing while $\bar{n}_c$ shows a moderate decrease towards unit, which means that particle production is approaching the limit of independent production (Poisson) in central collisions.

The obtained $\bar{N}$ and $\bar{n}_c$ are then averaged over reference-multiplicity-2 for each centrality bin, of which the boundaries are shown by mesh area in the bottom panel of Figure 2. Note by performing the analysis with the finest reference-multiplicity-2 bin, one avoids the additional fluctuation from a wide multiplicity bin which can significantly bias the variance.

The same procedure is applied on data from all energies and the final $\bar{N}$ and $\bar{n}_c$ as a function of collision energy is presented in Figure 3 and Figure 4, respectively. Not surprisingly the average total number of clans per event is found to increase either with decreasing centrality, or with increasing energy, indicating that more clans are produced when the system becomes larger and/or when there is more energy available for particle production. The average charged particles per clan shows little dependence on energy for peripheral collisions, which is expected if peripheral collisions are regarded as simple superpositions of independent binary collisions, thus they should show the same clustering property. From peripheral to central collisions, $\bar{n}_c$

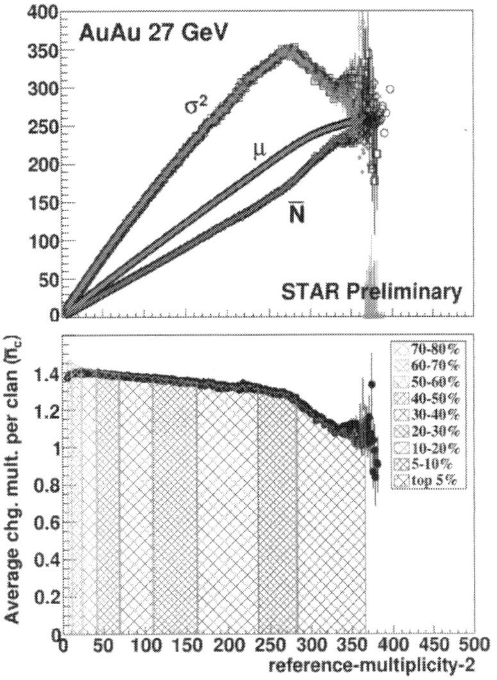

Figure 2. σ^2, μ, $\bar{N}$ (top panel) and $\bar{n}_c$ (bottom panel) as a function of reference-multiplicity-2 for Au+Au collisions at $\sqrt{s_{\rm NN}}$= 27 GeV. See text for details.

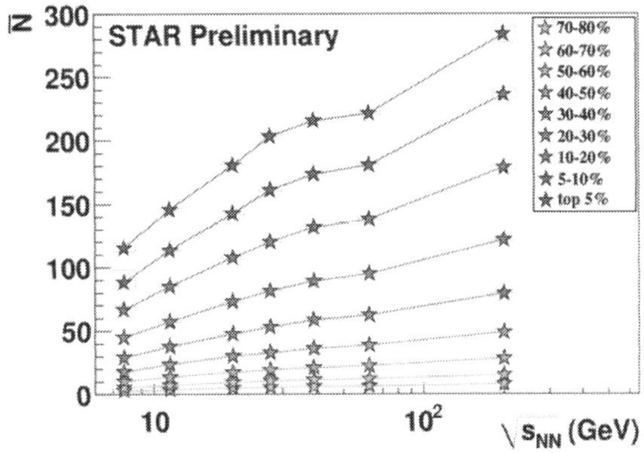

Figure 3. Average total number of clans per event $\bar{N}$ as a function of collision energy, for various centrality classes. Boxes represents systematic error resulted from variations in STAR's day-to-day operation during data-taking.

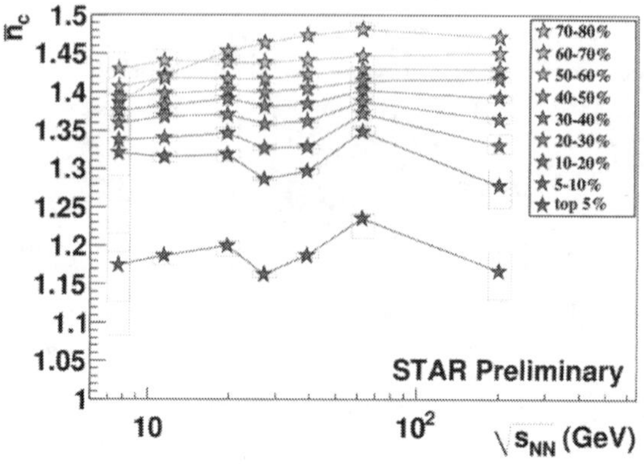

Figure 4. Average charged particles per clan $\bar{n}_c$ as a function of collision energy, for various centrality classes. Boxes represents systematic error resulted from variations in STAR's day-to-day operation during data-taking.

exhibits an energy dependence, with a minimum at 27 GeV and an enhancement at 62 GeV. Such dependence is seen in non-peripheral collisions and is strongest in most central collisions. If there is a phase transition, one would expect a disturbance in particle production which reveals itself through the change of clustering characteristics. The feature observed is a motivation for further studies in this direction. It is worth to point out that the dependence becomes less significant when the DCA cut is released to 2 cm or 3 cm, or when protons, contaminated by transported ones, are included. Both observations (not shown) indicate that the feature is prominent in primarily produced particles, for which the clan concept applies. The apparent enhancement at 62 GeV may be affected by worse beam condition at that particular energy and needs a further investigation.

Due to the change in operation condition, for example, temperature and air pressure of TPC, the efficiency is not a constant. Nevertheless, one can correct for efficiency and recover the efficiency-corrected $\bar{N}$ and $\bar{n}_c$ by applying the following two identities [13]:

$$\begin{aligned} \mu &= \frac{\mu'}{\epsilon} \\ p &= \frac{p'\epsilon}{1-p'(1-\epsilon)}, \end{aligned} \tag{7}$$

where μ' and p' are measured observables, and μ and p are efficiency-corrected ones. Here ϵ is the *detecting efficiency* in general meaning, which includes both the finite detector efficiency effect and the finite detector acceptance effect, ϵ = efficiency $\times$ acceptance. With that, the efficiency corrected $\bar{N}$ and $\bar{n}_c$ can be calculated,

$$\begin{aligned} \bar{N} &= -k\ln p = -k\ln[\frac{p'\epsilon}{1-p'(1-\epsilon)}] \\ \bar{n}_c &= \frac{\mu}{\bar{N}} = \frac{\mu'/\epsilon}{-k\ln[\frac{p'\epsilon}{1-p'(1-\epsilon)}]}. \end{aligned} \tag{8}$$

To study if the efficiency variation can cause such a feature, without loosing generality, assuming a 30% efficiency and 5% variation in year-to-year operation, the resulted variation in $\bar{n}_c$ is too small (comparable to symbol size in Figure 4) to explain the variation seen in data.

4. Summary

In this paper, STAR's measurement of average total number of clans per event, $\bar{N}$, and average charged particles per clan, $\bar{n}_c$, are studied for AuAu collisions at $\sqrt{s_{NN}}$= 7.7, 11.5, 19.6, 27, 39, 62.4 and 200 GeV, for a variety of centrality classes. $\bar{N}$ is found to increase with decreasing centrality and/or increasing energy, and $\bar{n}_c$ is found to decrease with decreasing centrality. Within the same centrality class, $\bar{n}_c$ exhibits a reduction between 19.6 GeV and 62 GeV, with the minimum around 27 GeV. The structure is not seen in peripheral collisions (50%-70%) but is visible in middle central collisions and, becomes prominent in central collisions (top 5%). The variation of mean number of particles per clan with energy cannot be explained by variation in detecting efficiency. Whether the observed structure is connected to a phase transition needs further investigations.

References

[1] Giovanni A and Van Hove L 1986 Z. Phys. C 30 391
[2] Giovanni A, Lupia S and Ugoccioni R 2002 Phys. Rev. D 65 094028
[3] Brambilla M, Giovannini A and Ugoccioni 2008 Physica A 387 1110
[4] Alner G J *et al.* [UA5 Collaboration] 1987 Phys. Rep. 154 247
[5] Arneodo M *et al.* [European Muon Collaboration] 1987 Z. Phys. C 35 335; 1987 Erratum-*ibid.*36, 512
[6] Adamus M *et al.*, [EHS/NA22 Collaboration] 1988 Z. Phys. C 37 215
[7] Abbott T *et al.*, [E-802 Collaboration] 1995 Phys. Rev. C 52 2663
[8] Bachler J *et al.*, [NA35 Collaboration] 1993 Z. Phys. C 57 541
[9] Adare A *et al.*, [PHENIX Collaboration] 2008 Phys. Rev. C 78 044902
[10] Aggarwal M. M. *et al.*, [STAR Collaboration] 2010 arXiv:1007.2613
[11] Bonner B. *et al.*, 2003 Nucl. Instr. and Meth. in Phys. Res. A 508 181; Llope W.J. (for the STAR Collaboration), 2012 Nucl. Instr. and Meth. in Phys. Res. A s661 s110-s113
[12] Anderson M. *et al.*, 2003 Nucl. Instr. and Meth. in Phys. Res. A 499 659
[13] Tang A. and Wang G. 2013 Phys. Rev. C 88 024905

30th Winter Workshop on Nuclear Dynamics (WWND2014) IOP Publishing
Journal of Physics: Conference Series **535** (2014) 012010 doi:10.1088/1742-6596/535/1/012010

Remarkable Features of Decaying Hagedorn States

M Beitel, K Gallmeister and C Greiner

Institut für Theoretische Physik, Goethe-Universität Frankfurt am Main,
Max-von-Laue-Str. 1, 60438 Frankfurt am Main, Germany

E-mail: beitel@th.physik.uni-frankfurt.de

Abstract. Hagedorn states (HS) are a tool to model the hadronization process which occurs in the phase transition phase between the quark gluon plasma (QGP) and the hadron resonance gas (HRG). Their abundance is believed to appear near the Hagedorn temperature T_H which in our understanding equals the critical temperature T_c. These hadron-like resonances are characterized by being very massive and by not being limited to quantum numbers of known hadrons. To generate a whole zoo of such new states we solve the covariantly formulated bootstrap equation by regarding energy conservation and conservation of the baryon number B, strangeness S and electric charge Q. To investigate their decay properties decay chain calculations of HS were conducted. One single (heavy) HS with certain quantum numbers decays by various two-body decay channels subsequently into final stable hadrons. Multiplicities of these stable hadrons, their ratios and their energy distributions are presented. Strikingly the final energy spectra of resulting hadrons show a thermal-like distribution with the characteristic Hagedorn temperature T_H. All hadronic properties like masses, spectral functions etc. are taken from the hadronic transport model Ultra Relativistic Quantum Molecular Dynamics (UrQMD).

1. Introduciton

Before the emergence of quantum chromodynamics (QCD) as the theory of strong interactions many phenomenological ideas came up trying to describe particle production in elementary but also in heavy ion collisions. The most important idea for this note goes back to R. Hagedorn [1] who proposed in 1965 that all particles found at that time and which would be found in the future belong to a common mass spectrum. This spectrum, better known as Hagedorn spectrum, exhibits the specific feature of being exponential in the infinite mass limit. The slope of Hagedorn spectrum's exponential part is solely determined by the Hagedorn temperature T_H. This temperature denotes the limiting temperature for hadronic matter since any partition function of a HRG with Hagedorn-like mass spectrum diverges as long as $T > T_H$. Above the Hagedorn temperature a new state of matter, namely the QGP, is assumed to be realized. How a phase transition from HRG to QGP and back exactly works is one of the most challenging problems of modern physics. One possible tool to investigate this phase transition is the application of HS which mainly contribute to the exponential part of the Hagedorn spectrum but also may appear in the 'hadronic' mass range too. The HS are created in multi-particle collisions most abundantly near T_H which in our understanding equals to the critical temperature T_c. Hagedorn states are color neutral objects which are allowed to have any quantum numbers as long as they are compatible to HS' mass. The appearance of HS in multi-particle collisions and their role was already discussed in [2, 3, 4, 5, 6]. In [7, 8, 9, 10, 11, 12]. the authors show that HS can also alter the occurrence of various phases from hadronic to deconfined partonic matter (first order, second order or crossover) improving QCD's equation of state as shown in [13, 14, 15]. The appearance of HS near T_c can explain, as shown in [4, 5, 6], the fast chemical equilibration of (multi-) strange baryons B and their anti-particles $\bar{B}$ at Relativistic Heavy Ion Collider (RHIC) energies considering HS as a kind of catalyst according to

$$\left(n_1\pi + n_2K + n_3\overline{K} \leftrightarrow\right) HS \leftrightarrow \overline{B} + B + X. \tag{1}$$

The dynamical evolution of this reaction is given through a set of coupled rate equations leading to a chemical equilibration time of about $t_{ch} \approx 5\,\mathrm{fm}/c$ for $B\bar{B}$-pairs. Without 'clustering' of pions and kaons to HS the same approach would result at least in $t_{ch} \approx 10\,\mathrm{fm}/c$ or more which is obviously too long. The inclusion of HS in a hadron resonance gas model provides also a lowering of the speed of sound, c_s and of the shear viscosity over entropy density ratio η/s at the phase transition and being in good agreement with lattice calculations [13, 14, 16, 17]. In addition, by comparing calculations with inclusion of HS to calculations without them, a significant lowering of the shear viscosity to entropy ratio, η/s, is observed [13, 18, 19, 16]. The inclusion of HS creates a minor dependence of the thermal fit parameters of particle ratios on the Hagedorn temperature, T_H, which is assumed to be equal to T_C [20]. The successful application of HS mentioned above calls to an implementation of them into the hadronic transport program UrQMD on the basis of $2 \leftrightarrow 1$-processes by regarding the principle of detailed balance. The role of HS on dynamical evolution of multiplicities, transport coefficients etc. will thus be investigated in future [21].

2. Model

The formulation of a whole zoo of HS will be provided as they will be created in binary collisions within the microscopic hadronic transport simulation program UrQMD [22]. Multiplicities (and their ratios) of stable hadrons stemming from cascading decay simulations of one single initial massive HS for different masses, radii and quantum number combinations are calculated. Additionally energy distribution of the decay products are examined and it is shown that all hadrons stemming from that cascade follow the Boltzmann distribution analogous alike a thermalized hadron resonance gas, although the final and freely moving hadrons are freed solely from the subsequent decay. The starting point of all calculations provided is the postulate of the statistical bootstrap model (SBM) stating that fireballs consist of fireballs which in turn consist of fireballs etc. . The mathematical formulation of this postulate leads to the well known bootstrap equation

$$\begin{aligned}\tau_{\vec{C}}(m) = \frac{R^3}{3\pi m}\sum_{\vec{C}_1,\vec{C}_2}\iint \mathrm{d}m_1\mathrm{d}m_2\,\tau_{\vec{C}_1}(m_1)\,m_1\tau_{\vec{C}_2}(m_2)\,m_2 \\ \times\, p_{cm}(m,m_1,m_2)\,\delta_{\vec{C},\vec{C}_1+\vec{C}_2}\end{aligned} \tag{2}$$

The functions $\tau_{\vec{C}_i}$ on the r.h.s. are spectral functions of two constituents which make up a HS with spectral function τ on the l.h.s. of Eq. (2). Strict conservation of total energy leads p_{cm} denoting the momenta of both constituent particles with masses m_1 and m_2 in the rest frame of made up HS with mass m,

$$p_{cm}(m,m_1,m_2) = \frac{1}{2m}\sqrt{\left(m^2 - m_1^2 - m_2^2\right)^2 - 4m_1^2m_2^2}, \tag{3}$$

as usual where charge conservation is assured by Kronecker's δ. The radius R denotes the size of created HS and is considered to be constant taking on some reasonable values which are discussed further below. Contrary to the well-known non-covariant bootstrap equation [23, 24], the expression here is formulated covariantly. In the general solution of Eq. (2), the number of constituents is theoretically infinite. The reason to consider two constituents case only is because HS will be implemented in hadron transport models like e.g. UrQMD as a whole zoo of new particles. In standard transport models maximally two particles in the incoming channel are allowed because the interaction probability is calculated on the basis of geometrical cross sections. On the other hand resonance decays in two hadrons are realized in UrQMD too making an implementation of further $(2 \leftrightarrow 1)$ processes, now involving HS, possible. For this kind of new processes the principle of detailed balance will strictly hold. The accepted error by the approximation of only two outgoing particles is roughly about 30%, which can be estimated by looking at the HS decay probability into n particles, $P(n) = (\ln 2)^{n-1} / (n-1)!$, yielding a probability for the decay into two particles of 69%, into three particles of 24% etc. [23]. Given the function $\tau(m)$ we proceed to the formulation of HS' total decay width

$$\Gamma_{\vec{C}}(m) = \frac{\sigma}{2\pi^2 \tau_{\vec{C}}(m)} \sum_{\vec{C}_1, \vec{C}_2} \iint dm_1 dm_2 \tau_{\vec{C}_1}(m_1) \tau_{\vec{C}_2}(m_2) \times p_{cm}^2(m, m_1, m_2)\, \delta_{\vec{C}, \vec{C}_1 + \vec{C}_2} \,. \tag{4}$$

This formula was derived by employing general formulae for cross section and decay width as given in [25] where for the (creation or fusion) cross section σ of HS only the $2 \to 1$ and for their decay width Γ only the $1 \to 2$ case were considered. Further we demanded the principle of detailed balance between creation and decay of HS to be valid which connects its creation with its decay properties in Eq. (4). The cross section will be considered simply as the geometrical value $\sigma = \pi R^2$. The partial decay width of a HS with mass m and charge vector $\vec{C}$ decaying into two particles with masses between m_1 and $dm_1 + m_1$, m_2 and $dm_2 + m_2$ and charges $\vec{C}_1$, $\vec{C}_2$ reads

$$\Delta\Gamma_{\vec{C}, \vec{C}_1, \vec{C}_2}(m, m_1, m_2) = \frac{\sigma}{2\pi^2} \frac{\Delta m_1 \tau_{\vec{C}_1}(m_1) \Delta m_2 \tau_{\vec{C}_2}(m_2)}{\tau_{\vec{C}}(m)} p_{cm}^2(m, m_1, m_2) \,. \tag{5}$$

The fractional two-body branching ratios $\mathcal{B}$ are just the ratio of partial and total decay widths, Eqs. (5) and (4),

$$\Delta\mathcal{B}_{\vec{C}, \vec{C}_1, \vec{C}_2}(m, m_1, m_2) = \frac{\Delta\Gamma_{\vec{C}, \vec{C}_1, \vec{C}_2}(m, m_1, m_2)}{\Gamma_{\vec{C}}(m)} . \tag{6}$$

3. Results

The bootstrap equation Eq. (2) in general is a highly non-linear integral equation of Volterra type which can be solved analytically for some special cases [26, 27]. The numerical solution of the given bootstrap equation for a meson-like, non-strange and electrically neutral (B=S=Q=0) Hagedorn spectrum for two different typical radii ($R_1 = 0.8\,\text{fm}, R_2 = 1.0\,\text{fm}$) is presented in Fig. 1. In the same figure also spectra for baryonic non-strange and electrically charged states ($B = 1, S = 0, Q = 1$) are shown. All Hagedorn spectra rise exponentially for masses $\geq 1.5\,\text{GeV}$ with different slopes for different radii, but for $m < 1.5\,\text{GeV}$ they all include and thus fit the 'hadronic' part of the spectrum. The slopes of the exponential part were determined by the fit function $\tau_{\text{fit}}(m) = Am^{-b} \exp(m/T_H)$, yielding the Hagedorn temperatures $T_H = 0.145\,\text{GeV}$ for $R = 1.0\,\text{fm}$ and $T_H = 0.162\,\text{GeV}$ for $R = 0.8\,\text{fm}$ both being rather independent on the chosen quantum number combinations. The total decay width of a Hagedorn state consists of three different contributions, where the first one considers only hadrons, the second hadrons and Hagedorn states, and the third one only HS in the outgoing channel as depicted in Fig. 2. The peak in the mass range of M_{HS} = 0-2 GeV on left part of Fig. 1 comes mainly from

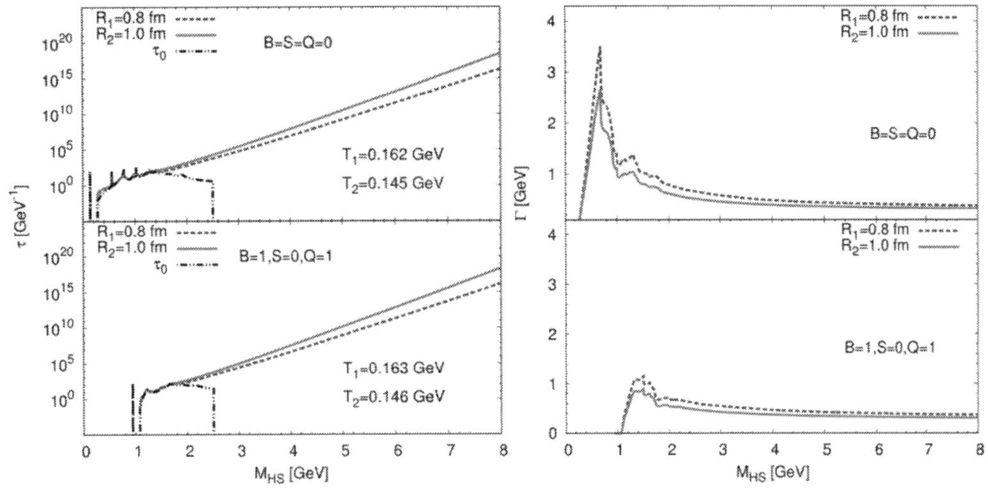

Figure 1. Meson-like ($B = S = Q = 0$) Hagedorn spectrum (upper left) and corresponding HS' total decay width (upper right) and baryonic ($B = 1, S = 0, Q = 1$) Hagedorn spectrum (lower left) and corresponding HS' total decay width (lower right) for two different radii. On the left figure additionally (fitted) Hagedorn temperatures are provided where the black line represents the sum of spectral functions of hadrons with the given quantum numbers.

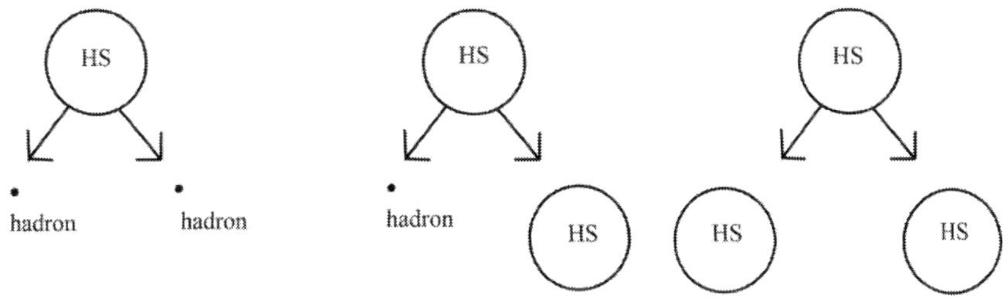

Figure 2. Decay of HS into two hadrons (left), decay into one hadron and one HS (center) and decay into two HS (right)

the first contribution, because in this mass range the phase space for pure hadronic decay is largest. The height of the peak depends on the number of hadronic pairs, whose quantum numbers all sum up to the quantum number of the Hagedorn state they are building up, being large for $B = S = Q = 0$ and rather small for $B = Q = 1, S = 0$. Another remarkable feature is that for both radii the total decay width tends to a constant value depending only on R for large masses. Having the numerous branching ratios Eq. (6) at hand, one is able to calculate hadronic multiplicities stemming from Hagedorn state decays. Here one starts with some initial heavy Hagedorn state, which decays subsequently down until hadrons are left only as shown in Fig. 3. Among those also non stable resonances might appear, which further undergo a hadronic feed down leaving one with light and stable hadrons with respect to the strong force like pions, kaons, etc. . All hadronic properties used here were taken from the transport model UrQMD [22]. Calculated multiplicities and their ratios for some uncharged ($B = S = Q = 0$) initial Hagedorn state are shown in Fig. 4 (upper part). One observes a linear dependence of all multiplicities on the initial Hagedorn state mass where the magnitude depends on the available phase space for each hadron. Thus in a decay of a charge neutral

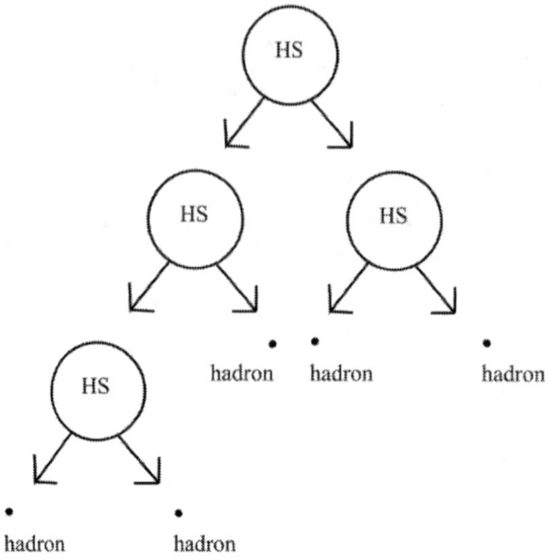

Figure 3. One possible decay chain of an initial heavy HS

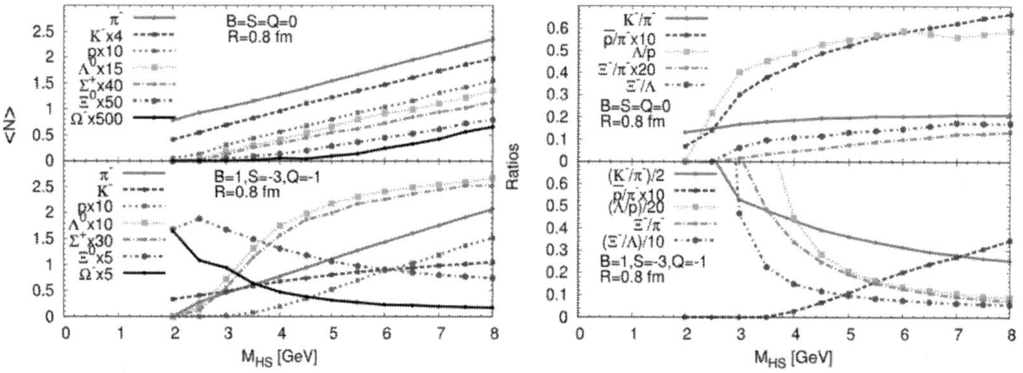

Figure 4. Hadronic multiplicities after a cascade decay of initial Hagedorn state with radius $R = 0.8\,\mathrm{fm}$ and $B = S = Q = 0$ (upper left) and $B = 1, S = -3, Q = -1$ (lower left). Corresponding multiplicity ratios are shown for $B = S = Q = 0$ (upper right) and $B = 1, S = -3, Q = -1$ (lower right). In both cases 'hadronic' feeddown is taken into account.

Hagedorn state π^- clearly expectedly dominate which have to be produced in pairs mostly with π^+ since exact charge conservation is enforced. Kaons, especially K^-, are even stronger suppressed not only of their larger mass but also due to the fact that they have to conserve both electric charge and strangeness. For the baryons presented the same argumentation holds since both have to conserve baryon number B and additionally electric charge Q for proton and strangeness S for Λ. For the multistrange hyperons Ξ^0 and Ω^- the production suppression is even stronger. This has to be contrasted with the results for a baryonic, multi-strange and electrically charged ($B = 1, S = -3, Q = -1$) Ω^--like Hagedorn state also shown in Fig. 4. Now the choice of Hagedorn state's initial quantum numbers is reflected in the preference of baryon production although they are much heavier than the presented mesons. Especially the abundance of hyperons (Ω^-, Ξ^0) compared to the case discussed before is striking since the easiest way to conserve the initial quantum numbers is the production of one $\Omega^-\pi^0$- or one $\Xi^0 K^-$ pair where on the other hand the phase space for all other hadrons with different quantum numbers is suppressed now. Hence exact conservation of quantum numbers always causes a competition between hadron's phase space and its quantum numbers. On the right part of Fig. 4 the corresponding multiplicity ratios for $R = 0.8\,\mathrm{fm}$ and same quantum number combinations are shown. A comparison of theoretical ratios with experimental results is also provided. Numerical values for the multiplicity ratios for Hagedorn state masses of 4 GeV and 8 GeV are listed in Tab. 1 and, for illustration, compared to experimental results for p-p- and Pb-Pb collisions at midrapidity, both measured by ALICE at LHC. The theoretical multiplicity ratios lie seemingly

	p-p	Pb-Pb	4 GeV	8 GeV
K^-/π^-	0.123(14)	0.149(16)	0.187	0.210
$\overline{p}/\pi^-$	0.053(6)	0.045(5)	0.043	0.066
Λ/π^-	0.032(4)	0.036(5)	0.021	0.038
$\Lambda/\overline{p}$	0.608(88)	0.78(12)	0.494	0.579
Ξ^-/π^-	0.003(1)	0.0050(6)	0.0023	0.0066
$\Omega^-/\pi^- \cdot 10^{-3}$	—	0.87(17)	0.086	0.560

Table 1. Comparison of particle multiplicity ratios from theory vs. p-p at $\sqrt{s_{NN}} = 0.9\,\mathrm{TeV}$ [28] and Pb-Pb at $\sqrt{s_{NN}} = 2.76\,\mathrm{TeV}$ [29, 30, 31], both from ALICE at LHC. Calculated values are listed for Hagedorn state masses of 4 GeV and 8 GeV. Numbers in brackets denote the error in the last digits of the multiplicity ratios.

close to the measured by ALICE, except for the very rare multi-strange baryon Ω^-. However, it has to be made clear, that the decay of HS alone is never assumed to describe the experimental data. The theoretical multiplicity ratios serve as a proof of reliability and reasonability of Hagedorn state's branching ratios defined in Eq. (6). This test is necessary before implementing them into a transport model. Beside multiplicities also the energy distribution of the decay products was examined. They are shown in Fig. 5 for an uncharged ($B = S = Q = 0$) Hagedorn state with initial mass $M_{HS} = 4\,\mathrm{GeV}$ and also $M_{HS} = 8\,\mathrm{GeV}$. The striking observations are that all particle species follow a Boltzmann-like distribution with the same slope independent on the initial Hagedorn state mass. Thus pions, kaons, protons spectra look alike stemming of a system with a temperature being $T_{th} = 0.162\,\mathrm{GeV}$. We remark that the final and freely moving hadrons are freed from the subsequent decays without any reintaractions. The particular finding is that the 'thermal' temperature T_{th} exactly equals the Hagedorn temperature T_H obtained from a fit in the left part of Fig. 1 for the case $R = 0.8\,\mathrm{fm}$. The Hagedorn temperature T_H was nothing but a slope parameter to fit the exponential part of the Hagedorn spectrum, where on the other hand T_{th} is the slope ('temperature') of the finally created hadrons. Starting with a bootstrap formula with no introduction of any notion of temperatures at all resulted in a 'thermalized' decay with a slope being the Hagedorn temperature.

4. Conclusion

A covariantly formulated bootstrap equation is presented which ensures energy and quantum number conservation. The solution of this bootstrap equation provides the Hagedorn spectra describing the 'hadronic' part adequately and being exponential for large masses. Given the Hagedorn spectra the total decay width formula for HS was presented which was obtained on the principle of detailed balance between their creation and their decay. The main characteristics of HS' total decay width are their large peaks for masses in the 'hadronic' range and roughly constant values for masses beyond that range. With partial and total decay width we were able to

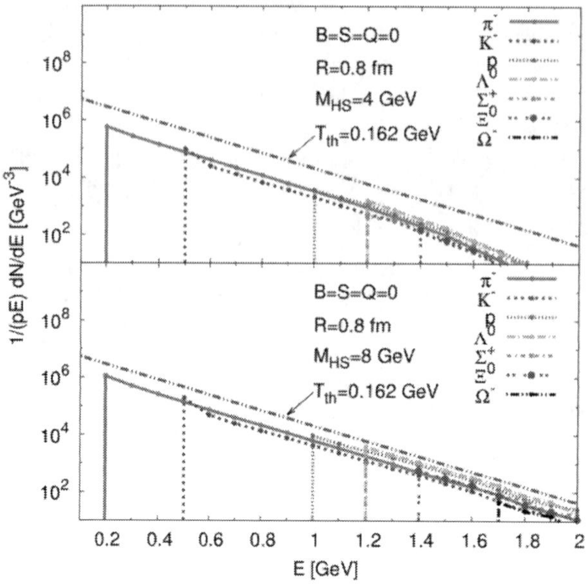

Figure 5. Energy spectra of hadrons stemming from cascade decay of charge neutral Hagedorn state with radius $R = 0.8\,\text{fm}$ and initial mass $M_{HS} = 4\,\text{GeV}$ and $M_{HS} = 8\,\text{GeV}$.

define HS' branching ratios needed for decay simulations. As a specific case we considered decay chains of one single initial Hagedorn state cascading down by various two (intermediate) particle decay channels until stable hadrons were left only. The multiplicity ratios of those stable hadrons where compared to Pb-Pb ALICE data at LHC to check the reasonability and reliability of HS' theoretical branching ratios. Also the energy distribution of those hadrons were examined for two initial HS' masses. The striking findings were that all hadrons stemming from such a decay chain, without any reinteractions, are first all thermal and second exhibit the same temperature. The particular feature of this 'thermal' temperature is that it equals the Hagedorn temperature gained from a fit of Hagedorn spectrum's exponential part. Summarizing, such a finding gives fresh insight into the microscopic and thermal-like hadronization in ultrarelativistic e^+-e^-- (see eg. [32]), hadron-hadron-, and also especially in heavy ion collisions: An implementation of the presented Hagedorn state decays in addition to their production mechanisms into the transport approach UrQMD offers a new venue for allowing strongly interacting hadronic multiparticle collisions in a consistent scheme being important in the vicinity of the deconfinement transition by creating and decaying more exotic HS. Understanding faster thermalization and chemical equilibration, but also microscopic transport properties can be thoroughly investigated in future [21].

5. Acknowledgements

The authors acknowledges discussions with K. Bugaev. This work was supported by the Bundesministerium für Bildung und Forschung (BMBF), the HGS-HIRe and the Helmholtz International Center for FAIR within the framework of the LOEWE program launched by the State of Hesse. Numerical computations have been performed at the Center for Scientific Computing (CSC).

References

[1] Hagedorn R 1965 *Nuovo Cim. Suppl.* **3** 147–186
[2] Greiner C and Leupold S 2001 *J. Phys. G* **27** L95–L102
[3] Greiner C, Koch-Steinheimer P, Liu F M, Shovkovy I A and Stoecker H 2005 *J. Phys. G* **31** S725–S732
[4] Noronha-Hostler J, Greiner C and Shovkovy I A 2008 *Phys. Rev. Lett.* **100** 252301
[5] Noronha-Hostler J, Greiner C and Shovkovy I 2010 *J. Phys. G* **37** 094017
[6] Noronha-Hostler J, Beitel M, Greiner C and Shovkovy I 2010 *Phys. Rev. C* **81** 054909
[7] Moretto L, Bugaev K, Elliott J and Phair L 2006 *Europhys. Lett.* **76** 402–408
[8] Zakout I, Greiner C and Schaffner-Bielich J 2007 *Nucl. Phys. A* **781** 150–200
[9] Zakout I and Greiner C 2008 *Phys. Rev. C* **78** 034916
[10] Ferroni L and Koch V 2009 *Phys. Rev. C* **79** 034905
[11] Bugaev K, Petrov V and Zinovjev G 2009 *Phys. Rev. C* **79** 054913
[12] Ivanytskyi A, Bugaev K, Sorin A and Zinovjev G 2012 *Phys. Rev. E* **86** 061107
[13] Noronha-Hostler J, Noronha J and Greiner C 2009 *Phys. Rev. Lett.* **103** 172302
[14] Majumder A and Muller B 2010 *Phys. Rev. Lett.* **105** 252002
[15] Karsch F, Redlich K and Tawfik A 2003 *Eur.Phys.J.* **C29** 549–556
[16] Noronha-Hostler J, Noronha J and Greiner C 2012 *Phys. Rev. C* **86** 024913
[17] Jakovac A 2013 *Phys. Rev. D* **88** 065012
[18] Gorenstein M, Hauer M and Moroz O 2008 *Phys. Rev. C* **77** 024911
[19] Itakura K, Morimatsu O and Otomo H 2008 *J. Phys. G* **35** 104149
[20] Noronha-Hostler J, Ahmad H, Noronha J and Greiner C 2010 *Phys. Rev. C* **82** 024913
[21] Beitel M, Gallmeister K and Greiner C 2014 in preparation
[22] Bass S, Belkacem M, Bleicher M, Brandstetter M, Bravina L *et al.* 1998 *Prog. Part. Nucl. Phys.* **41** 255–369
[23] Frautschi S C 1971 *Phys. Rev. D* **3** 2821–2834
[24] Hamer C and Frautschi S C 1971 *Phys. Rev. D* **4** 2125–2137
[25] Beringer J *et al.* (Particle Data Group) 2012 *Phys. Rev. D* **86**(1) 010001
[26] Yellin J 1973 *Nucl. Phys. B* **52** 583–594
[27] Hagedorn R and Montvay I 1973 *Nucl. Phys. B* **59** 45–86
[28] Aamodt K *et al.* (ALICE Collaboration) 2011 *Eur. Phys. J. C* **71** 1594
[29] Abelev B *et al.* (ALICE Collaboration) 2013 *Phys. Rev. C* **88** 044910
[30] Abelev B B *et al.* (ALICE Collaboration) 2013 *Phys. Rev. Lett.* **111** 222301
[31] Abelev B B *et al.* (ALICE Collaboration) 2014 *Phys. Lett. B* **728** 216–227
[32] Becattini F 1996 *Z. Phys. C* **69** 485–492

Journal of Physics: Conference Series **535** (2014) 012011 doi:10.1088/1742-6596/535/1/012011

Double ratio of charmonia in $p+\text{Pb}$ collisions at $\sqrt{s_{\text{NN}}} = 5.02$ TeV

Yunpeng Liu[1], **Che Ming Ko**[1,2], **Taesoo Song**[1,3]

[1] Cyclotron Institute, Texas A&M University, College Station, Texas 77843, USA
[2] Department of Physics and Astronomy, Texas A&M University, College Station, Texas 77843, USA
[3] Frankfurt Institute for Advanced Studies, J. W. Goethe University, 60438 Frankfurt am Main, Germany

E-mail: yliu@comp.tamu.edu

Abstract. Based on a kinetic description of J/ψ dissociation and production in an expanding quark-gluon plasma that is described by a 2+1 dimensional ideal hydrodynamics, we have studied the double ratio $R_{p\text{Pb}}(\psi')/R_{p\text{Pb}}(J/\psi)$ of charmonia in p+Pb collisions at $\sqrt{s_{\text{NN}}} = 5.02$ TeV by including not only the cold nuclear matter effects but also the hot nuclear matter effects. We find that the double ratio of prompt charmonia is significantly suppressed in the most central collisions as a result of the hot nuclear matter effects.

Charmonia suppression in heavy ion collisions was first suggested in Ref. [1] based on the consideration of color screening in quark-gluon plasma (QGP). The anomalous suppression of J/ψ production, besides the normal suppression due to the cold nuclear matter effects, observed in Pb+Pb collisions at the Super Proton Synchrotron (SPS) seemed to confirm this prediction and thus suggested that a QGP was produced in these collisions [2]. Since then, there have been extensive experimental and theoretical studies on J/ψ production and suppression in both elementary and nuclear collisions. In particular,

- Baseline study of charmonia production in $p+p$ collisions has been carried out using the color singlet model (CSM) [3] and/or the color octet model (COM) [4].
- Cold nuclear matter effects, including the shadowing effect, the Cronin effect and the nuclear absorption, have been studied in p+A and d+A collisions.
- Hot medium effects, including the color screening, gluon dissociation, quasi-free scattering dissociation, regeneration and so forth, have been studied in $A+A$ collisions.

However, with the high beam energy at the LHC, the initial energy density in p+Pb collisions can be sufficient large for the formation of a QGP. It is then of interest to study the hot medium effects on quarkonia production in these collisions and compare them to those due to the cold nuclear matter [5, 6, 7, 8, 9, 10, 11, 12].

In Ref. [13], we have carried out such a study by using a 2+1 dimensional ideal hydrodynamic model to describe the bulk dynamics of $p+\text{Pb}$ collisions at $\sqrt{s_{\text{NN}}} = 5.02$ TeV, i.e., based on the equation

$$\partial_\mu T^{\mu\nu} = 0 \tag{1}$$

with the boost invariant condition. In the above, $T^{\mu\nu} = (\epsilon + p)u^\mu u^\nu - g^{\mu\nu}p$ is the energy-momentum tensor in terms of the energy density ϵ, pressure p, and four velocity u. For the equation of state $\epsilon(p)$, we have used that of Ref. [14] based on an ideal gas of massless partons for the QGP and massive hadrons for the hadronic matter, together with a bag constant which leads to a critical temperature $T_c = 165$ MeV [15, 16, 17] at zero baryon chemical potential. For the initial energy density at time $\tau_0 = 0.6$ fm/c and its rapidity dependence, it is obtained from a multiple phase transport (AMPT) model [18]. Selecting the most central 10% collisions according to the yield of partons in AMPT, the maximum initial energy density is found at rapidity $y = -2$.

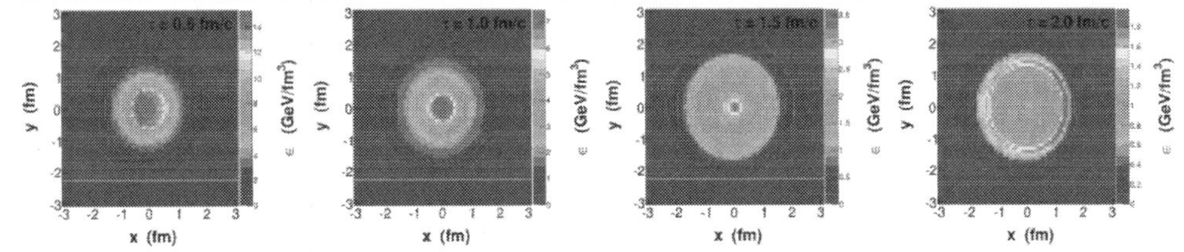

Figure 1. (Color online) Energy density ϵ at rapidity $y = -2$ in the most central 10% $p+$Pb collisions at $\sqrt{s_{\rm NN}} = 5.02$ TeV as a function of the transverse coordinate x and y relative to the center of the proton at different proper time τ.

In Fig. 1, we show the energy density distribution in the transverse plane at rapidity $y = -2$ for different proper time τ. The origin of the coordinates is located at the center of the proton, and the center of the lead lies on the negative x-axis. The initial energy density is roughly isotropic in the transverse plane with a slightly larger value on the lead side. The maximum value is $\epsilon_0 \approx 15$ GeV/fm^3, which corresponds to a temperature $T_0 \approx 290$ MeV.

Since the energy density is large and the volume is small compared with those in A+A collisions at lower energies, the system expands fast, leading to a quick decreas of the energy density at the center. At time $\tau = 2.0$ fm/c, the energy density at the center is even smaller than that in the peripheral region, and the maximum energy density is at the radius $r \approx 1.5$ fm. The value of the energy density on the positive axis is slightly larger than that at the negative axis. Since the center of the QGP produced in p+Pb collisions hadronizes earlier, its spacial topology differs from that in Pb+Pb collisions[1].

For the time evolution of the distribution function $f(\mathbf{x}, \mathbf{p}, t)$ of J/ψ in the phase space of coordinate $\mathbf{x}$ and momentum $\mathbf{p}$, we use the transport equation

$$\partial_t f + \mathbf{v} \cdot \nabla f = -\alpha f + \beta, \tag{2}$$

where $\mathbf{v}$ is the velocity of the J/ψ, and α and β are the dissociation and the regeneration rate, respectively. The dissociation rate

$$\alpha(T, u, p) = \frac{N_g}{E} \int \frac{d\mathbf{k}}{(2\pi)^3 E_g} k \cdot p f_g(T, u, k) \sigma_D. \tag{3}$$

takes into account the gluon dissociation process $J/\psi + g \to c + \bar{c}$ with the cross section σ_D [19, 20, 14]. The regeneration rate takes into account the inverse process $c + \bar{c} \to J/\psi + g$, whose cross section is related to that of the gluon dissociation process by the detailed balance

[1] The Euler characteristic χ changes from 2 to 4 in central p+Pb collisions.

relation, and can be calculated using the distribution functions of charm and anticharm quarks. For simplicity, we take the charm quark distribution to be in thermal equilibrium, i.e.,

$$f_c(\mathbf{x}, \mathbf{q}_c) = \rho_c(\mathbf{x}) \frac{N}{e^{q_c \cdot u/T} + 1}, \tag{4}$$

where ρ_c is the number density of charm quarks and satisfies the conservation equation $\partial_t \rho_c + \nabla \cdot (\rho_c \mathbf{v}_m) = 0$ with $\mathbf{v}_m$ being the velocity of the medium, and $N \equiv \left[(2\pi)^{-3} \int d\mathbf{q}_c \left(e^{q_c \cdot u/T} + 1\right)^{-1}\right]^{-1}$ is the normalization factor for the Fermi distribution. For anticharm quarks, their distribution function $f_{\bar{c}}$ has a similar form. We also include a velocity-dependent dissociation temperature T_d to describe the color screening effect [21].

Different from Pb+Pb collisions at the LHC, at most one pair of charm and anticharm quarks are produced in a p+Pb collision, and they are also not likely to reach thermal equilibrium . These effects are included via a canonical enhancement factor $C_{ce} = 1 + 1/(dN_c^{\rm dir}/dy)$ in the evaluation of the equilibrium number of J/ψ, as in the statistical model [22, 23, 24], and the relaxation reduction factor $r = 1 - \exp(-\tau/\tau_r)$ [25] with the charm quark relaxation time $\tau_r = 7$ fm/c [26, 27]. More details can be found in Ref. [13]. We note that these effects can be more accurately studied in a transport model [28].

For the initial distribution of J/ψ, it is obtained from the Glauber model with the inclusion of initial-state cold nuclear matter effects and a J/ψ production cross section of $d\sigma_{pp}/dy = 5.68$ μb for pp collisions at 5.02 TeV. The latter is obtained from interpolating the experimental results at lower ($\sqrt{s_{\rm NN}} = 2.76$ TeV) [29] and higher ($\sqrt{s_{\rm NN}} = 7$ TeV) [30] energies with a power-law form and an average transverse momentum square $\langle p_T^2 \rangle = 9.5$ GeV^2 at middle rapidity. As to the rapidity distribution, it is assumed to be Gaussian with parameters taken from Refs. [31, 32]. The density distribution of initially produced J/ψ in space is assumed to be proportional to the thickness of a uniform solid sphere of radius $r = 0.8$ fm as that for the proton.

For the cold nuclear matter effects on J/ψ and charm quarks, they are assumed to be the same and are taken from Refs. [33, 20], which then lead to a suppression in proton rapidity and an enhancement in Pb rapidity.

The above treatment for the ground state J/ψ is generalized to include its excited states χ_c and ψ' with different dissociation temperatures T_D and cross sections [34, 35] as in Ref. [21]. For the initial abundance of these excited states, they are determined from that of J/ψ by using the empirically known feed-down contributions to J/ψ production in $p+p$ collisions, i.e., using the proportion 6 : 3 : 1 for direct J/ψ, feed-down from χ_c and from ψ' [36]. They are further assumed to subject to the same cold nuclear matter effects as the J/ψ.

For the contribution to J/ψ production from the decay of regenerated χ_c and ψ', they are included with the branch ratios from Ref. [37].

In Fig. 2, we show the double ratio of prompt charmonia as a function of rapidity. It is seen that the double ratio at rapidity from −4 to 0 is smaller than unit due to stronger suppression of ψ' in the hot medium. In the most central 10% collisions, it is smaller than 0.1, while in collisions with centrality 40%-60%, it is between 0.3 and 0.4. Since the formation time of J/ψ and ψ' is much longer than that for the proton and Pb to pass through each other, the cold nuclear matter effects are not expected to differ much between the J/ψ and ψ'.

In experiments, the inclusive double ratio is easier to measure. According to a recent study [38], the contribution of B meson decay increases the double ratio in heavy ion collisions. The corresponding result for p+Pb collisions is shown in Fig. 3. The qualitative effect of B meson decay is the same as that in Pb+Pb collisions [38], and the double ratio is above 0.4 in the whole rapidity range.

In summary, we have studied the double ratio of quarkonia in p+Pb collisions at $\sqrt{s_{\rm NN}} = 5.02$ TeV in a transport approach based on a 2+1 dimensional ideal hydrodynamic model and found

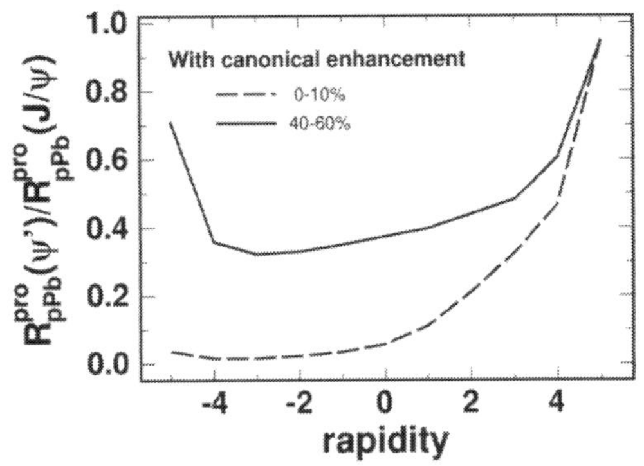

Figure 2. Double ratio $R_{\rm pPb}^{\rm pro}(\psi')/R_{\rm pPb}^{\rm pro}(J/\psi)$ of prompt charmonia in $p+{\rm Pb}$ collisions at $\sqrt{s_{\rm NN}} = 5.02$ TeV as a function of rapidity for centralities 0-10% and 40-60%.

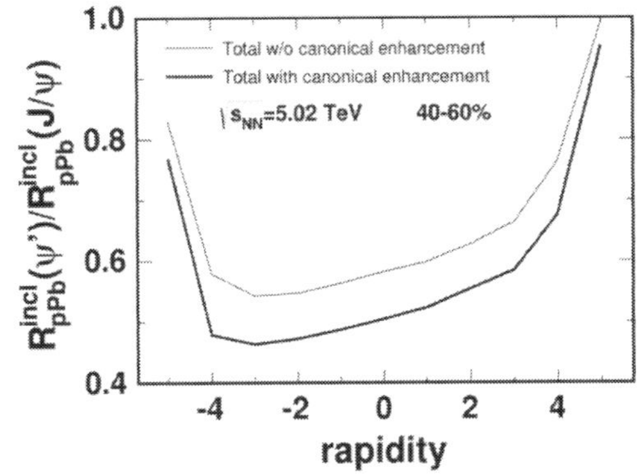

Figure 3. Double ratio $R_{\rm pPb}^{\rm incl}(\psi')/R_{\rm pPb}^{\rm incl}(J/\psi)$ of inclusive charmonia in $p+{\rm Pb}$ collisions at $\sqrt{s_{\rm NN}} = 5.02$ TeV as a function of rapidity for centrality 40-60%.

that both the prompt and inclusive double ratios are smaller than one, especially for the prompt one in the most central 10% collisions. This is not expected from the cold nuclear matter effects as they exist before J/ψ and ψ' are formed. Measuring the double ratio of charmonia thus provides the possibility to study the hot medium effects in p+Pb collisions at the LHC.

Acknowledgments

This work was supported by the U.S. National Science Foundation under Grant No. PHY-1068572, the US Department of Energy under Contract No. DE-FG02-10ER41682, and the Welch Foundation under Grant No. A-1358.

References

[1] Matsui T and Satz H 1986 *Phys. Lett.* **B178** 416
[2] Gonin M *et al.* (NA50) 1996 *Nucl. Phys.* **A610** 404c–417c
[3] Chang C H 1980 *Nucl. Phys.* **B172** 425–434
[4] Bodwin G T, Braaten E and Lepage G P 1995 *Phys. Rev.* **D51** 1125–1171 (*Preprint* hep-ph/9407339)
[5] Gerschel C and Hufner J 1988 *Phys. Lett.* **B207** 253–256

[6] Gousset T and Pirner H 1996 *Phys. Lett.* **B375** 349–354 (*Preprint* hep-ph/9601242)
[7] McLerran L D and Venugopalan R 1999 *Phys. Rev.* **D59** 094002 (*Preprint* hep-ph/9809427)
[8] Kopeliovich B and Zakharov B 1991 *Phys. Rev.* **D44** 3466–3472
[9] Kopeliovich B, Nemchick J, Nikolaev N N and Zakharov B 1994 *Phys. Lett.* **B324** 469–476 (*Preprint* hep-ph/9311237)
[10] Kopeliovich B 2003 *Phys. Rev.* **C68** 044906 (*Preprint* nucl-th/0306044)
[11] Iancu E and Venugopalan R 2003 (*Preprint* hep-ph/0303204)
[12] Ferreiro E, Fleuret F, Lansberg J and Rakotozafindrabe A 2009 *Phys. Lett.* **B680** 50–55 (*Preprint* 0809.4684)
[13] Liu Y, Ko C M and Song T 2014 *Phys. Lett.* **B728** 437–442 (*Preprint* 1309.5113)
[14] Liu Y p, Qu Z, Xu N and Zhuang P f 2009 *Phys. Lett.* **B678** 72–76 (*Preprint* 0901.2757)
[15] Aoki Y, Fodor Z, Katz S and Szabo K 2006 *Phys. Lett.* **B643** 46–54 (*Preprint* hep-lat/0609068)
[16] Borsanyi S *et al.* (Wuppertal-Budapest Collaboration) 2010 *JHEP* **1009** 073 (*Preprint* 1005.3508)
[17] Aoki Y, Endrodi G, Fodor Z, Katz S and Szabo K 2006 *Nature* **443** 675–678 (*Preprint* hep-lat/0611014)
[18] Lin Z W, Ko C M, Li B A, Zhang B and Pal S 2005 *Phys. Rev.* **C72** 064901 (*Preprint* nucl-th/0411110)
[19] Peskin M E 1979 *Nucl. Phys.* **B156** 365
[20] Manceau L (ALICE) 2013 (*Preprint* 1307.3098)
[21] Liu Y, Xu N and Zhuang P 2013 *Phys. Lett.* **B724** 73–76 (*Preprint* 1210.7449)
[22] Andronic A, Braun-Munzinger P, Redlich K and Stachel J 2007 *Nucl. Phys.* **A789** 334–356 (*Preprint* nucl-th/0611023)
[23] Kostyuk A 2005 (*Preprint* nucl-th/0502005)
[24] Liu Y, Greiner C and Kostyuk A 2013 *Phys. Rev.* **C87** 014910 (*Preprint* 1207.2366)
[25] Grandchamp L and Rapp R 2002 *Nucl. Phys.* **A709** 415–439 (*Preprint* hep-ph/0205305)
[26] Zhao X and Rapp R 2008 *Phys. Lett.* **B664** 253–257 (*Preprint* 0712.2407)
[27] Song T, Han K C and Ko C M 2012 *Phys. Rev.* **C85** 054905 (*Preprint* 1203.2964)
[28] Liu Y, Ko C M and Li F 2014 (*Preprint* 1406.6648)
[29] Abelev B *et al.* (ALICE Collaboration) 2012 *Phys. Lett.* **B718** 295–306 (*Preprint* 1203.3641)
[30] Aamodt K *et al.* (ALICE Collaboration) 2011 *Phys. Lett.* **B704** 442–455 (*Preprint* 1105.0380)
[31] Bossu F, del Valle Z C, de Falco A, Gagliardi M, Grigoryan S *et al.* 2011 (*Preprint* 1103.2394)
[32] Liu Y, Qu Z, Xu N and Zhuang P 2010 *J. Phys.* **G37** 075110 (*Preprint* 0907.2723)
[33] Albacete J L *et al.* 2013 *Int. J. Mod. Phys. E Vol.* **22** 1330007 (*Preprint* 1301.3395)
[34] Wang X N and Yuan F 2002 *Phys. Lett.* **B540** 62–67 (*Preprint* nucl-th/0202018)
[35] Arleo F, Gossiaux P, Gousset T and Aichelin J 2002 *Phys. Rev.* **D65** 014005 (*Preprint* hep-ph/0102095)
[36] Zoccoli A *et al.* (HERA-B Collaboration) 2005 *Eur. Phys. J.* **C43** 179–186
[37] Beringer J *et al.* (Particle Data Group) 2012 *Phys. Rev.* **D86** 010001
[38] Chen B, Liu Y, Zhou K and Zhuang P 2013 *Phys. Lett.* **B726** 725–728 (*Preprint* 1306.5032)

Journal of Physics: Conference Series 535 (2014) 012012 doi:10.1088/1742-6596/535/1/012012

Multiple Parton Interactions with ALICE: from pp to p–Pb

Andreas Morsch
for the ALICE Collaboration
CERN, 1211 Geneva, Switzerland
E-mail: andreas.morsch@cern.ch

Abstract. The study of multiplicity dependent di-hadron angular correlations allows us to assess the contribution of multiple-parton interactions to particle production. We will review these measurements in pp and p–Pb collisions with the ALICE detector at the LHC and discuss the results in the context of centrality determination and other multiplicity dependent observables in p–Pb.

1. Introduction

The concept of multiple parton–parton interactions (MPI) provides the theoretical basis to understand global event properties of non-diffractive, minimum bias pp collisions. The model also allows for a straightforward interpretation of the fact that at high $\sqrt{s}$ the leading order cross-section for $2 \rightarrow 2$ parton scatterings with momentum transfer $Q > Q_{\rm min} \gg \Lambda_{\rm QCD}$ exceeds the total pp cross-section at a range of $Q_{\rm min}$-values where perturbative QCD is applicable [1]. At LHC energies, this happens already at $Q_{\rm min} \approx 4$ GeV/c [2, 3, 4]. In a naive factorisation approach the mean number of scatterings per event $\langle n_{\rm hard} \rangle$ is equal to the ratio of the hard and the total cross-section, $\sigma_{\rm hard}/\sigma_{\rm tot}$. In more realistic phenomenological models like PYTHIA6 [5], PYTHIA8 [6] and HERWIG++ [4], the colour screening effect regularises the divergence ($\propto 1/p_{\rm T}^4$) of the hard cross-section at low $p_{\rm T}$ (< 2 GeV/c). Moreover, being limited by energy and momentum conservation $n_{\rm hard}$ can not reach arbitrarily large values. In particular, these models implement a pp impact parameter dependence which explains the so called jet-pedestal effect in the underlying event of hard collisions [7].

The underlying event is conventionally defined as the region transverse to the leading jet or particle in an event, $\pi/3 < |\Delta\varphi| < 2\pi/3$, where $\Delta\varphi$ is the azimuthal distance to the leading object. Figure 1 shows for pp collisions at $\sqrt{s} = 7$ TeV the measured charged particle density in the transverse region as a function of the leading particle transverse momentum ($p_{\rm T}$) [8]. The particle density exhibits a steep rise at low $p_{\rm T}$ until $p_{\rm T} \approx 4$ GeV/c where a plateau is reached. In the framework of MPI, events with a higher leading particle $p_{\rm T}$ correspond to more central pp collisions in which the probability for additional uncorrelated hard scatterings is enhanced. This effect is also well known in nucleus–nucleus collisions and described by the proportionality of yield from hard processes to the nuclear overlap function. For example, in Pb–Pb collisions about half of yield of particles or jets from hard processes is found in the 10% most central collisions. Consequently selecting high-$p_{\rm T}$ particles one selects also preferentially more central collisions until a maximal centrality bias is reached.

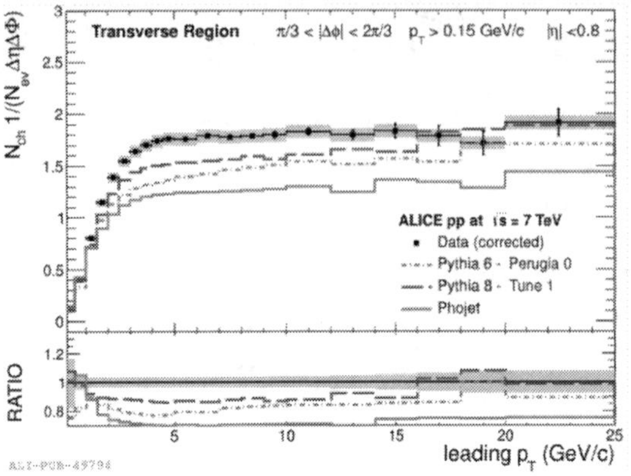

Figure 1. Number density in the transverse region of leading particles as the function of the leading particle p_T for pp collisions at $\sqrt{s} = 7$ TeV [8].

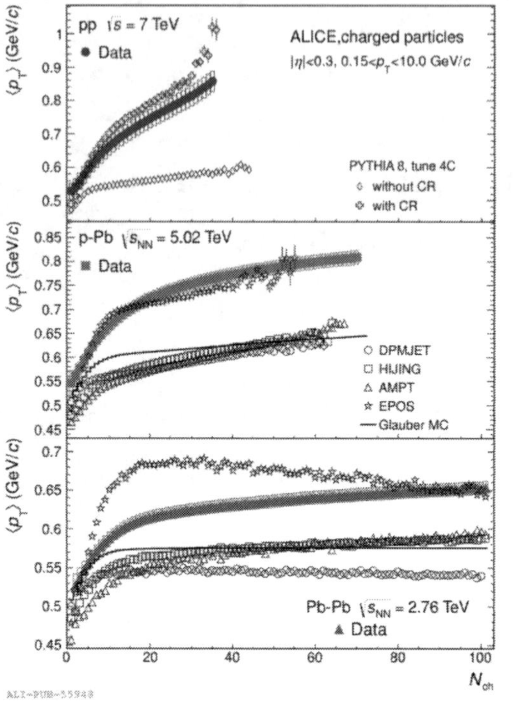

Figure 2. The mean particle transverse momentum as a function of multiplicity in $|\eta| < 0.3$ for pp collisions at $\sqrt{s} = 7$ TeV (upper panel), p–Pb collisions at $\sqrt{s_{NN}} = 5.02$ TeV (middle panel) and Pb–Pb collisions at $\sqrt{s_{NN}} = 2.76$ TeV (lower panel) [11].

Coherence effects between multiple scatterings, called colour re-connections (CR), have been introduced in order to explain the steep rise of the particle mean p_T as a function of the event multiplicity N_{ch} observed in pp collision [9, 10] . Figure 2 (upper panel) shows the measured $\langle p_T \rangle (N_{ch})$ compared to the results from PYTHIA8 simulations with and without CR [11]. The model without CR clearly fails to describe the data, whereas by including this effect the agreement is quite good. The EPOS LHC Monte Carlo [12] generator which includes collective hydrodynamic flow for small systems like pp, is also in good agreement with $\langle p_T \rangle (N_{ch})$ and the underlying event measurements at the LHC.

At LHC energies, the large number of initial hard parton–parton scatterings is a common feature of high-multiplicity pp, p–Pb and Pb–Pb collisions. In the MPI model, high-multiplicity pp collisions arise from low-impact parameter collisions and statistical upward fluctuations of the number of MPIs per event. Depending on the position and width of the pseudorapidity (η) window $\Delta\eta$ in which multiplicity is counted these events are also expected to contain harder than average partonic collisions and with partons fragmenting into a larger than average number of hadrons (fragmentation bias). In Pb–Pb, the mean number of initial parton–parton scatterings is determined almost entirely by the collision centrality. Additional biases are weak. Proton-lead collisions can be expected to lie in between these two cases. In models that treat p–Pb collisions as independent p–N collisions, the number of parton–parton scatterings is expected to be determined by the p–Pb and p–N centralities. Therefore a detailed insight into MPI related effects in pp collisions is needed to understand pp as a reference for p–Pb.

2. The structure of high multiplicity pp collisions

Two-Particle Azimuthal Correlations

Two-Particle azimuthal correlations represent a powerful tool to understand the origin of high-multiplicity pp collisions. Such studies involve measuring the distributions of the relative azimuthal angle $\Delta\varphi$ between pairs of particles: a "trigger" particle in a certain transverse momentum $p_{\mathrm{T,trig}}$-interval and an "associated" particle in a $p_{\mathrm{T,assoc}}$-interval. In these correlations, the fragmentation products of parton-parton scatterings manifest themselves as characteristic near-side ($\Delta\varphi = 0$) peak and away-side ($\Delta\varphi = \pi$) ridge structures. The number of correlated particles per trigger particle is defined as the yield of particles in the peaks over the uncorrelated background which is constant in $\Delta\varphi$. In models describing high-multiplicity pp events as superpositions of multiple scatterings this normalised yield is expected to be constant or to increase as a function of multiplicity. In the case where a soft component in the p_{T}-range of the trigger becomes dominant at high multiplicities, the yield per trigger-particle decreases.

In pp collisions at LHC energies, an increase of the near- and away-side yields as a function of the charged particle multiplicity is observed (see Fig. 3) [13]. This rise is well described by recent PYTHIA6 tunes and PYTHIA8 and in these models it is due to the increase of the number of MPIs and the increase of the average Q^2 of the collisions. The number of observed trigger particles depends on the number of initial partons and the number of fragments per parton. Since the latter rises with multiplicity, a non linear increase of the number of trigger particles with multiplicity is observed. Hence, the number of trigger particles is not a good measure of the number of MPIs contributing to the event multiplicity.

In order to reduce the sensitivity to parton fragmentation, ALICE studies the number of uncorrelated seeds ($N_{\mathrm{uncorrelated\,seeds}}$), defined as the ratio of the number of trigger particles to the sum of near and away-side yields per trigger particle plus one. This ratio increases almost linearly with multiplicity and indicates an onset of a saturation at the highest multiplicities 4. In PYTHIA, $N_{\mathrm{uncorrelated\,seeds}}$ is proportional to the number of MPIs and the saturation is related to a steep drop of the probability distribution for the number of MPIs (for example by 4 orders of magnitude between 12 and 24 for PYTHIA6 tune Perugia-0).

Sphericity

Another way to characterise the structure of high-multiplicity pp events is the transverse sphericity. This event-by-event observable can vary between 0 (jet like event) and 1 (particle distribution isotropic in azimuth). In data this observable shows a strong rise as a function of multiplicity until it reaches a plateau value (Fig. 5) [14]. The rise is well reproduced by Monte Carlo event generators. However, at the highest multiplicities they exhibit a turnover to lower sphericity values indicating a stronger (higher p_{T}) jet component as compared to data.

Heavy Flavor yields

An important test of the MPI origin of high multiplicity events is the study of heavy flavour yields as a function of multiplicity. Heavy quarks (c, b) are created in hard processes with a minimum momentum transfer of $Q > 2m_{\mathrm{Q}} \gg \Lambda_{\mathrm{QCD}}$ and they are very rarely created during hadronization of a different parton species. Hence, heavy flavour hadrons are ideal tags for hard processes down to zero p_{T}. Under the assumption that the number of MPIs is proportional to the hard cross section and that the soft particle multiplicity scales with the number of MPIs one expects that the yield from any hard sub-process increases with multiplicity. In ALICE, this behaviour as been verified for D-meson and J/ψ production (see Fig. 6).

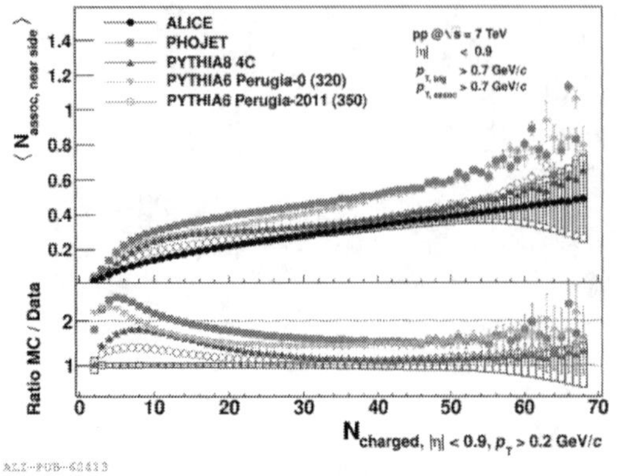

Figure 3. Number of associated particles per trigger particle on the nearside from two-particle azimuthal correlations with $p_{\rm T,trig}, p_{\rm T,assoc} > 0.7$ GeV/c in pp collisions at $\sqrt{s} = 7$ TeV [13].

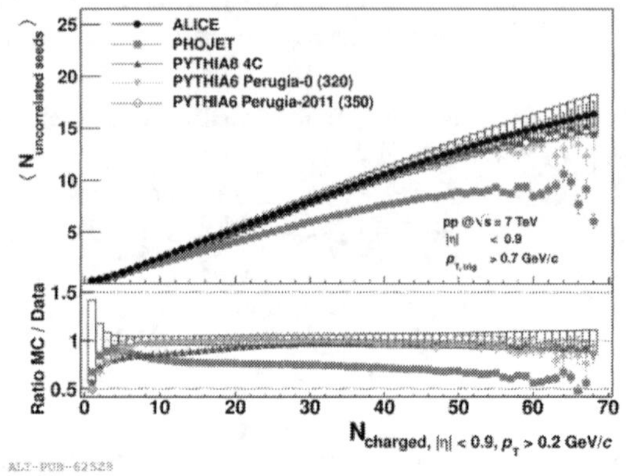

Figure 4. Number of uncorrelated seeds as defined in the text from two-particle azimuthal correlations with $p_{\rm T,trig}, p_{\rm T,assoc} > 0.7$ GeV/c in pp collisions at $\sqrt{s} = 7$ TeV [13].

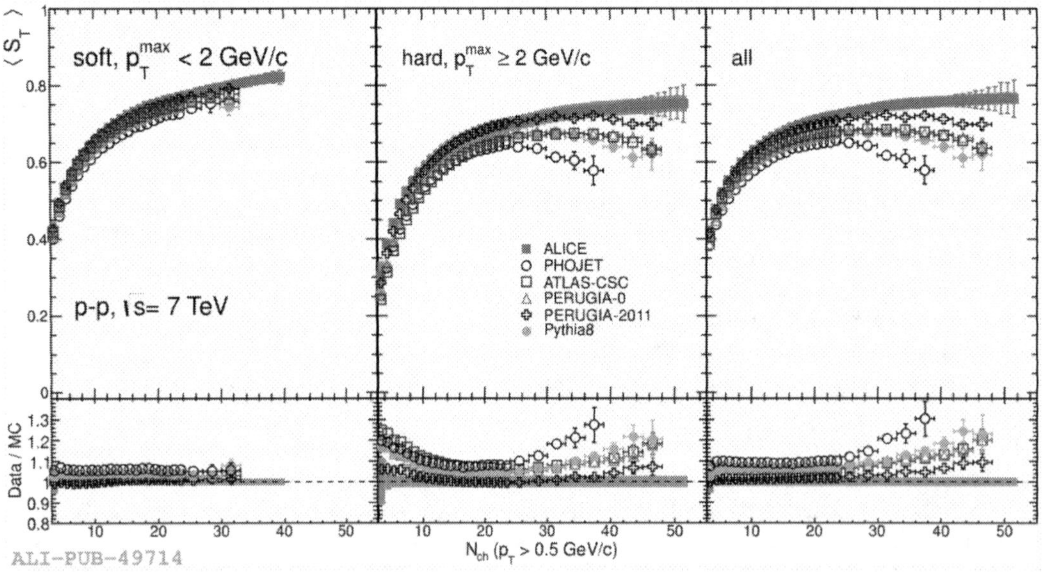

Figure 5. The transverse sphericity as a function of multiplicity in pp collisions at $\sqrt{s} = 7$ TeV for soft events defined as events not containing any particle with $p_{\rm T} > 2$ GeV/c (left panel), hard events defined as events containing at least one particle with $p_{\rm T} > 2$ GeV/c (middle panel) and all events (right panel) [14].

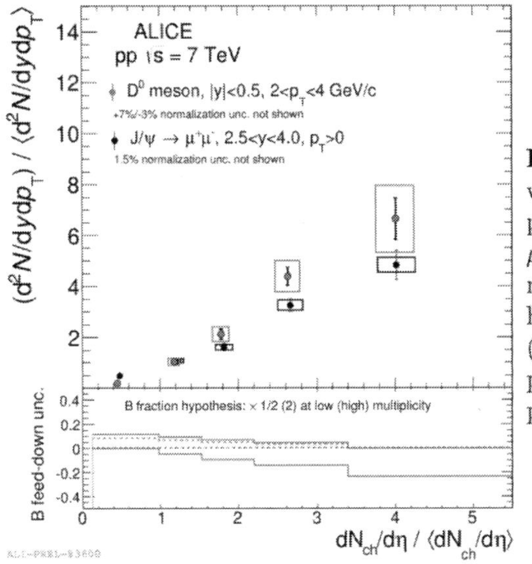

Figure 6. D^0-meson self normalised yields vs multiplicity in $2 < p_T < 4$ GeV/c in pp collisions at 7 TeV compared to J/psi $\to \mu^+\mu^-$. Multiplicity was measured via the number of tracklets (segments formed by two hits in different layers of the pixel detector (SPD) compatible with the reconstructed primary vertex) and is expressed as charged particle density normalised to its mean value.

3. MPIs in p–A

In p–A collisions the number of MPIs is proportional to the number of binary nucleon–nucleon collisions ($N_{\rm coll}$). These MPIs overlap in a transverse collision area similar to the one in pp. However, in comparison to pp, much larger number of MPIs can be reached. It is plausible that effects related to MPIs observed in pp are also relevant in p–A collisions. For example effects due to colour reconnections could be enhanced in p–A due to the larger number of MPIs.

Using the same factorisation approach as in pp and assuming that p–A collisions can be described as an independent superposition of pp collisions one can write for the mean number of MPIs in p–A:

$$\langle n_{\rm hard}\rangle_{\rm pA} = \langle N_{\rm coll}\rangle_{\rm MB}\langle n_{\rm hard}\rangle_{\rm pp} \tag{1}$$

This implies that in general, particle yields from hard processes $Y_{\rm hard}$ normalised by the number of binary collisions scale like

$$\frac{Y_{\rm hard}}{\langle N_{\rm coll}^{\rm cent}\rangle} \propto \frac{\langle n_{\rm hard}\rangle_{\rm pN}}{\langle n_{\rm hard}\rangle_{\rm pp}} \tag{2}$$

For centrality integrated p–A collisions the ratio $\langle n_{\rm hard}\rangle({\rm pN})/\langle n_{\rm hard}\rangle({\rm pp})$ is unity. However it is important to note that in this framework it can deviate from unity in centrality event classes. This fact has to be taken into account when studying cold nuclear effects comparing p–A results to the the pp baseline. Two examples, the nuclear modification factor and $\langle p_T\rangle$ as a function of multiplicity are discussed in the following sections.

MPIs and centrality selection in p–A

In general, centrality is defined via estimators that depend monotonically on the number of nucleon-nucleon collisions, e.g. multiplicity and summed energy in a certain pseudo-rapidity range. ALICE employs various sub-detector systems covering disjunct pseudo-rapidity ranges to estimate centrality. $CL1$ is the number of clusters counted in the 2nd layer of the Silicon Pixel Detector covering $|\eta| < 1.4$. $V0A$ and $V0C$ are the summed amplitudes measured by a pair of scintillator arrays covering $2.8 < \eta < 5.1$ and $-3.7 < \eta < -1.7$, respectively. Centrality classes are defined as percentiles of the multiplicity/summed-amplitude distributions.

Journal of Physics: Conference Series **535** (2014) 012012 doi:10.1088/1742-6596/535/1/012012

In order to study the influence of the centrality selection on MPIs in a coherent superposition of p–N collisions we coupled the PYTHIA6 event generator to a p–Pb Glauber MC calculation (G-PYTHIA). For each MC Glauber event PYTHIA is used N_{coll} times to generate N_{coll} independent pp collisions. The resulting charged particle multiplicity distribution in the above mentioned η ranges are used to define centrality classes similar to the ones used in data [15]. Figure 7 shows the average number of hard scatterings ($\langle n_{\text{hard}} \rangle$) per binary collision as a function of centrality. In the 20% most central collisions $\langle n_{\text{hard}} \rangle$ per N_{coll} is higher than the centrality averaged value and in the most peripheral collisions this ratio is far below this average. As discussed above this bias has consequences for the binary scaling of hard processes. Figure 8 shows the nuclear modification factor $Q_{\text{pPb}}^{\text{CL1}}$ which is the ratio of the inclusive charged particle spectrum measured in p–Pb to the pp reference scaled by N_{coll}. ALICE does not use the conventional notation R_{pPb} to indicate the biased nature of the observable. At high p_{T} (> 6 GeV/c), the centrality dependence of $Q_{\text{pPb}}^{\text{CL1}}$ reflects the bias on $\langle n_{\text{hard}} \rangle$. In addition $Q_{\text{pPb}}^{\text{CL1}}$ rises with p_{T}. This is the consequence of an additional autocorrelation bias. Hard processes contribute themselves to the multiplicity. This contribution increases with the hardness of the scattering. Hence pp events containing a hard scattering are classified more central than an average collision. At high p_{T} the G-PYTHIA model of independent collisions describes the data well. Global particle production scales with the number of participants and as we will see later in the low- and intermediate p_{T} region effects reminiscent of collective behaviour in A–A are observed. Hence, as expected our simple model does not describe the low- and intermediate p_{T} region. One important observation in this region that will be discussed in the next section is the increase of the average transverse momentum $\langle p_{\text{T}} \rangle$ with multiplicity.

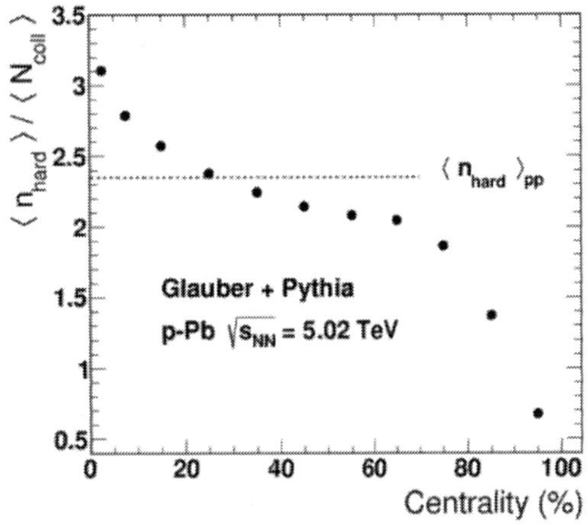

Figure 7. Average number of hard collisions per binary p–nucleon collision as a function of centrality from the G-PYTHIA calculation described in the text.

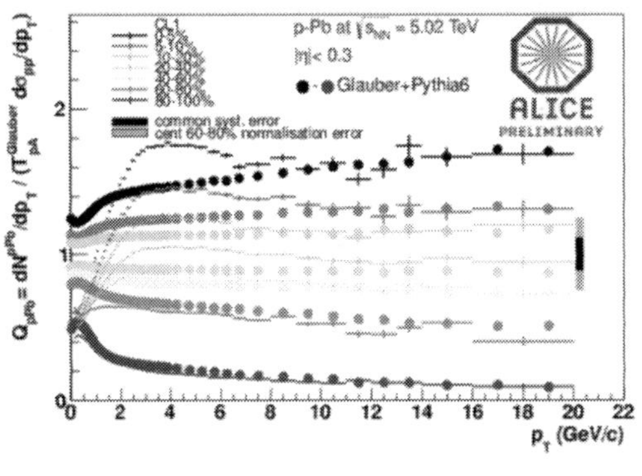

Figure 8. Q_{pPb} for the CL1 centrality estimator compared to the G-PYTHIA calculation as described in the text for p–Pb collisions at $\sqrt{s_{\text{NN}}} = 5.020$ TeV .

Multiplicity dependent measurements in p–Pb

Figure 2 shows the average transverse momentum as a function of multiplicity for pp collisions at $\sqrt{s} = 7$ TeV, p–Pb at $\sqrt{s_{\text{NN}}} = 5.02$ TeV and Pb–Pb at $\sqrt{s_{\text{NN}}} = 2.76$ TeV [11]. From measurements of, $\langle p_{\text{T}} \rangle (N_{\text{ch}})$ in pp at several energies we expect that the $\sqrt{s}$ -dependence of this observable is weak and, hence, we assume the results for the three collision systems can be

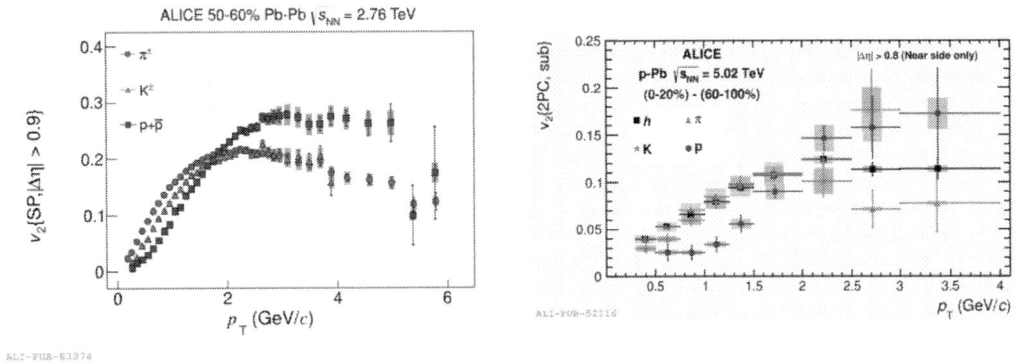

Figure 9. For hadrons, pions, kaons and protons: (left panel) $v_2(p_T)$ for mid-central Pb–Pb collisions [24] and (right panel) $v_2\{2PC, sub\}(p_T)$ from 2-particle correlations in the 0–20% multiplicity class after subtraction of the correlation from the 60–100% multiplicity class [26].

directly compared. With respect to Pb–Pb, in p–Pb, $\langle p_T \rangle$ shows a much stronger increase with multiplicity following the pp data up to $N_{ch} = 14$. Note that multiplicities around 14 correspond to typical p–Pb collision, whereas pp collisions at this multiplicity are already strongly biased ($N_{ch} > 14$ corresponds to 50% (10%) of the p–Pb (pp) cross-section). As discussed in the introduction, in pp this rise cannot be explained by an independent superposition of multiple parton scatterings and can be attributed to the effect of colour re-connections between strings. Similar to the model described above one can try to describe the rise of $\langle p_T \rangle$ in p–Pb collisions by a superposition of parton scatterings from an incoherent superposition of p–nucleon collisions [16, 11] (see 2 middle panel, solid line). The observed rise of $\langle p_T \rangle$ is significantly stronger than expected suggesting that also in this case effects from interactions between MPIs like colour re-connections are at work. The EPOS generator which includes hydrodynamic collective flow can reproduce the p–Pb data, however, it fails to describe peripheral Pb–Pb collisions. ALICE has also measured the multiplicity dependence of $\langle p_T \rangle$ for identified particles (π, K, p, Λ). Here, a clear mass ordering $\langle p_T \rangle_\Lambda > \langle p_T \rangle_p > \langle p_T \rangle_K > \langle p_T \rangle_\pi$ is observed [18], which is an indication for collective expansion with a common velocity field. The same kind of mass ordering is also qualitatively expected from colour re-connections [17].

In heavy-ion collisions, the increase of $\langle p_T \rangle$ and its mass ordering find their natural explanation in the collective radial expansion of the system [19]. This picture can be tested in a blast-wave framework with a simultaneous fit to all particles for each multiplicity bin [20]. From the fit one obtains the mean radial velocity $\langle \beta \rangle$ and kinetic freeze-out temperature T_{kin}. It is found that the trend of T_{kin}-$\langle \beta \rangle$ as a function of multiplicity is very similar in p–Pb and Pb–Pb collisions [18]. The same trend, albeit at a 30% smaller T_{kin}, can be also reproduced with pp collisions simulated by PYTHIA8 when colour reconnections are included.

Two-particle angular correlations in p–Pb
The strongest evidence for collective effects in p–Pb results from the study of triggered 2-particle angular correlations in the azimuthal ($\Delta\varphi$) and pseudo-rapidity ($\Delta\eta$) differences. These correlations show symmetric ridge like-structures elongated in $\Delta\eta$ at the near-side ($\Delta\varphi = 0$) and away-side ($\Delta\varphi = \pi$) of the trigger particle [21, 22, 23]. They are very similar to the momentum anisotropy observed in Pb–Pb [24], where the effect is attributed to collectivity (flow) (Fig. 9) [26]. Whereas these correlations can be explained by hydrodynamic flow or models based on the Color Glass Condensate [25], the above mentioned MPI Monte Carlo models based on colour re-connections are not yet able to describe this observation.

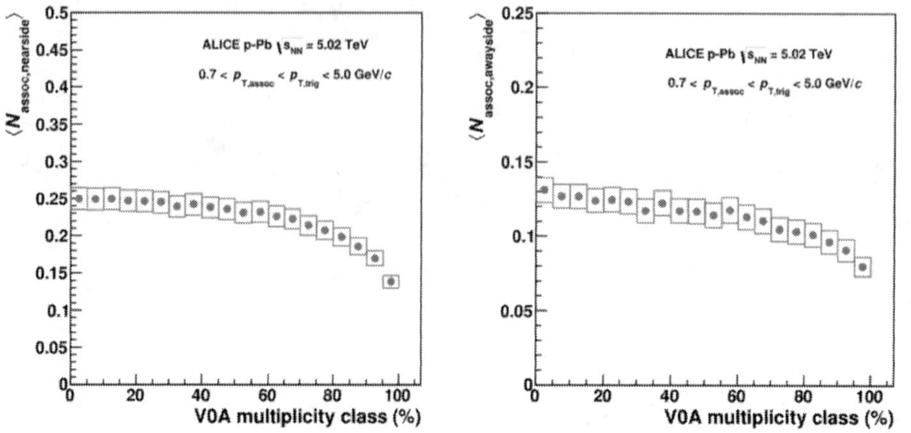

Figure 10. Near-side (left) and away-side (right) per-trigger yield as a function of V0A multiplicity class in p–Pb collisions at $\sqrt{s_{\rm NN}} = 5.02$ TeV with $0.7 < p_{\rm T,assoc}, p_{\rm T,trig} < 5.0$ GeV/c [27].

Whereas $\langle p_{\rm T}\rangle$ and long-range correlations are sensitive to collective effects, possible local changes to parton fragmentation can be studied by jet-like short range correlations in $\Delta\eta - \Delta\varphi$. Figure 10 shows the near-side and away-side per trigger yields as a function of the V0A multiplicity class in p–Pb collisions at $\sqrt{s_{\rm NN}} = 5.02$ TeV [27]. Contrary to the strong event activity dependence observed for $\langle p_{\rm T}\rangle$ and for the long-range correlations, the yields measured in jet-like correlations stay unmodified over a wide range of V0A multiplicity percentile. Only for the lowest multiplicities ($> 70\%$) a significant decrease of the yields is observed. It can be expected that this observation puts new constraints on models implementing coherence effects between MPI or collective hadronization.

3.1. Summary

ALICE has explored the rich phenomenology of MPI in pp collisions. In particular, the study of the multiplicity dependence of jet-like azimuthal two-particle correlations shows in a quite direct way that, in agreement with Monte Carlo simulations, high multiplicity pp events can be understood as a superposition of semi-hard scatterings.

Studies of observables as a functions of multiplicity have been extended to p–Pb. Phenomena which are in nucleus–nucleus well established as the consequence of collective hydrodynamic flow are observed. The observed increase of $\langle p_{\rm T}\rangle$ and its mass ordering has been also observed in pp where it can be attributed to colour re-connections between strings formed by multiple parton-parton interactions. It can be speculated that this effect plays also an important role in p–Pb where the MPIs overlap in a transverse region similar to pp. The EPOS Monte Carlo generator which includes collective fragmentation effects via hydrodynamic flow for small collision systems is in agreement with the pp and p–Pb data. Hydrodynamic flow is also a natural explanation for the symmetric double-ridge structure observed in p–Pb. With the aim to get more insight into the possible influence of these effects on parton fragmentation, ALICE has studied jet-like angular correlations at low $p_{\rm T}$ as a function of multiplicity in p–Pb. Surprisingly the fragmentation properties expressed as the yield associated to a trigger particles stays constant over a wide range of multiplicities. Future model comparisons will show whether these findings can be used to further constrain coherent fragmentation effects.

References

[1] Sjöstrand T and van Zijl M 1987 *Phys. Rev.* D **36** 2019
[2] Sjostrand T and Skands P Z 2004 *JHEP* **0403** 053
[3] Sjostrand T and Skands P Z 2005 *Eur. Phys. J.* C **39** 129
[4] Bahr M, Butterworth J M and Seymour M H 2009 *JHEP* **0901** 065
[5] Sjostrand T, Mrenna S and Skands P Z 2006 JHEP **0605** 026
[6] Sjostrand T, Mrenna S and Skands P Z 2008 *Comput. Phys. Commun.* **178** 852
[7] Sjostrand T and van Zijl M 1987 *Phys. Lett.* B **188** 149
[8] Abelev B B *et al.* (ALICE Collaboration) 2012 *JHEP* **1207** 116
[9] Skands P Z and Wicke D 2007 2007 *Eur. Phys. J.* C **52** 133
[10] Gieseke S, Rohr C and Siodmok A, 2012 *Eur. Phys. J.* C **72** 2225
[11] Abelev B B *et al* (ALICE collaboration) 2013 *Phys. Lett.* B **727** 371-380
[12] Pierog T *et al.* arXiv:1306.0121 [hep-ph].
[13] Abelev B B *et al.* (ALICE Collaboration) 2013 *JHEP* **1309** 049
[14] Abelev B *et al.* (ALICE Collaboration) 2012 *Eur. Phys. J.* C **72** 2124
[15] Morsch A (ALICE Collaboration) 2014 *J. Phys. Conf. Ser.* **509** 012021
[16] Bzdak A and Skokov V *Phys. Lett.* B **726** (2013) 408
[17] A. Ortiz Velasquez *et al. Phys. Rev. Lett.* **111** (2013) 4, 042001
[18] Abelev B B *et al* (ALICE Collaboration) 2014 *Phys. Lett.* B **728** 25-38
[19] Heinz U W 2004 Concepts of heavy ion physics, CERN-2004-001-D
[20] Schnedermann E, Sollfrank J and Heinz U W 1993 *Phys. Rev.* C **48** 2462
[21] Chatrchyan S (CMS Collaboration) *et al* 2013 *Phys. Lett.* B **718** 795
[22] Abelev B *et al* (ALICE Collaboration) 2013 *Phys. Lett.* B **719** 29-41
[23] Aad G *et al* (ATLAS Collaboration) 2013 *Phys. Rev. Lett.* **110** 182302
[24] Abelev B B *et al.* (ALICE Collaboration) 2014 arXiv:1405.4632 [nucl-ex]
[25] Dusling K and Venugopalan R 2013 *Phys. Rev.* D **87** 5 054014
[26] Abelev B B *et al.* (ALICE Collaboration) 2013 *Phys. Lett.* B **726** 164-177
[27] Abelev B B *et al.* (ALICE Collaboration) 2014 arXiv:1406.5463 [nucl-ex].

Anisotropic flows and the shear viscosity of the QGP within a transport approach

S. Plumari[a,b], **G.L. Guardo**[a,b], **A. Puglisi**[b], **F. Scardina**[a,b] **and V. Greco**[a,b]

[a] *Department of Physics and Astronomy, University of Catania, Via S. Sofia 64, I-95125 Catania (Italy)*
[b] *Laboratorio Nazionale del Sud, INFN-LNS, Via S. Sofia 63, I-95125 Catania (Italy)*

E-mail: salvatore.plumari@hotmail.it

Abstract. We study the build up of elliptic flow v_2 and high order harmonics v_n within a transport approach at fixed shear viscosity to entropy density ratio η/s and with initial state fluctuations. In particular we study the effect of a temperature dependent η/s for two different beam energies: RHIC for Au+Au at $\sqrt{s} = 200\,GeV$ and LHC for $Pb + Pb$ at $\sqrt{s} = 2.76\,TeV$. We find that for the two different beam energies considered the suppression of the elliptic flow and of higher harmonics $v_3(p_T)$ and $v_4(p_T)$ due to the viscosity of the medium have different contributions coming from the cross over or QGP phase. Moreover, we discuss the correlation between the initial spatial anisotropies ϵ_n and flow coefficients v_n. We observe that the elliptic flow v_2 is strongly correlated with initial eccentricity ϵ_2. While higher harmonics v_3 and v_4 are weakly correlated to their asymmetry measure in coordinate space ϵ_3 and ϵ_4.

1. Introduction

The experimental results accumulated in these years in the ultra relativistic heavy ion collisions (uRHICs) first in the experiments conducted at RHIC and more recently at the LHC has shown that the elliptic flow $v_2 = \langle cos(2\,\varphi_p)\rangle = \langle (p_x^2 - p_y^2)/(p_x^2 + p_y^2)\rangle$, is the largest ever observed in these uRHICs [1, 2]. The elliptic flow is a measurement of the momentum anisotropy of the emitted particles and it is an observable that encodes information about the transport properties of the matter created in these collisions. In particular it is an observable sensitive to the shear viscosity to entropy density ratio η/s. Theoretical calculations within viscous hydrodynamics [3, 4] and in the recent years also calculation performed within transport approach [5, 6, 7] have shown that this large value of v_2 is consistent with a matter with a very low shear viscosity to entropy density ratio η/s close to the conjectured lower bound for a strongly interacting system, with $\eta/s = 1/4\pi$ [8]. Moreover, both calculations have shown that the v_2 depends sensitively on η/s. Furthermore, in the recent years the possibility to measure experimentally the event-by-event angular distribution of emitted particle has made possible the measurement of higher harmonics $v_n = \langle cos(n\,\varphi_p)\rangle$ for $n > 2$. The comparison between event-by-event viscous hydrodynamical calculations and the experimental results for v_n seems to confirm a finite but not too large value of $4\pi\eta/s \sim 1 - 3$ [9, 10]. However, small value of $\eta/s \sim 1/4\pi$ is not an evidence of the creation of a QGP phase. A phenomenological estimation of its temperature dependence could give information if the matter created in these collisions undergoes a phase transition [11, 12].

Information about a temperature dependence of η/s can be achieved studying the $v_2(p_T)$ and the high order harmonic $v_n(p_T)$ in wider range of energies. Similar studies have been performed using a transport approach but only for the elliptic flow [13].

There are several theoretical indications that η/s should have a particular behavior with the temperature[11, 12, 14, 15, 16, 17]. As an example in Fig.(1) it is shown a collection of theoretical results about the temperature dependence of η/s. Fig.(1) shows that in general η/s should have a typical behavior of phase transition with a minimum close to the critical temperature T_C [11, 12]. On one hand at low temperature estimations of η/s performed in

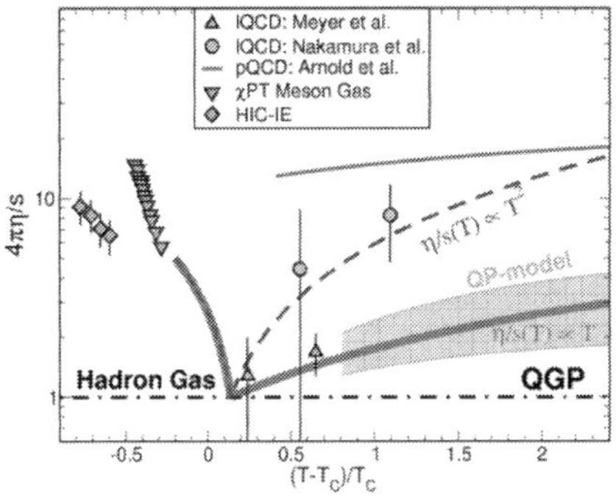

Figure 1. Different parametrizations for η/s as a function of the temperature. The orange area refers to the quasi-particle model predictions for η/s [18]. The different lines indicate different possible T dependencies while green line represents the result for the pQCD calculation [19]. Symbols are as in the legend. See the text for more details.

the chiral perturbation theory for a meson gas [14, 15], have shown that in general η/s is a decreasing function with the temperature, see down-triangles in Fig.(1). Similar results for η/s have been extrapolated from heavy-ion collisions at intermediate energies, see HIC-IE diamonds in Fig.(1). On the other hand at higher temperature $T > T_c$ lQCD calculation have shown but with large error bars that in general η/s becomes an increasing function with the temperature [16], see up-triangles and circles in Fig.(1). In general due to the large error bars in the lQCD results for η/s it is not possible to infer a clear temperature dependence in the QGP phase. The analysis at different energies of $v_2(p_T)$ and the extension to high order harmonics $v_n(p_T)$ can give further information about the T dependence of η/s. In this proceeding we will show some new results on $v_n(p_T)$ for the two different systems $Au + Au$ at $\sqrt{s} = 200\,GeV$ and $Pb + Pb$ at $\sqrt{s} = 2.76\,TeV$.

2. Kinetic approach at fixed shear viscosity to entropy density ratio

In this section we introduce a transport approach in which it is possible to fix the η/s ratio. This approach provides a tool to estimate the viscosity of the plasma valid in a wider range of η/s and p_T than hydrodynamical framework. We perform such simulations using a relativistic transport code developed in these years to perform studies of the dynamics of heavy-ion collisions at both RHIC and LHC energies [5, 7, 20, 21, 22]. The evolution of the phase-space distribution function $f(x, p, t)$ is given by solving the Relativistic Boltzmann Transport (RBT) equation:

$$p^\mu\, \partial_\mu f(x,p) = \int d\Gamma_2 d\Gamma_{1'} d\Gamma_{2'} (f_{1'} f_{2'} - f f_2) |\mathcal{M}|^2 \delta^4(p + p_2 - p_{1'} - p_{2'})\,, \qquad (1)$$

where $d\Gamma_k = d^3p_k/2E_k(2\pi)^3$ and $\mathcal{M}$ denotes the transition amplitude for the elastic processes which is directly linked to the differential cross section $|\mathcal{M}|^2 = 16\pi\, s\,(s - 4M^2) d\sigma/dt$ with s the Mandelstam invariant. Numerically we solve the RBT equation using the so called test particle method and the collision integral is solved by using Monte Carlo methods based on the stochastic interpretation of transition amplitude [23, 5, 21].

In order to study the dynamical evolution of the fireball with a certain $\eta/s(T)$ we determine locally in space and time the total cross section σ_{tot} needed to have the wanted local viscosity.

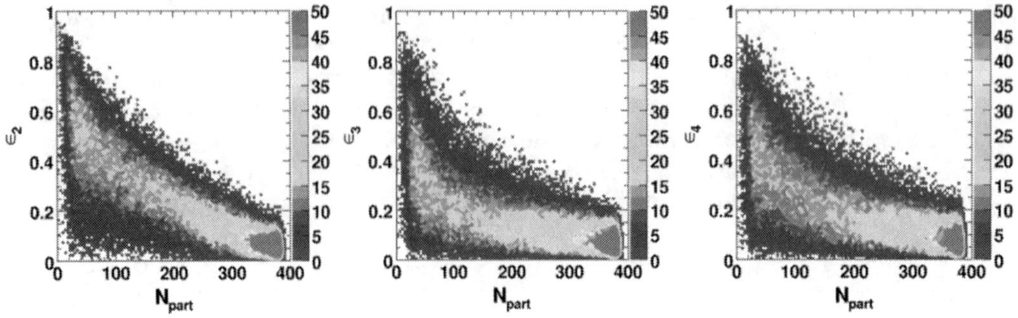

Figure 2. In the left, middle and right panel it is shown the coefficient ϵ_n as a function of the participant nucleons N_{part} respectively for $n = 2, 3$ and 4.

In the Chapmann-Enskog theory and for a pQCD inspired cross section, typically used in parton cascade approaches [24, 25, 5, 26, 27, 23, 6], with $d\sigma/dt \sim \alpha_s^2/(t - m_D^2)^2$, the η/s is given by the following expression:

$$\eta/s = \frac{1}{15}\langle p\rangle\,\tau_\eta = \frac{1}{15}\frac{\langle p\rangle}{g(a)\sigma_{tot}\rho}, \qquad (2)$$

where $a = m_D/2T$, with m_D being the screening mass regulating the angular dependence of the cross section, while $g(a)$ is the proper function accounting for the pertinent relaxation time $\tau_\eta^{-1} = g(a)\sigma_{tot}\rho$ associated to the shear transport coefficient and it is given by:

$$g(a) = \frac{1}{50}\int dy y^6 \left[(y^2+\frac{1}{3})K_3(2y) - yK_2(2y)\right] h\left(\frac{a^2}{y^2}\right), \qquad (3)$$

where K_n-s are the Bessel functions and the function h relate the transport cross section to the total cross section $\sigma_{tr}(s) = \sigma_{tot}\, h(m_D^2/s)$ with $h(\zeta) = 4\zeta(1+\zeta)[(2\zeta+1)ln(1+1/\zeta) - 2]$. Therefore the total cross section is evaluated locally by mean the Eq.(2).

3. Initial conditions

Our simulations start at the initial time $\tau_0 = 0.6 fm/c$. The initial conditions are longitudinal boost invariant. In the following discussion, we will consider two different types of initial conditions in coordinate space. One of them consist in a fixed distribution in transverse plane by using the standard Glauber model as used in previous works, see [5, 7, 20, 13]. The second one consist of an initial profile changing event by event that will be described below.

In order to generate an event by event initial profile we use the Monte-Carlo Glauber model to obtain the initial density distribution for each event. Woods-Saxon distribution is used to sample the nucleons randomly in the two colliding nucleus. In this way we generate a discrete distribution for these nucleons. We employ a geometrical method to determine if the two nucleons one for the nucleus A and the other one for the nucleus B are colliding: two nucleons collide each other if the relative distance in the transverse plane is $d_T \leq \sqrt{\sigma_{NN}/\pi}$ where σ_{NN} is the nucleon-nucleon cross section. The calculation of N_{coll} and N_{part} is given by counting the number of collisions and the number of participating nucleons for each event. In this way the position of the participating nucleons will fluctuate from event-to-event. The discrete distribution for the nucleons is converted to a smooth one by assuming for each nucleon a gaussian distribution centered in the nucleon position. Finally, the density in the transverse

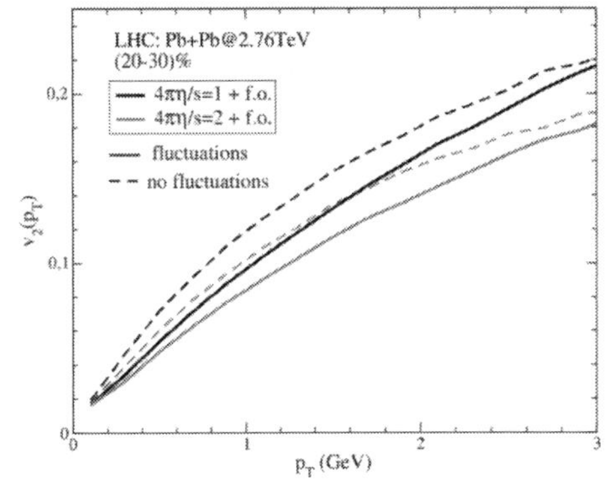

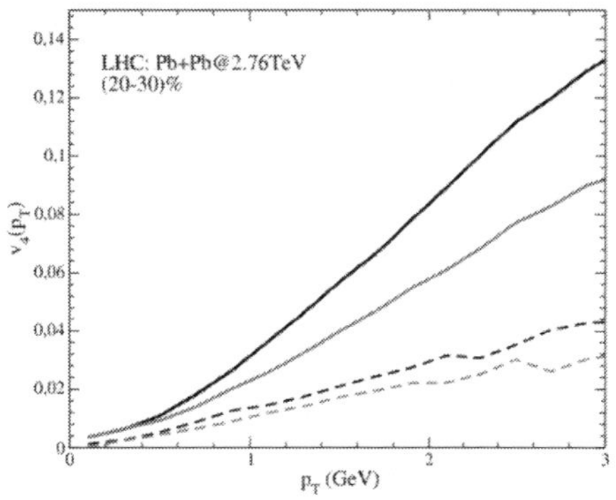

Figure 3. Left: differential elliptic flow $v_2(p_T)$ at midrapidity and for $(20-30\%)$ centrality. The black and red solid lines refer to the case with fluctuations and for $4\pi\eta/s = 1 + f.o.$ and $4\pi\eta/s = 2 + f.o.$ respectively. Term f.o. means the increase of η/s at lower temperature. The black and red dashed lines are for the case without fluctuations in the initial geometry. Right: differential $v_4(p_T)$ at midrapidity and for $(20-30\%)$ centrality with the same legend as in the left panel.

plane is given by the following sum

$$\rho_T(x,y) \propto \sum_{i=1}^{N_{part}} \exp\left[-\frac{(x-x_i)^2+(y-y_i)^2}{2\sigma_{xy}^2}\right] \tag{4}$$

where σ_{xy} is the Gaussian width and in the following calculations it has been fixed to $\sigma_{xy} = 0.5\,fm$. Given $\rho_T(x,y)$ we can calculate the initial anisotropy in coordinate space. The initial anisotropy is quantified in terms of the following coefficients ϵ_n:

$$\epsilon_n = \sqrt{< r_T^n \cos(n\phi) >^2 + < r_T^n \sin(n\phi) >^2} / < r_T^2 > \tag{5}$$

where $r_T = \sqrt{x^2+y^2}$ and $\phi = arctan(y/x)$ are the polar coordinate in the transverse plane. In the following discussion we will concentrate on the fireball produced in $Au + Au$ at $\sqrt{s_{NN}} = 200\,GeV$ and $Pb + Pb$ at $\sqrt{s_{NN}} = 2.76\,TeV$. For the nucleon-nucleon cross section we have used $\sigma_{NN} = 42\,mb$ and $70\,mb$ respectively. Fig.(2) shows the ϵ_2, ϵ_3 and ϵ_4 as a function of the number of participant N_{part}.

For the initialization in momentum space at RHIC (LHC) energies we have considered for partons with transverse momentum $p_T \leq p_0 = 2\,GeV$ $(3\,GeV)$ a thermalized spectrum in the transverse plane with a maximum initial temperature $T_0 = 340\,MeV$ $(510\,MeV)$. While for partons with $p_T > p_0$ we have assumed the spectrum of non-quenched minijets according to standard NLO-pQCD calculations with a power law shape. The transverse momentum of the particles is distributed uniformly in the azimuthal angle.

4. Effects of η/s on the $v_n(p_T)$

In this section we discuss the results for the final $v_n(p_T)$ corresponding to the different initial configurations. Differential flow coefficients $v_n(p_T)$ are very sensitive observables to the transport properties of the medium like the ratio η/s and they can give us more informations about the matter created in the heavy ion collisions. In the following discussion with $v_n(p_T)$ we mean the root mean square $\sqrt{< v_n^2 >}$ as it has been done in experimental data using the event plane

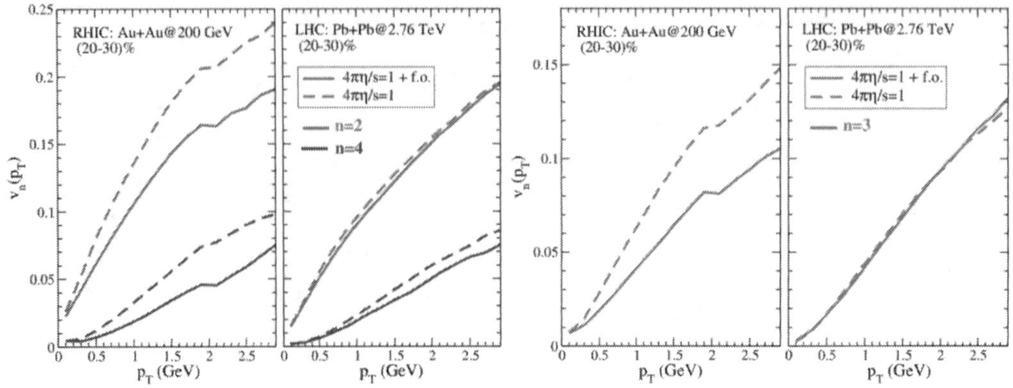

Figure 4. Left panel: differential $v_2(p_T)$ and $v_4(p_T)$ green and blue lines respectively at midrapidity and for $(20-30)\%$ collision centrality. The comparison is between the two systems: $Au + Au$ at $\sqrt{s} = 200\,GeV$ and $Pb + Pb$ at $\sqrt{s} = 2.76\,TeV$. The dashed lines refer to the case with a constant $\eta/s = (4\pi)^{-1}$ during all the evolution while the solid lines to the case with $\eta/s = (4\pi)^{-1}$ at higher temperature and with an increasing η/s ratio at lower temperature. Right panel: differential $v_3(p_T)$ red lines with the same legend as in the left panel.

method. In the left panel of Fig.(3) we compare the differential $v_2(p_T)$ obtained with an initial state fluctuations (solid lines) as discussed in the previous section with the one obtained for an event by event averaged initial profile (dashed lines). These results are for the system $Pb + Pb$ at $\sqrt{s} = 2.76\,TeV$ and for $(20-30\%)$ centrality. For this collision centrality we have simulated $N_{event} = 500$ events. We have checked the convergence of our results for the v_2, v_3 and v_4 with lattice spacing of the calculation grid and with the number of test particles. As shown in the left panel of Fig.(3) the fluctuations in the initial geometry reduce the elliptic flow of about 10%. In general the introduction of the fluctuations in the initial geometry play the role to generate the higher order harmonics in particular the odd harmonics which were absent in the averaged initial configuration by symmetry. Different behavior were observed for the 4-th harmonic as shown in right panel of Fig.(3)where in this case we have that the role of the fluctuation is to increase the $v_4(p_T)$ by a factor 2. This is due to the fact that ϵ_4 is almost completely generated by the fluctuations. Moreover, comparing the black and red solid lines with the dashed ones we observe that now the v_4 becomes more sensitive to the value of the viscosity.

In order to study the role of the η/s ratio on the build up of the elliptic flow $v_2(p_T)$ and on the high order harmonics $v_n(p_T)$ we have performed two different calculations: one with a constant $4\pi\eta/s = 1$ during all the evolution of the system and the other one with $4\pi\eta/s = 1$ at higher temperature in the QGP phase and an increasing η/s in the cross over region towards the estimated value for hadronic matter $4\pi\eta/s \approx 6$ as depicted in Fig.(1). Such an increase of η/s at lower temperature allows for a smooth realistic realization of the kinetic freeze-out. This last kind of calculation is labeled by f.o.. This is because at lower temperature, according to the formula Eq.(2), to account for the increase of η/s towards the estimated value for the hadronic matter, the total cross section decrease.

In the left panel of Fig.(4) it is shown the elliptic flow $v_2(p_T)$ and the $v_4(p_T)$ (green and blue lines) at mid rapidity for $(20-30)\%$ centrality for both RHIC $Au + Au$ at $\sqrt{s} = 200\,GeV$ (left panel) and LHC $Pb+Pb$ at $\sqrt{s} = 2.76\,TeV$ (right panel). In general in agreement with what has been obtained in hydrodynamical calculations, the increase of the viscosity of the medium has the effect to reduce the v_2 and v_4. As we can see at RHIC energies comparing the blue dashed

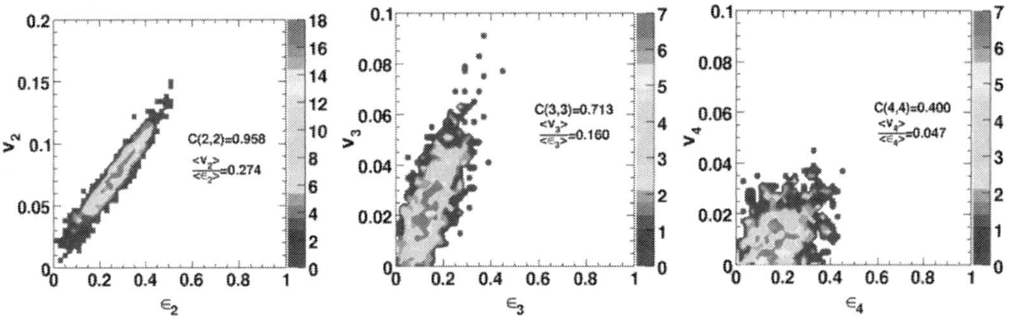

Figure 5. In the left, middle and right panel it is shown v_n and ϵ_n at mid rapidity and for $(20-30)\,\%$ respectively for $n = 2, 3$ and 4. In these calculations we have fixed $\eta/s = 1/(4\pi)$ during all the evolution of the system.

lines with the solid ones, in the left panel of Fig.(4), the $v_2(p_T)$ is sensitive to the increase of the η/s at lower temperature close to the cross over region. In particular the effect is a reduction of the elliptic flow of about 17%. A similar trend were observed for the 4-th harmonic $v_4(p_T)$ where we have a reduction due to the increase of η/s at lower temperature but the effect in this case is larger then the previous one of about 30%. The different sensitivity to the η/s can be attributed to their different formation time, $t_{v_4} > t_{v_2}$ [26]. On the other hand at LHC energies, left panel of Fig.(4), the scenario is different, the elliptic flow is almost unaffected by the increase of η/s ratio at low temperature (in the hadronic phase) as we can see comparing the green dashed line with the solid one. Instead we observe that the increase of η/s at lower temperature has a more sensitive effect on the $v_4(p_T)$ with a reduction of about 10%, see blue solid and dashed lines. Again this different sensitivity to the η/s in the cross over region between v_2 and v_4 at LHC are consistent with the results obtained at RHIC energies and related to the different formation time of the harmonics. The greater sensitivity at RHIC energies of both v_2 and v_4 to the η/s at low temperature is related to the different life time of the fireball. In fact the life time of the fireball at LHC is greater than that at RHIC, 8-10 fm/c at LHC against 4-5 fm/c at RHIC [28]. In general this means that at RHIC energies the v_n have not enough time to fully develop in the QGP phase. While at LHC energies we have that the v_n develops almost completely in the QGP phase and therefore it is less sensitive to the dynamics at lower T. In the right panel of Fig.(4) it is shown the triangular flow $v_3(p_T)$ (red lines) at mid rapidity for $(20-30)\%$ centrality and for both RHIC $Au+Au$ at $\sqrt{s} = 200\,GeV$ (left panel) and LHC $Pb+Pb$ at $\sqrt{s} = 2.76\,TeV$ (right panel). In agreement with what has been obtained for the even harmonics v_2 and v_4, we observe at RHIC energies a reduction of $v_3(p_T)$ due to the increase of the η/s at low temperature with a reduction of about 25%, while at LHC it is almost insensitive to the change of η/s in the cross over region.

5. Correlations between v_n and ϵ_n

In the recent years, the correlation between integrated v_2 and high order harmonics v_3, v_4 with the initial asymmetry in coordinate space ϵ_2, ϵ_3 and ϵ_4 has been studied in the event-by-event viscous hydrodynamics framework [29, 30, 31]. In general it has been shown that the elliptic flow is strongly correlated with initial eccentricity while a weaker correlation has been found for higher harmonics v_3, v_4 with ϵ_3 and ϵ_4. In this section we study these correlations within a transport approach with initial state fluctuation and for fixed η/s. In particular the results shown in this section are for an $\eta/s = 1/(4\pi)$ during all the evolution of the fireball.

In the left hand side of Fig.(5), we have shown the final integrated elliptic flow $< v_2 >$ as a function of the corresponding initial eccentricity $< \epsilon_2 >$ for each event. We observe that for $20-30\%$ of centrality the elliptic flow results are strongly correlated with the initial eccentricity. This is consistent with a correlation coefficient close to 1. The correlation coefficient is given by:

$$C(y,x) = \frac{\sum_i (x_i - \langle x \rangle)(y_i - \langle y \rangle)}{\sqrt{\sum_i (x_i - \langle x \rangle)^2 \sum_i (y_i - \langle y \rangle)^2}} \quad (6)$$

For the plot in the left panel of Fig.(5)we get $C(2,2) \approx 0.96$. In the middle and right panel of Fig.(5), we have shown the results obtained for the higher harmonics $< v_3 >$ and $< v_4 >$ respectively and for $20-30\,\%$ centrality. We observe that the correlation between triangular flow $< v_3 >$ and its initial $< \epsilon_3 >$ is weaker then that between $< v_2 >$ and $< \epsilon_2 >$ with a correlation coefficient $C(3,3) \approx 0.71$. Furthermore for the fourth harmonic flow $< v_4 >$ we observe an uncorrelation with its initial $< \epsilon_4 >$ as shown by a correlation coefficient close to zero. One explanation for this no linear correlation between $< v_4 >$ and $< \epsilon_4 >$ is that the for final $< v_4 >$ there is also a correlation with the initial $< \epsilon_2 >$. In [31] has been shown it possible to have a good linear correlation between $< v_4 >$ and a linear combination of the initial $< \epsilon_2 >$ and $< \epsilon_4 >$.

6. Conclusions

We have investigated within an event-by-event transport approach at fixed η/s, the build up of $v_2(p_t)$ and higher harmonics $v_3(p_T)$ and $v_4(p_T)$ for two different beam energies: RHIC for $Au + Au$ at $\sqrt{s} = 200\,GeV$ and LHC for $Pb + Pb$ at $\sqrt{s} = 2.76\,TeV$. We have found that at RHIC the $v_n(p_T)$ are more sensitive to the value of η/s at low temperature and in particular this sensitivity increase with the order of the harmonics. At LHC we get a different effect, the $v_n(p_T)$ are not so sensitive to the increase of η/s at low temperature. Finally, we have studied also the relation between v_n and ϵ_n for $n = 2, 3$ and 4 in ultra relativistic heavy-ion collisions within an event-by-event relativistic transport approach. We have shown that the second and third harmonic flows are correlated to their initial eccentricities, a result similar to that has been obtained in the recent years in viscous hydrodynamics.

7. Acknowledgments

V.Greco, F. Scardina and G.L. Guardo acknowledge the support of the ERC-StG Grant under the QGPDyn project.

References

[1] Adams J *et al.* (STAR Collaboration) 2005 *Nucl.Phys.* **A 757** 102–183 (*Preprint* nucl-ex/0501009)
[2] Aamodt K *et al.* (ALICE Collaboration) 2010 *Phys.Rev.Lett.* **105** 252302 (*Preprint* 1011.3914)
[3] Romatschke P and Romatschke U 2007 *Phys.Rev.Lett.* **99** 172301 (*Preprint* 0706.1522)
[4] Song H and Heinz U W 2008 *Phys.Rev.* **C 78** 024902 (*Preprint* 0805.1756)
[5] Ferini G, Colonna M, Di Toro M and Greco V 2009 *Phys.Lett.* **B 670** 325–329 (*Preprint* 0805.4814)
[6] Xu Z and Greiner C 2009 *Phys.Rev.* **C 79** 014904 (*Preprint* 0811.2940)
[7] Plumari S and Greco V 2012 *AIP Conf.Proc.* **1422** 56–61 (*Preprint* 1110.2383)
[8] Kovtun P, Son D and Starinets A 2005 *Phys.Rev.Lett.* **94** 111601 (*Preprint* hep-th/0405231)
[9] Schenke B, Jeon S and Gale C 2012 *Phys.Rev.* **C 85** 024901 (*Preprint* 1109.6289)
[10] Gale C, Jeon S, Schenke B, Tribedy P and Venugopalan R 2013 *Phys.Rev.Lett.* **110** 012302 (*Preprint* 1209.6330)
[11] Csernai L P, Kapusta J and McLerran L D 2006 *Phys.Rev.Lett.* **97** 152303 (*Preprint* nucl-th/0604032)
[12] Lacey R A, Ajitanand N, Alexander J, Chung P, Holzmann W *et al.* 2007 *Phys.Rev.Lett.* **98** 092301 (*Preprint* nucl-ex/0609025)
[13] Plumari S, Greco V and Csernai L 2013 (*Preprint* 1304.6566)
[14] Prakash M, Prakash M, Venugopalan R and Welke G 1993 *Phys.Rept.* **227** 321–366
[15] Chen J W, Li Y H, Liu Y F and Nakano E 2007 *Phys.Rev.* **D 76** 114011 (*Preprint* hep-ph/0703230)
[16] Meyer H B 2007 *Phys.Rev.* **D 76** 101701 (*Preprint* 0704.1801)

[17] Das S K and Alam J e 2011 *Phys.Rev.* **D 83** 114011 (*Preprint* 1011.4181)
[18] Plumari S, Alberico W M, Greco V and Ratti C 2011 *Phys.Rev.* **D 84** 094004 (*Preprint* 1103.5611)
[19] Arnold P B, Moore G D and Yaffe L G 2003 *JHEP* **0305** 051 (*Preprint* hep-ph/0302165)
[20] Plumari S, Puglisi A, Colonna M, Scardina F and Greco V 2013 *J.Phys.Conf.Ser.* **420** 012029 (*Preprint* 1209.0601)
[21] Plumari S, Puglisi A, Scardina F and Greco V 2012 *Phys.Rev.* **C 86** 054902 (*Preprint* 1208.0481)
[22] Ruggieri M, Scardina F, Plumari S and Greco V 2013 *Phys.Lett.* **B 727** 177–181 (*Preprint* 1303.3178)
[23] Xu Z and Greiner C 2005 *Phys.Rev.* **C 71** 064901 (*Preprint* hep-ph/0406278)
[24] Zhang B, Gyulassy M and Ko C M 1999 *Phys.Lett.* **B 455** 45–48 (*Preprint* nucl-th/9902016)
[25] Molnar D and Gyulassy M 2002 *Nucl.Phys.* **A 697** 495–520 (*Preprint* nucl-th/0104073)
[26] Greco V, Colonna M, Di Toro M and Ferini G 2008 *Prog.Part.Nucl.Phys.* (*Preprint* 0811.3170)
[27] Plumari S, Baran V, Di Toro M, Ferini G and Greco V 2010 *Phys.Lett.* **B 689** 18–22 (*Preprint* 1001.2736)
[28] Aamodt K *et al.* (ALICE Collaboration) 2011 *Phys.Lett.* **B 696** 328–337 (*Preprint* 1012.4035)
[29] Niemi H, Denicol G, Holopainen H and Huovinen P 2013 *Phys.Rev.* **C 87** 054901 (*Preprint* 1212.1008)
[30] Chaudhuri A, Haque M R, Roy V and Mohanty B 2013 *Phys.Rev.* **C 87** 034907 (*Preprint* 1211.2040)
[31] Gardim F G, Grassi F, Luzum M and Ollitrault J Y 2012 *Phys.Rev.* **C 85** 024908 (*Preprint* 1111.6538)

Journal of Physics: Conference Series **535** (2014) 012014 doi:10.1088/1742-6596/535/1/012014

Transverse Flow of Gluon Fields in Heavy Ion Collision

Guangyao Chen and Rainer . Fries

Cyclotron Institute and Department of Physics and Astronomy, Texas A&M University, College Station, TX 77843, USA

E-mail: glorychern@gmail.com

Abstract. We describe the dynamics of initial gluon elds in heavy ion collision using a formal recursive solution of the Yang Mills equations and solving for the energy momentum tensor analytically in a boost-invariant setup. We generalize the original McLerran-Venugopalan (MV) model in order to allow for realistic nuclear pro les. This leads to a transverse flow of gluon elds. This flow pattern is inherited by the quark gluon plasma fluid after thermalization. Its most interesting aspect is a rapidity-odd flow component. We show that this rapidity-odd flow does not break boost invariance and that it emerges naturally from the Yang Mills equations. It leads to directed flow of particles and introduces angular momentum to the system.

1. Introduction

Experiments at the Relativistic Heavy Ion Collider (RHIC) and the Large Hadron Collider (LHC) provide opportunities to study a deconfined phase of CD matter, uark Gluon Plasma (GP) [1, 2, 3]. In high energy nuclear collisions, the wave functions of incident nuclei are dominated by wee partons. High occupation numbers permit a quasi-classical description of the gluon fields in the nuclei before collisions called Color Glass Condensate (CGC), which should be valid up to a time scale inversely proportional to the saturation scale Q_s. Predictions from CGC can serve as initial conditions for a further hydrodynamic evolution of the bulk matter created in high energy nuclear collision [4, 5, 6].

There have been many efforts devoted to solving the ang Mills equations for the gluon field in the forward light cone after the collision, both numerically and analytically [7, 8, 9]. Here we use a formal recursive solution of the ang-Mills equations found by Fries et al. [7] to calculate the energy momentum tensor. In a generalized version of the McLerran-Venugopalan (MV) model [10, 11], which allows for transverse dynamics (MVTD), we calculate the flow of energy in transverse direction and find a rapidity-odd flow term which will eventually lead to directed flow of particles. These results had first been reported in [12, 13].

This article is organized as follows. First we briefly review the MV model in section 2. Then, in section 3 we show that the rapidity-odd flow term is a natural consequence of Gauss' Law in an electrodynamic analogue. We compute the Poynting vector of the gluon field in the forward light cone in the MVTD framework in section 4. Finally we discuss the phenomenological consequence of flowing gluon fields in heavy ion collision in section 5.

2. Gluon Fields in the MV Model

In the McLerran-Venugopalan model, the simplest incarnation of CGC, partons with large B orken-x moving along the light cone can be treated as a $SU(3)$-current $J^\mu = \delta^{\mu+}\delta(x^-)\rho_{\underline{a}}t^{\underline{a}}$ in light cone gauge, which generates soft gluons through the ang Mills equations $D_\mu F^{\mu\nu} = J^\nu$. The local charge density $\rho_{\underline{a}}(\vec{r})$ is stochastic with vanishing mean because of local color neutrality, but the variance is finite. A Gaussian distribution with variance μ is used to describe $\rho_{\underline{a}}(\vec{r})$ in the MV model,

$$\langle \rho_{\underline{a}}(\vec{r}_1)\rho_{\underline{b}}(\vec{r}_2)\rangle = \frac{g^2}{N_c^2-1}\delta_{\underline{ab}}\mu(\vec{r}_1)\delta^2(\vec{r}_1-\vec{r}_2)\,. \tag{1}$$

In our notation, μ is the average, *color-summed*, squared density of charges in the transverse plane.

A recursive solution [7] has been found for the ang-Mills equations [14] of the gluon field after collisions of two charges. The chromo-electric and chromo-magnetic fields $\vec{E}$ and $\vec{B}$ immediately after the collision of nuclei can then be calculated.

In the MV model, most of the physical observables are related to the gluon distribution function of a nucleus in light cone gauge [15]$(i,j=1,2)$

$$\langle A^i_{\underline{a}}(\vec{r})A^j_{\underline{b}}(\vec{r})\rangle = \frac{g^2}{8\pi(N_c^2-1)}\delta_{\underline{ab}}\delta^{ij}\mu(\vec{r})\ln\frac{Q^2}{m^2}\,, \tag{2}$$

where Q is a V cutoff and m is an infrared scale that can be introduced as an effective gluon mass. In the original calculation [15], μ was taken to be a constant area density, which is not a realistic approximation for nuclei. We have checked that the CD dynamics at larger distances can be separated from color glass physics if variations of $\mu(\vec{r})$ are small on the IR length scale $\sim 1/m$, i.e.

$$m^{-1}|\nabla\mu(\vec{r})| \ll \mu(\vec{r})\,. \tag{3}$$

Many well-known results of the MV model, in particular Eq. (2), hold with corrections appearing as an expansion in gradients $m^{-1}\nabla^i\mu(\vec{r})$. Details of this calculation will be reported elsewhere [16].

The longitudinal chromo-electric and -magnetic fields after the collision for small times τ can be written in terms of purely transverse gauge fields $A_1^i = A^i_{\underline{a},1}(\vec{r})t^{\underline{a}}$ and $A_2^i = A^i_{\underline{a},2}(\vec{r})t^{\underline{a}}$ of the nuclei before the collision, in their respective light cone gauge [7, 16, 17],

$$E^3 = \left(1+\frac{\tau^2}{4}D^2\right)E_0 + \mathcal{O}(\tau^4) \tag{4}$$

$$B^3 = \left(1+\frac{\tau^2}{4}D^2\right)B_0 + \mathcal{O}(\tau^4)\,, \tag{5}$$

where $D^2 = D^iD^i$ is the square of the covariant derivative with respect to gauge field $A_1^i + A_2^i$. $E_0 = ig\left[A_1^j, A_2^j\right]$, $B_0 = ig\epsilon^{jk}\left[A_1^j, A_2^k\right]$ are the longitudinal fields at $\tau = 0$, which are consistent with boundary conditions. Transverse fields do not receive contribution from zeroth order in τ, i.e. they vanish at $\tau = 0$. At order τ, the transverse fields can be expressed in terms of the initial longitudinal fields as [7, 16, 17]

$$E^i = -\frac{\tau}{2}\left(\sinh\eta\, D^iE_0 + \cosh\eta\,\epsilon^{ij}D^jB_0\right) \tag{6}$$

$$B^i = -\frac{\tau}{2}\left(\sinh\eta D^iB_0 - \cosh\eta\,\epsilon^{ij}D^jE_0\right)\,. \tag{7}$$

The transverse fields will receive corrections at order τ^3 and higher, but here we focus on the early time behavior. Higher order (in τ) contributions are discussed in [16, 18] .

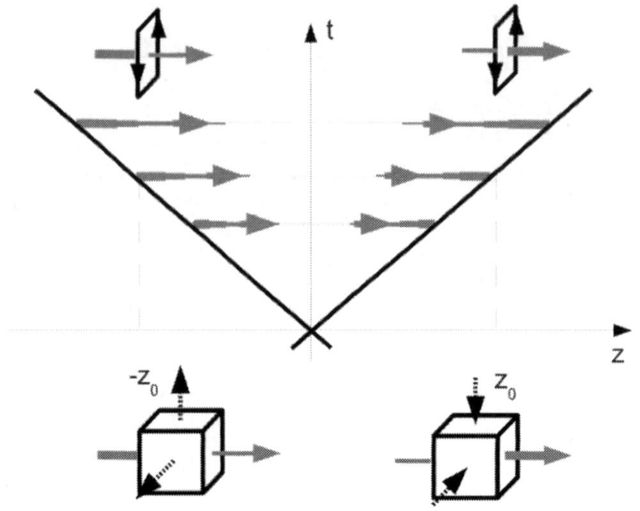

Figure 1. Rapidity-even transverse fields from Ampère's and Faraday's Laws and rapidity-odd transverse fields from Gauss' Law illustrated by transverse fields at $z = z_0$ and $z = -z_0$. Initial longitudinal fields are indicated by solid grey arrows, thickness reflects field strength. Refer to [13] for details.

Now the energy momentum tensor can be calculated from chromo-electric and -magnetic fields straightforwardly, e.g. the initial energy density is $\epsilon_0 = T^{00}(\tau = 0) = (E_0^2 + B_0^2)/2$. We are particularly interested in the Poynting vector T^{0i} which represents the initial transverse energy flux. The leading behavior of T^{0i} at small τ can be decomposed into two terms $T^{0i} = S_+^i + S_-^i$ where

$$S_+^i = \frac{\tau}{2} \cosh \eta \left(E_0 D^i E_0 + B_0 D^i B_0 \right) \tag{8}$$

is even in η, and,

$$S_-^i = \frac{\tau}{2} \sinh \eta \, \epsilon^{ij} \left(E_0 D^j B_0 - B_0 D^j E_0 \right) \tag{9}$$

is odd in η.

We would like to emphasize that the energy momentum tensor with such a Poynting vector is boost invariant despite the rapidity-odd flow term. Here boost invariance is defined for the full tensor $T^{\mu\nu}$ as

$$\Lambda_\mu^{\mu'}(y) T_{\mu'\nu'}(\eta) \Lambda_\nu^{\nu'}(y) = T_{\mu\nu}(\eta + y)\,, \tag{10}$$

where $\Lambda(y)$ is the Lorenz boost tensor with rapidity y along the z-axis. We will show in the next section that the rapidity-odd field is a natural consequence of the field equations.

3. An Electrodynamic Analogue

An electrodynamics problem with equivalent boundary conditions can provide an intuitive interpretation of the origin of the η-odd flow term. In the forward light cone the free electromagnetic fields obey the homogenous Maxwell Equations $\partial_\mu F^{\mu\nu} = 0$. We require equivalent boundary conditions on the light cone $\tau = 0$, i.e., $\vec{E}(\tau = 0, \vec{r}) = E_0(\vec{r})\vec{e}_z$, $\vec{B}(\tau = 0, \vec{r}) = B_0(\vec{r})\vec{e}_z$. In other words, the initial fields are again purely longitudinal. In principle this abelian problem can be solved analytically [14, 16], but we explore the solution order by order in powers of τ as we did in the case of CD for comparison.

The longitudinal fields in the abelian case can be obtained immediately from the CD solutions,

$$E^3 = \left(1 + \frac{t^2 - z^2}{4}\nabla^2\right) E_0 \tag{11}$$

$$B^3 = \left(1 + \frac{t^2 - z^2}{4}\nabla^2\right) B_0 \,, \tag{12}$$

while the transverse fields are

$$E^i = \frac{z}{2}\nabla^i E_0 + \frac{t}{2}\epsilon^{ij}\nabla^j B_0 \tag{13}$$

$$B^i = \frac{z}{2}\nabla^i B_0 - \frac{t}{2}\epsilon^{ij}\nabla^j E_0 \,, \tag{14}$$

for small times τ, i.e. $t^2 \approx z^2$. Now we can check these solutions against Gauss', Ampère's and Faraday's Laws.

It is easy to see that longitudinal fields E^3 and B^3 are decreasing away from the light cone $t^2 = z^2$. At the same transverse point, the electric (magnetic) flux through a transverse square area of size a^2 at fixed points $z = z_0 > 0$ and $z = -z_0$ have the same initial values $E_0 a^2$ ($B_0 a^2$) and diminish at the same rates $\nabla^2 E_0 a^2 t/2$ ($\nabla^2 B_0 a^2 t/2$). This reduction induces magnetic (electric) fields curling with the same chirality around the longitudinal fields at these two points, see Fig. 1.

On the other hand we consider two cubes of volume a^3. One cube has two sides located at z_0 and $z = z_0 - a$ along the z axis, and the total electric (magnetic) flux out of the box due to the longitudinal field is $-z_0 a^3 \nabla^2 E_0/2 > 0$ ($-z_0 a^3 \nabla^2 B_0/2 > 0$). The other cube has two sides located at $-z_0$ and $z = -z_0 + a$, and the net flux of longitudinal field of this cube has the opposite sign. From Gauss' Law we will have transverse electric (magnetic) fields with opposite signs induced at z_0 and $z = -z_0$.

To summarize, the transverse fields contain η-odd terms due to Gauss' Law and η-even terms due to Ampère's and Faraday's Laws. It should not be surprising that such transverse fields naturally give rise to an η-odd flow of energy as well.

4. Averaging ver Field Configurations

The expectation value of the initial energy density can be calculated from gluon distribution functions after averaging over source configurations [19],

$$\varepsilon_0(\vec{r}) = \frac{g^6}{32\pi^2}\frac{N_c}{N_c^2 - 1}\ln^2\frac{Q^2}{m^2}\mu_1(\vec{r})\mu_2(\vec{r}) \,. \tag{15}$$

The η-even flow term of the Poynting vector becomes,

$$S^i_+ = -\frac{\tau}{2}\cosh\eta\frac{g^6}{32\pi^2}\frac{N_c}{N_c^2 - 1}\ln^2\frac{Q^2}{\hat{m}^2}\nabla^i\left(\mu_1\mu_2\right) = -\frac{\tau}{2}\cosh\eta\nabla^i\epsilon_0 \,. \tag{16}$$

It mimics hydrodynamic flow by following the gradient of the energy density. The rapidity-odd flow is

$$S^i_- = -\frac{\tau}{2}\sinh\eta\frac{g^6}{32\pi^2}\frac{N_c}{N_c^2 - 1}\ln^2\frac{Q^2}{m^2}\left(\mu_2\nabla^i\mu_1 - \mu_1\nabla^i\mu_2\right) \,, \tag{17}$$

where the expectation value of two gluon fields in light cone gauge with one derivative has been used [16],

$$\left\langle \left(\partial^k A^i_{\underline{a}}\right)(\vec{x}_\perp) A^j_{\underline{b}}(\vec{x}_\perp)\right\rangle = \frac{g^2}{16\pi(N_c^2 - 1)}\delta_{\underline{ab}}\ln\frac{Q^2}{m^2}\nabla^l\mu(\vec{x}_\perp)\left(\delta^{jl}\delta^{ik} - \delta^{il}\delta^{jk} - \delta^{kl}\delta^{ij}\right) \,. \tag{18}$$

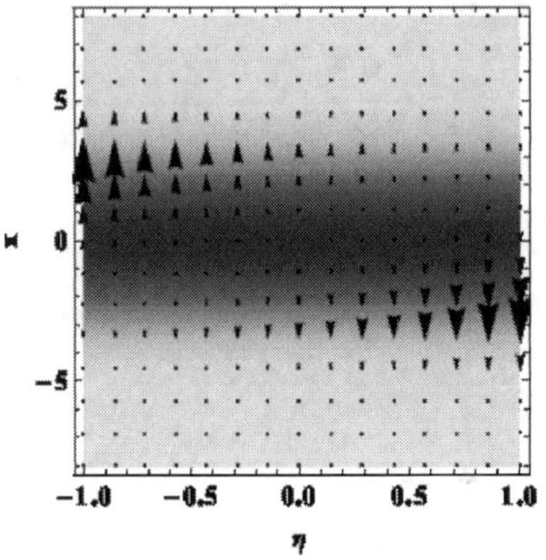

Figure 2. Flow field V^i (black arrows) and energy density ϵ_0 (shading) in the $\eta - x$-plane defined by $y = 0$ for Au+Au collisions at $b = 6$ fm. The flow will lead to a tilted fireball.

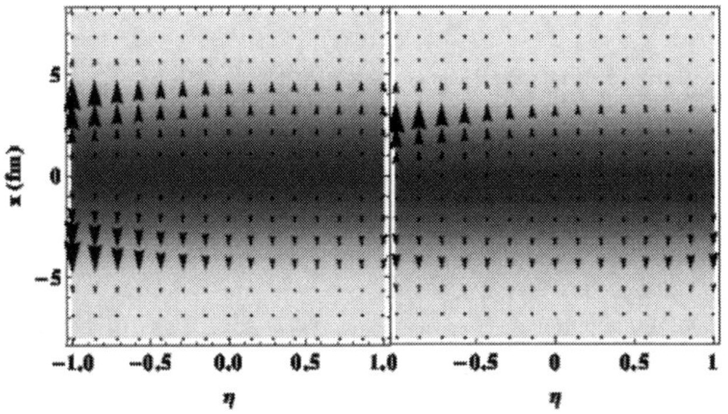

Figure 3. Flow field V^i for Au+Cu collisions in the $\eta - x$-plane. Left: $b = 0$ fm. Right: $b = 2$ fm.

. Discussion of Results

It is useful to normalize the initial average transverse energy flow by the average initial energy density,

$$V^i = \frac{T^{0i}}{\epsilon_0} = -\frac{\tau}{2}\left(\cosh\eta\frac{\nabla^i(\mu_1\mu_2)}{\mu_1\mu_2} + \sinh\eta\frac{\mu_2\nabla^i\mu_1 - \mu_1\nabla^i\mu_2}{\mu_1\mu_2}\right) \tag{19}$$

which is independent of the V cutoff and the IR regulator. It is a proxy for the velocity field if the system were to decay rapidly into thermalized particles. sing Woods-Saxon profiles for incident nuclei, V^i has been calculated in several situations [13].

Fig. 2 shows the average flow field V^i for the collision of two gold nuclei at impact parameter $b = 6$ fm in $\eta - x$-plane. The nucleus centered at $x = 3$ fm travels in the positive η-direction. The fireball shows a rotation pattern as if the initial gluon flux tubes preferred to expand in the wake of spectator nucleons.

Fig. 3 shows the average flow fields V^i for Au+Cu collisions in the $\eta - x$-plane at impact two parameters: $b = 0$ fm and $b = 2$ fm. The gold nucleus is traveling in the positive η direction. The flow field leads to a more prominent expansion on the copper side. The flow pattern becomes even more involved for Au+Cu collisions at finite impact parameter [13].

The flow of energy implies a finite angular momentum with magnitude comparable to that found in other calculations [20, 21]. Even without detailed knowledge of the thermalization mechanism, it is still reasonable to expect such a flow pattern to survive in a hydrodynamic phase after (partial) thermalization because of energy and momentum conservation. Eventually it should contribute to directed flow of particles observed at RHIC and LHC [22, 23, 24]. The matching to ideal hydrodynamic initial conditions has been worked out in [25]. A matching to 3+1D viscous hydrodynamics initial conditions will provide a tilted fireball similar to the fire streak model [26]. Our calculation suggests that color glass could be the microscopic mechanism to transfer angular momentum to the fireball. A detailed comparison to experimental data will require hydrodynamic simulations which are ongoing.

Acknowledgments
This pro ect was supported by the .S. National Science Foundation through CAREER grant PH -0847538, and by the ET Collaboration and DOE grant DE-FG02-10ER41682. We thank oseph Kapusta and ang Li who contributed to this pro ect.

References

[1] K. Adcox *et al.* [PHENIX Collaboration], Nucl. Phys. A **757**, 184 (2005)
[2] J. Adams *et al.* [STAR Collaboration], Nucl. Phys. A **757**, 102 (2005)
[3] B. Muller, J. Schukraft and B. Wyslouch, Ann. Rev. Nucl. Part. Sci. **62**, 361 (2012)
[4] P. F. Kolb and U. W. Heinz, in *Quark Gluon Plasma 3*, ed. R. C. Hwa et al., World Scienti c (2003), preprint nucl-th 0305084
[5] M. Luzum and P. Romatschke, Phys. Rev. C **78**, 034915 (2008) [Erratum-ibid. C **79**, 039903 (2009)]
[6] B. Schenke, S. Jeon and C. Gale, Phys. Rev. C **82**, 014903 (2010)
[7] R. J. Fries, J. I. Kapusta and Y. Li, preprint arXiv:nucl-th 0604054
[8] T. Lappi and L. McLerran, Nucl. Phys. A **772**, 200 (2006)
[9] J. Vredevoogd and S. Pratt, Phys. Rev. C **79**, 044915 (2009)
[10] L. D. McLerran and R. Venugopalan, Phys. Rev. D **49**, 3352 (1994)
[11] L. D. McLerran and R. Venugopalan, Phys. Rev. D **49**, 2233 (1994)
[12] G. Chen and R. J. Fries, J. Phys. Conf. Ser. **446**, 012021 (2013)
[13] G. Chen and R. J. Fries, Phys. Lett. B **723**, 417 (2013)
[14] A. Kovner, L. D. McLerran and H. Weigert, Phys. Rev. D **52**, 3809 (1995) Phys. Rev. D **52**, 6231 (1995)
[15] J. Jalilian-Marian, A. Kovner, L. D. McLerran and H. Weigert, Phys. Rev. D **55**, 5414 (1997)
[16] G. Chen, R. J. Fries, J. I. Kapusta and Y. Li, in preparation
[17] H. Fu ii, K. Fukushima and Y. Hidaka, Phys. Rev. C **79**, 024909 (2009)
[18] R. J. Fries, G. Chen, proceeding of this workshop.
[19] T. Lappi, Phys. Lett. B **643**, 11 (2006)
[20] . -T. Liang and X. -N. Wang, Phys. Rev. Lett. **94**, 102301 (2005) [Erratum-ibid. **96**, 039901 (2006)]
[21] V. Vovchenko, D. Anchishkin and L. P. Csernai, Phys. Rev. C **88**, no. 1, 014901 (2013)
[22] B. I. Abelev *et al.* [STAR Collaboration], Phys. Rev. Lett. **101**, 252301 (2008)
[23] J. Adams *et al.* [STAR Collaboration], Phys. Rev. C **73**, 034903 (2006)
[24] B. Abelev *et al.* [ALICE Collaboration], Phys. Rev. Lett. **111**, no. 23, 232302 (2013)
[25] R. J. Fries, J. I. Kapusta and Y. Li, Nucl. Phys. A **774**, 861 (2006)
[26] J. Gosset, J. I. Kapusta and G. D. Westfall, Phys. Rev. C **18**, 844 (1978)

Journal of Physics: Conference Series **535** (2014) 012015 doi:10.1088/1742-6596/535/1/012015

Centrality categorization and application to physics effects in high-energy $d + A$ collisions

J.D. Orjuela Koop for the PHENIX Collaboration

Department of Physics, University of Colorado Boulder. 390 UCB Boulder, CO 80309-0390

E-mail: `javier.orjuelakoop@colorado.edu`

Abstract. The study of numerous physics effects in small collision systems requires a careful characterization of the event geometry. In particular, many such phenomena have a strong dependence on the impact parameter of the collision. We describe the methodology utilized by PHENIX to select centrality classes in d+Au collisions via cuts on charge deposited at backward (Au-going) rapidity. The measured charge can be mapped to other geometric quantites using a Monte Carlo Glauber model. We also describe how autocorrelations between the process of interest and the backward rapidity charge introduce bias effects that alter the measurement of centrality-dependent invariant yields. Our framework provides a method to compute correction factors to account for such effects. We discuss their calculation and validation using the HIJING Monte Carlo Generator. It is found that centrality bias correction factors are small and slightly p_T dependent for d+Au collisions at $\sqrt{s_{NN}} = 200$ GeV, yet an order of magnitude larger and strongly p_T dependent for p+Pb collisions at $\sqrt{s_{NN}} = 5.02$ TeV. The implications of such corrections are discussed for selected physics observables and effects.

1. Introduction

Small collision systems constitute an excellent ground for the study of a wide range of physics effects. Such effects include gluon saturation, color neutralization [1], the modification of parton distribution functions in nuclei [2][3] and—more recently—the open question of collective flow in small systems [4, 5, 6, 7].

Nonetheless, a prerequisite to any such study is to characterize the event geometry in terms of quantities such as the number of binary collisions N_{coll}, the number of participants N_{part}, and the impact parameter (or centrality) of the event. Not only are these values needed to calculate important quantities such as nuclear modification factors, allowing us to compare particle yields in $p+p$ and $p(d)$+A events, but they also play a role in analyzing strongly geometry-dependent phenomena such as collective flow.

Customarily, high-energy experiments at RHIC and the LHC have determined event geometry by correlating geometric quantities with far-backward rapidity multiplicity as a physical observable. However, this correlation is biased when additional conditions are imposed on the event—such as the production of particles at midrapidity—leading to miscalculated centrality-dependent invariant particle yields.

In these proceedings, we discuss centrality categorization in d+Au events, following Ref. [8]. We begin by describing the methodology used by the PHENIX experiment to categorize event centrality and determine other geometric quantities from backward rapidity multiplicity measurements. We then discuss how bias effects arise for events with particle production at midrapidity. Afterwards, we examine the computation of bias correction factors and study their centrality and transverse momentum dependence at RHIC and LHC energies. Finally, we discuss the implications of such corrections for selected physics observables and effects.

2. Centrality Categorization in the PHENIX Experiment

The PHENIX experiment utlilizes two types of detectors to determine geometry in d+A events. Namely, a pair of beam-beam counters (BBCs) spanning the pseudorapidity range $3.0 < |\eta| < 3.9$, and zero-degree calorimeters (ZDCs) covering $|\eta| < 6$ [9, 10]. The BBCs measure charge deposited in the Au-going direction, which can be mapped to geometric quantities following a standard Monte Carlo Glauber approach. On the other hand, the ZDCs measure the energy of spectator neutrons, which can be used to validate our geometry characterization approach.

We now describe the Monte Carlo Glauber approach [11] used to map the charge deposited in the Au-going BBC to geometric quantities, such as $N_{\rm coll}$, $N_{\rm part}$, and centrality. d+Au events are simulated by randomly sampling the positions of nucleons in the deuteron from the Hulthén wavefunction [12]

$$\psi_d(r_{pn}) = \left(\frac{\alpha\beta(\alpha+\beta)}{2\pi(\alpha-\beta)^2}\right)^{1/2} \frac{\exp(-\alpha r_{pn}) - \exp(-\beta r_{pn})}{r_{pn}} \tag{1}$$

where r_{pn} is the inter-nucleon distance, with $\alpha = 0.228$ fm^{-1} and $\beta = 1.18$ fm^{-1}. For the gold nucleus, we sample from a Woods-Saxon distribution

$$\rho(r) = \frac{\rho_0}{1 + \exp[(r-R)/a]} \tag{2}$$

with parameters $R = 6.38$ fm and $a = 0.54$ fm. We assume a collision between nucleons to occur when they come within a distance of $\sqrt{\sigma_{NN}/\pi}$ of each other, with $\sigma_{NN} = 42$ mb. We are then able to count the number of wounded nucleons in each nucleus, thus obtaining $N_{\rm coll}$ and $N_{\rm part}$.

Once these quantities have been determined, we map them to experimental observables assuming summed charge in the Au-going BBC to be directly proportional to $N_{\rm coll}$, with fluctuations described by a negative binomial distribution (NBD) [13, 14]:

$$NBD(x;\mu,\kappa) = \left(1+\frac{\mu}{\kappa}\right)\frac{(\kappa+x-1)!}{x!(\kappa-1)!}\left(\frac{\mu}{\mu+\kappa}\right)^x. \tag{3}$$

The parameters μ and κ are found by fitting the experimental BBC charge distribution. We obtain the charge deposited in the BBC by folding the Glauber $N_{\rm coll}$ distribution, $Gl(n)$, with the NBD fluctuation response

$$P(x) = \sum_{n=1}^{N_{\rm binary}} Gl(n) \times NBD(x; n\mu, n\kappa). \tag{4}$$

Centrality classes can then be established by slicing the multiplicity distribution into quantiles, as shown in the upper panel of Figure 1. We observe a good agreement between real data and the Glauber calculation, as shown in the bottom panel of Figure 1. The observed deviation from unity is due to trigger inefficiency for low charge deposition. The minimum bias trigger consists in the simultaneous detection of at least one particle in both arms of the BBC.

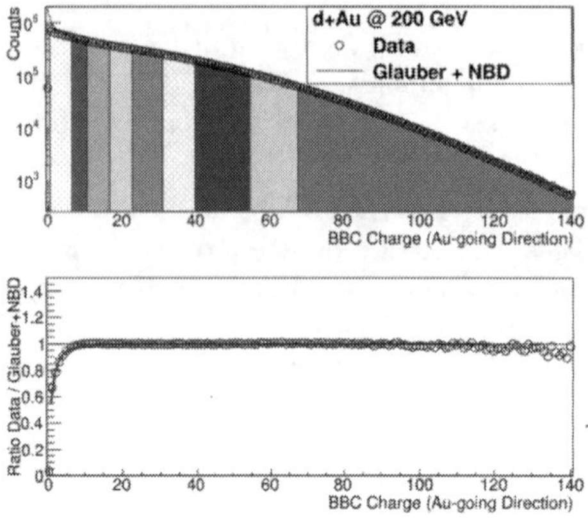

Figure 1: (upper) BBC Au-going charge distribution from d+Au events. Open circles represent real data. Glauber+NBD calculation shown as a histogram. (lower) Ratio of actual data to Glauber calculation results.

3. Bias Factor Corrections

In the previous section we showed how BBC multiplicity can be mapped to geometric quantities using standard Monte Carlo Glauber+NBD techniques. Once event geometry has been determined, we can calculate invariant particle yields for any given centrality selection. However, the mapping between multiplicity and geometry is biased for events where particles are produced at midrapidity, leading to the miscalculation of invariant yields. In this section we investigate the origin of the bias and how it can be corrected.

To understand the origin of the autocorrelation bias, let's consider $p+p$ collisions at 200 GeV, with a total inelastic cross section of 42 mb. In such case, measurements indicate that the PHENIX minimum bias trigger will fire $52 \pm 4\%$ of the time. However, this efficiency increases to $75 \pm 3\%$ for events with charge particle production at midrapidity as a result of their higher multiplicity.

Of the total inelastic $p+p$ cross section, we find that 28 of the 42 mb come from non-diffractive collisions, with the remaining 14 mb from single and double-diffractive events [15]. Nonetheless, since diffractive events result in particle production very close to the beam line, the minimum bias trigger will be biased towards non-diffractive collisions, where particle production at midrapidty is greatest. As a result, measured invariant yields will be biased towards a larger value.

Since nucleus-nucleus collisions effectively comprise multiple $p+p$ events, the bias effect described above will be present in d+Au events as well. In such case, there will exist a bias towards higher charge deposition in the BBC and, hence, towards larger centrality. Let's now examine the implications for centrality-dependent yields. In the case of peripheral events, there will be a deficit of midrapidity particles as they migrate to a more central categorization, resulting in smaller measured invariant yields that need to be corrected up through multiplication by an appropriate numerical factor. On the other hand, a midrapidity particle surplus in central events requires invariant yields to be corrected down.

We use the Glauber+NBD approach to compute numerical correction factors for each centrality class. We simulate d+Au events as comprising N individual $p+p$ collisions. Of these, we take just one to be biased toward higher multiplicity and higher trigger efficiency, with the other $N-1$ remaining unaffected. Biased collisions are modeled with a greater NBD contribution to BBC multiplicity in the Au-going direction, implemented by scaling the μ and

κ parameters of the NBD by 1.55 ± 0.23, as measured in $p+p$ data. We then take the ratio of invariant yields with and without the above bias, obtaining the correction factors shown in Table 1.

Table 1: Bias correction factors and different geometrical quantities for d+Au events at 200 GeV, as calculated with a Glauber+NBD approach.

Centrality	**0%-20%**	**20%-40%**	**40%-60%**	**60%-88%**
Bias Factor	0.94 ± 0.01	1.00 ± 0.01	1.03 ± 0.02	1.03 ± 0.006
$\langle N_{\text{coll}} \rangle$	15.1 ± 1.0	10.2 ± 0.7	6.6 ± 0.4	3.2 ± 0.2
$\langle N_{\text{part}} \rangle$	15.2 ± 0.6	11.1 ± 0.6	7.8 ± 0.4	4.3 ± 0.2

4. HIJING Study of the Bias Factors

The previous section described the calculation of bias correction factors for d+Au events based on the scaling of BBC multiplicity by a factor of 1.55. However, experimental data indicate a p_T dependence of this effect, as shown in the right panel of Figure 2. Our Glauber+NBD model does not allow for the study of this dependence. Therefore, we turn to the HIJING Monte Carlo generator [16] to examine the implications of this p_T dependence on our bias correction method, noting that any results will be model-dependent and not to be applied to experimental data.

4.1. Bias Factor Corrections for d+Au at 200 GeV

We assess the validity of HIJING for the study at hand by simulating p+p events at $\sqrt{s_{NN}} = 200$ GeV. We find the simulated minimum bias trigger efficiency to be 48%, compared to the measured value of $52 \pm 4\%$. It is also found that the presence of midapidity particles (i.e., $|\eta| < 0.35$) with $p_T > 1$ GeV increases BBC multiplicity by a factor of 1.62, compared with the previously quoted value of 1.55 from data. Furthermore, Figure 2 illustrates how HIJING qualitatively reproduces the p_T dependence of BBC multiplicity scaling observed in data.

In order to calculate bias factor correction factors in d+Au collisions, we define 4 centrality classes (0-20%, 20-40%, 40-60% and 60-88%) using the simulated BBC multiplicity distribution. We then calculate the particle yield at midrapidity for each centrality, thus obtaining *measured* yields where the effect of the autocorrelation bias is expected to be present.

We can isolate bias effects by determining the N_{coll} distribution for each centrality and then sorting events into centrality classes by matching these N_{coll} distributions. When particle yields are calculated in this manner, we refer to them as *truth* yields. Therefore, the ratio of *truth* to *measured* yields is exactly the desired bias correction factor. Table 2 shows the mean bias factors computed with HIJING for midrapidity particles with $p_T > 1$ GeV, where a good agreement is seen between HIJING and Glauber+NBD values.

The HIJING bias correction factors are shown as a function of p_T in left panel of Figure 3. We observe a modest p_T dependence of the bias factors, varying by less than 10% across p_T. The bias factors are seen to slightly rise in central events, and to slightly drop in peripheral events, in a manner consistent with expectations.

4.2. Bias Factor Corrections for p+Pb at 5.02 TeV

Given recent data from the LHC regarding p+Pb collisions at $\sqrt{s_{NN}} = 5.02$ TeV, we apply our HIJING methodology to this collision system, using charged particle multiplicity in the Pb-going direction within $-4.9 < \eta < -3.1$ to determine centrality.

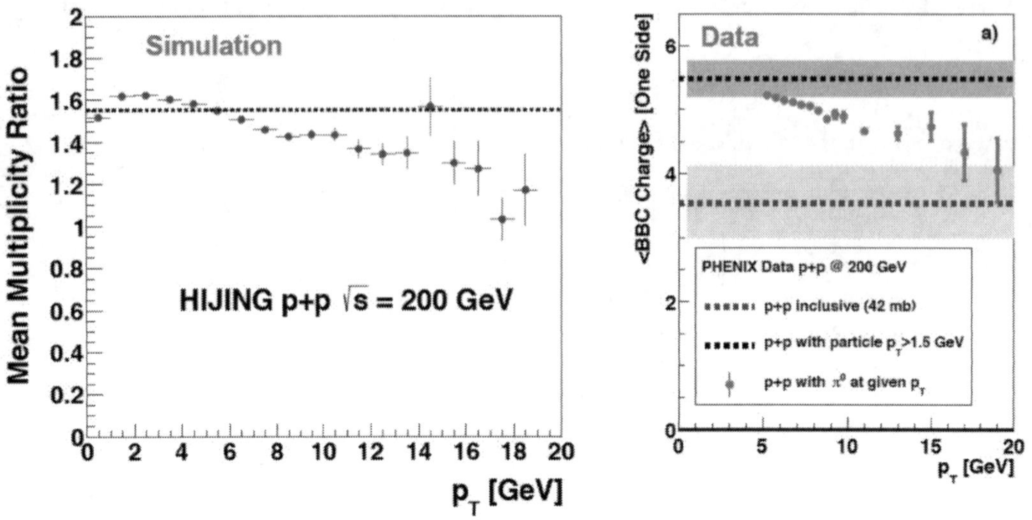

Figure 2: (left) Simulated ratio of Au-going BBC multiplicity for $p + p$ events with particle production at midrapidity to all inelastic collisions at $\sqrt{s_{NN}} = 200$ GeV. The dashed line at 1.55 indicates the measured increase in BBC multiplicity from data. (right) Measured charge deposition in the Au-going BBC as a function of midrapidity particle p_T for $p + p$ events at $\sqrt{s_{\mathrm{NN}}} = 200$ GeV.

Table 2: Mean bias factor corrections as a function of p_T for each centrality as calculated with HIJING.

Centrality	Mean Bias Factor Correction $1 \leq p_T < 5$	Mean Bias Factor Correction $5 \leq p_T < 10$	Mean Bias Factor Correction $10 \leq p_T < 15$	Mean Bias Factor Correction $15 \leq p_T < 20$
0-20%	0.951 ± 0.001	0.962 ± 0.001	1.000 ± 0.005	1.038 ± 0.020
20-40%	0.996 ± 0.001	1.008 ± 0.001	1.010 ± 0.006	0.996 ± 0.021
40-60%	1.010 ± 0.001	1.022 ± 0.001	1.019 ± 0.007	1.005 ± 0.025
60-88%	1.030 ± 0.001	1.026 ± 0.001	0.999 ± 0.008	0.991 ± 0.030

As before, we begin by simulating $p + p$ events at $\sqrt{s_{NN}} = 5.02$ TeV. We find that the presence of a particle at midrapidity increases multiplicity in the Pb-going direction by 1.67.

We calculate bias correction factors for 5 centrality categories (0-20%, 20-40%, 40-60%, 60-80% and 80-100%). The results, as a function of p_T are shown in Figure 3. We observe very large and p_T dependent correction factors, where multi-parton interactions play a substantial role [17]. These bias factors can be contrasted with ALICE data [18], as shown in Figure 4. Bearing in mind that Q_{pPb} scales as the inverse of the bias factors, we observe a good qualitative agreement between our HIJING results and the ALICE Q_{pPb}.

5. Application to Selected Physics Observables

In this section we present selected geometry-dependent physics observables available for study in d+Au collisions.

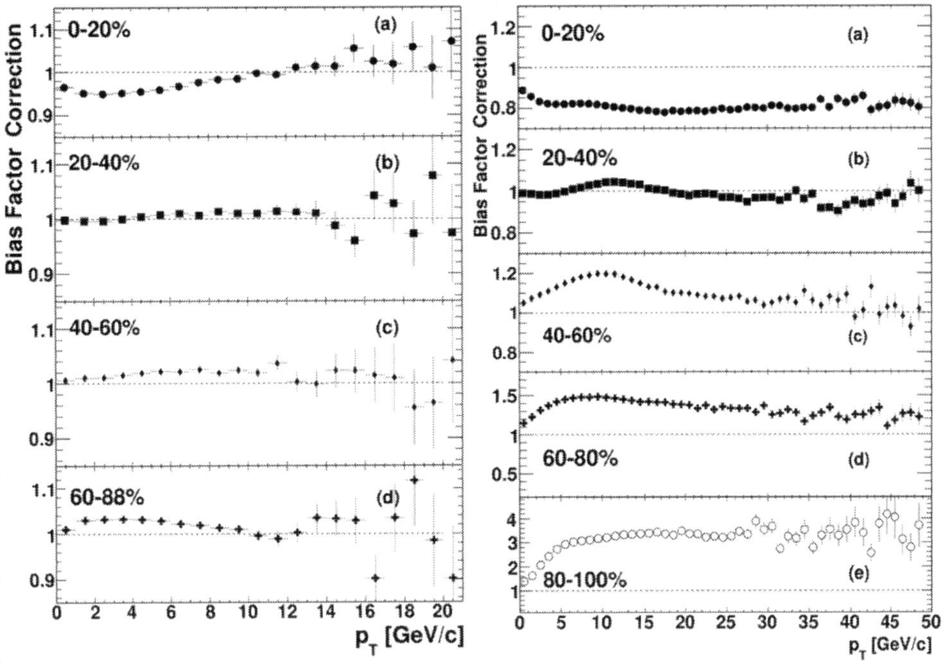

Figure 3: Bias factor corrections for (left) d+Au events at $\sqrt{s_{NN}} = 200$ GeV and (right) p+Pb events at $\sqrt{s_{NN}} = 5.02$ TeV.

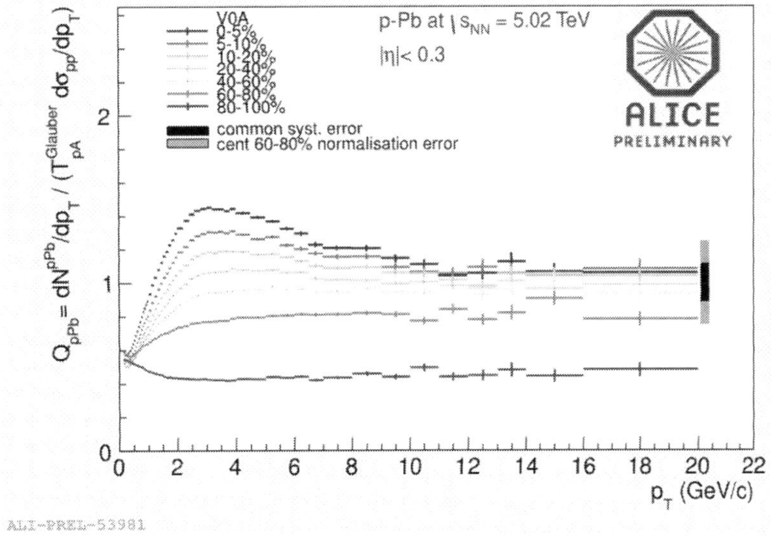

Figure 4: $Q_{pPb}(p_T)$ for p+Pb at $\sqrt{s_{NN}} = 200$ GeV for different centrality classes in the ALICE experiment [18].

5.1. Transverse Energy, Charged Particle Production and Bjorken Energy Density

Transverse energy and charged particle production are examples of important centrality-dependent quantities in the study of heavy-ion collisions [19].

Figure 5 illustrates transverse energy and charged particle production in PHENIX for d+Au and $Au + Au$ at $\sqrt{s_{NN}} = 200$ GeV as a function of centralitity (characterized by the number of participants $N_{\rm part}$). We observe a linear scaling of both quantities with $N_{\rm part}$, as well as the continuity of the curve from central d+Au to peripheral $Au + Au$ events.

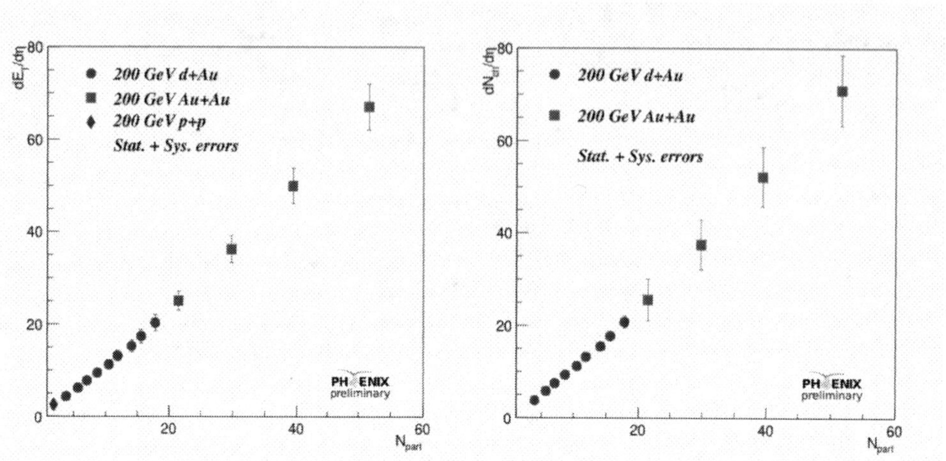

Figure 5: (left) Transverse energy production in d+Au and $Au + Au$ at $\sqrt{s_{NN}} = 200$ GeV as a function of centrality. (right) charged particle production in d+Au and $Au+Au$ at $\sqrt{s_{NN}} = 200$ GeV as a function of centrality.

The ratio of transverse energy to the overlap area between colliding nuclei corresponds to the Bjorken energy density,

$$\epsilon_{Bj}\tau = \frac{1}{\pi R^2}\frac{dE_T}{d\eta}. \tag{5}$$

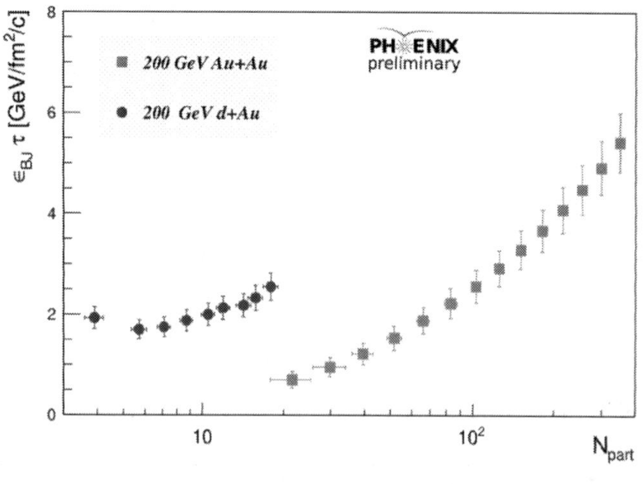

Figure 6: Bjorken energy density as a function of centrality for d+Au and $Au + Au$ events at $\sqrt{s_{NN}} = 200$ GeV.

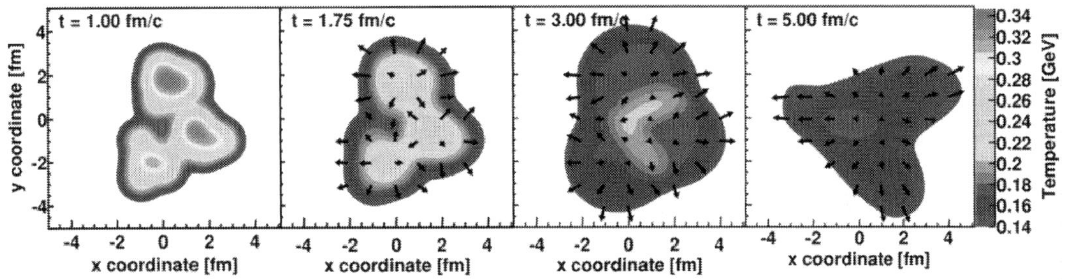

Figure 7: Time evolution of a ^{3}He + Au event from initial to final state. Color is proportional to temperature, while arrows represent velocity of individual fluid cells. Notice how initial state spatial triangularity gives rise to final state triangularity in momentum space [20].

Figure 6 plots the centrality dependence of the Bjorken energy density for d+Au and $Au + Au$ at $\sqrt{s_{NN}} = 200$ GeV. In this case, a discontinuity in the curve is observed as we go from central d+Au to peripheral $Au + Au$. Since transverse energy production is seen to be continuous, this indicates that the overlap area is a centrality-dependent geometric quantity that can be calculated with our Glauber+NBD model.

5.2. Collective Flow in ^{3}He+Au Collisions

Recent data from RHIC (d+Au) and LHC (p+Pb) provide compelling evidence for collective flow in small systems [4, 5, 6, 7]. However, there is currently no consensus to discriminate between the effects of initial-state spatial anisotropies and those from the subsequent evolution of the medium.

^{3}He+Au collisions have been proposed as a test system whose intrinsic triangular geometry can be leveraged to disentangle initial state effects from viscous damping through comparison with systems with one and two initial hotspots [20].

The same Glauber+NBD approach used to categorize centrality in d+Au can be used to generate the initial event geometry in ^{3}He+Au events. Just as the Hulthén wavefunction for the deuteron is used as input for the Glauber model, a study of ^{3}He+Au requires a description of the helium wavefunction [21]. This initial geometry is then propagated with viscous hydrodynamic code [22]. As shown in Figure 7, it is observed that intrinsic initial state triangularity survives as significant final state triangular flow, even with viscous damping. This example serves to illustrate the wide applicability of our Glauber+NBD approach to characterize event geometry and compare theoretical results to experimental findings.

6. Summary

We have described the methodology used by the PHENIX experiment to categorize geometry in d+Au collisions. The method is subject to inherent bias effects arising for events with charged particle production at midrapidity, which alter the measurement of centrality-dependent invariant yields. Our framework allows for the calculation of bias correction factors to account for such effects. Simulations with the HIJING generator indicate that bias correction factors are small for d+Au events at $\sqrt{s_{NN}} = 200$ GeV, but large and substantially pT-dependent for p+Pb at $\sqrt{s_{NN}} = 5.02$ TeV.

References

[1] A. Accardi et al., arXiv:1212.1701.
[2] K. Eskola, H. Paukkunen, and C. Salgado, J. High Energy Phys. **04**, 065 (2009)
[3] I. Helenius, K. J. Eskola, H. Honkanen, and C. A. Salgado, J. High Energy Phys. **07**, 073 (2012)
[4] A. Adare et al.(PHENIX Collaboration), Phys.Rev.Lett. **111**, 212301 (2013)
[5] G. Aad et al. (ATLAS Collaboration), Phys. Rev. Lett. **110**, 182302 (2013).
[6] B. Abelev et al. (ALICE Collaboration), Phys. Lett. B **719**, 29 (2013).
[7] S. Chatrchyan et al. (CMS Collaboration), Phys. Lett. B **718**, 795 (2013).
[8] A. Adare et al. (PHENIX Collaboration), arXiv:1310.4793 (to be published).
[9] M. Allen et al. (PHENIX Collaboration), Nucl. Instrum. Methods A **499**, 549 (2003).
[10] C. Adler et al. (STAR Collaboration), Nucl. Instrum. Methods A **499**, 433 (2003).
[11] M. L. Miller, K. Reygers, S. J. Sanders, and P. Steinberg, Ann. Rev. Nucl. Part. Sci. **57**, 205 (2007).
[12] L. Hulthen and M. Sagawara, Handbuch der Physik **39**, 14 (1957)
[13] A. Giovannini and L. Van Hove, Z. Phys. C **30**, 391 (1986)
[14] T. Abbott et al. (E802 Collaboration), Phys. Rev. C **52**,2663 (1995).
[15] S. S. Adler et al. (PHENIX Collaboration), Phys. Rev. C **77**, 014905 (2008).
[16] M. Gyulassy and X.-N. Wang, Comput. Phys. Commun. **83**, 307 (1994).
[17] M. H. Seymour and A. Siodmok, arXiv:1307.5015.
[18] A. Morsch (ALICE Collaboration), arXiv:1309.5525.
[19] S.S. Adler et al. (PHENIX Collaboration).Phys.Rev.C **71**:034908 (2005)
[20] J.L. Nagle et al., arxiv:1312.4565.
[21] J. Carlson and R. Schiavilla, Rev. Mod. Phys. **70**, 743 (1998)
[22] M. Luzum and P. Romatschke, Phys.Rev.C **78**, 034915 (2008), 0804.4015.

Journal of Physics: Conference Series **535** (2014) 012016 doi:10.1088/1742-6596/535/1/012016

Equation of state, fluctuations and other recent results from LQCD

Stefan Krieg[a,b]
for the Wuppertal-Budapest collaboration
[a]Department of Physics, University of Wuppertal,
Gaußstr. 20, D-42119, Germany
[b]Forschungszentrum Jülich, Jülich, D-52425, Germany

E-mail: s.krieg@fz-juelich.de

Abstract. We report on a continuum extrapolated result [1] for the equation of state (EoS) of QCD with $N_f = 2 + 1$ dynamical quark flavors and discuss preliminary results obtained with an addition dynamical charm quark ($N_f = 2 + 1 + 1$). Recent results on fluctuations of conserved charges (ratios of generalized susceptibilites) are reviewed and an update on crosscheck calculations using other complementary lattice actions is given. For all our final results, the systematics are controlled, quark masses are set to their physical values, and the continuum limit is taken using at least three lattice spacings corresponding to temporal extents up to $N_t = 16$.

1. Full result for the $N_f = 2 + 1$ equation of state

The rapid transition from the quark-gluon-plasma 'phase'[1] to the hadronic phase in the early universe and the QCD phase diagram are subjects of intense study in present heavy-ion experiments (LHC@CERN, RHIC@BNL, and the upcoming FAIR@GSI). This transition can be studied in a systematic way in Lattice QCD (for recent reviews see, e.g., [3, 4, 5]). The equation of state (EoS) of QCD, (i.e, the pressure p, energy density ϵ, trace anomaly $I = \epsilon - 3p$, entropy $s = (\epsilon + p)/T$, and the speed of sound $c_s^2 = dp/d\epsilon$ as functions of the temperature) has been determined by several lattice groups, however, a full result was, until recently, still lacking. Our of ref. [15] constituted a full result at three characteristic temperatures, which we have now extended to the full temperature range in [1] and made available electronically [2].

Our calculation is based on a tree-level Symanzik improved gauge action with 2-step stout-link improved staggered fermions. The precise definition of the action can be found in ref. [16], its advantageous scaling properties are discussed in ref. [8]. In particular, while it approaches the continuum value of the Stefan-Boltzmann limit in the infinite temperature limit $T \to \infty$ slower than actions with p4 or Naik terms (the latter is an additional fermionic term in the asqtad and hisq actions), it behaves monotonous and reaches the asymptotic a^2 behavior quite "early". Extrapolations from moderate temporal extents, e.g., using $N_t \geq 8$, allow for a smooth continuum extrapolation and provide an accuracy on the percent level, the typical accuracy one aims to reach. Additionally, applying simple tree-level improvement factors for the bulk

[1] Since this transition is a cross-over [6], this use of the term 'phase' is somewhat abusive, and indicates only the dominant degrees of freedom.

thermodynamic observables brings the individual data points for the different N_t very close to the continuum limit.

This points to an important advancement of the calculation described here over our previous results of [15]: we now include a large range of $N_t = 12$ data points and one $N_t = 16$ data point located at the peak position. Previously, we only had $N_t = 12$ results at three characteristic temperature values available. The $T \to 0$ limit is difficult due to taste-breaking effects, but is crucial since the renormalization is done at zero temperature, i.e. $p(T{=}0){=}0$. A mismatch at T=0 leads to a shift in the whole EoS. Previously, we calculated the difference in the pressure between the physical theory and its counterpart with 720 MeV heavy pions at a selected temperature (100 MeV), where the latter theory has practically zero pressure, and we, therefore, get $p(T = 100$ MeV) in the physical theory with the desired normalization. The difference of this result and the prediction by the Hadron Resonance Gas model (HRG) was then included in the systematical error. With our increased range of temporal extents, we now can use five lattice spacings to fix the additive term in the pressure, arriving at a complete agreement with the hadron resonance gas model at low temperatures. We also improved the precision on our line of constant physics (LCP, see ref. [1]), and used two different methods to set the scale (based on the w_0 scale [18] or on f_k) in order to control the systematical error related to scale setting.

These two different scale setting procedures entered into our 'histogram' method [19] used to estimate systematical errors, along with a range of other fit methods, each of which is an in principle completely valid approach. We then calculated the goodness of fit Q and weights based on the Akaike information criterion AICc [20, 21] and looked at the unweighted or weighted (based on Q or AICc) distribution of the results. The median is the central value, whereas the central region containing 68% of all the possible methods gives an estimate on the systematic uncertainties. This procedure provides very conservative errors. Here, we had four basic types of continuum extrapolation methods (with or without tree level improvement for the pressure and with a^2 alone or a^2 and a^4 discretization effects) and two continuum extrapolation ranges (including or excluding the coarsest lattice $N_t{=}6$ in the analysis). We used seven ways to determine the subtraction term at T=0 (subtracting directly at the same gauge coupling β or interpolating between the β values with various orders of interpolation functions), and the aforementioned two scale procedures. Finally, we had eight options to determine the final trace anomaly by choosing among various spline functions, giving altogether $4{\cdot}2{\cdot}7{\cdot}2{\cdot}8{=}896$ methods. Note that using either an AICc or Q based distribution changed the result only by a tiny fraction of the systematic uncertainty. Furthermore, the unweighted distribution always delivered consistent results within systematical errors.

The systematic error procedure clearly demonstrates the robustness of our final result. Even in the case of applying or not applying tree level improvement, where the data points at finite lattice spacing change considerably, the agreement between the continuum extrapolated results, and hence the contribution to the systematic error, is on the few percent level.

The continuum extrapolated trace anomaly is shown in Figure 1.

The pressure is obtained via integration from the trace anomaly, see Figure 2 (left) together with the predictions of the hadron resonance gas (HRG) model at low temperatures. There is a perfect agreement with HRG in the hadronic phase. The energy and entropy densities as well as the speed of sound are shown in the right panel of Figure 2.

2. Update on the $N_f = 2+1+1$ Equation of State

So far, the equation of state is known only in 2+1 flavor QCD. The contribution from the sea charm quarks most likely matter at least for $T > 300-400$ MeV (for an illustration, see Figure 3).

The $N_f = 2+1$ lattice results of the previous section agree with the HRG at low temperatures and are correct for the small to medium temperatures, and, as is shown in Figure 3, at

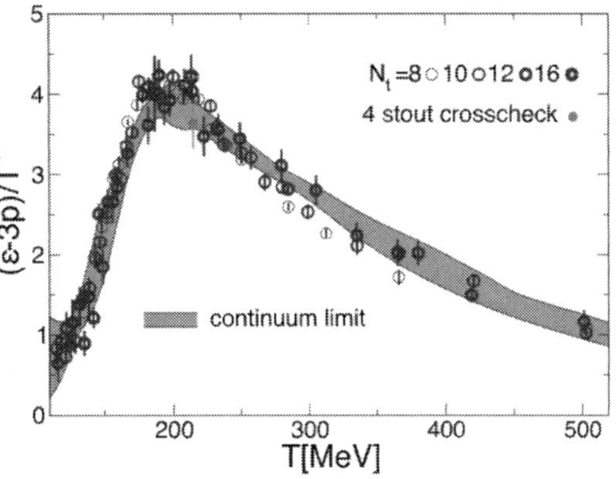

Figure 1. the trace anomaly as a function of the temperature. The continuum extrapolated result with total errors is given by the shaded band. Also shown is a cross-check point computed in the continuum limit with our new and different lattice action at $T = 214$ MeV, indicated by a smaller filled red point, which serves as a crosscheck on the peak's hight.

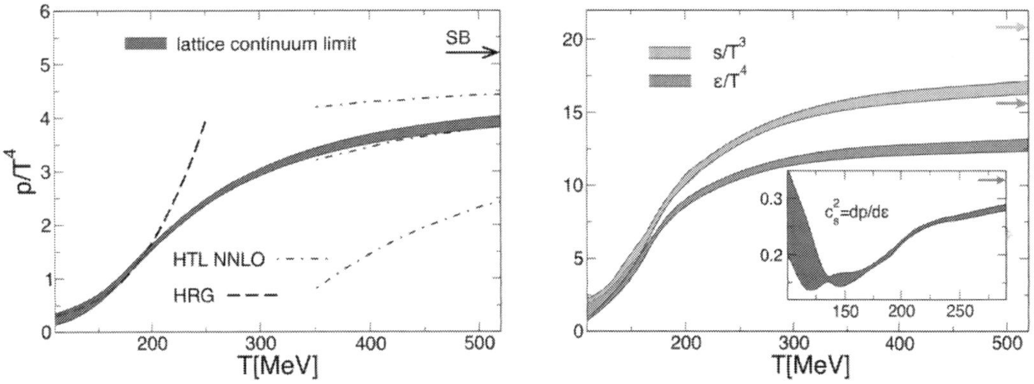

Figure 2. *Left:* continuum extrapolated result for the pressure with $N_f = 2{+}1$ flavors, including the HRG prediction and a comparison to the NNLO Hard Thermal Loop result of ref. [28] at high temperatures (renormalization scales $\mu = \pi T$, $2\pi T$ or $4\pi T$). *Right:* entropy and energy density. The insert shows the speed of sound.

temperatures of about 1 GeV perturbative results become sufficiently precise. Therefore, we need to calculate the EoS with a dynamical charm only for the remaining temperatures in the region of ≈ 300 MeV $< T <\approx 1000$ MeV.

As mentioned in the previous section, we are using a new lattice action for these calculations. More precisely, where our $N_f = 2 + 1$ calculation used an action with 2 levels of stout gauge link averaging in the coupling of the fermions to the gauge fields, we increased this to 4 levels with a smearing parameter of $\rho = 0.125$, for further details see ([1]). The crosscheck point shown in Figure 1 was computed using this new action. Since it perfectly agrees with the $N_f = 2 + 1$ results, even though it was computed using a dynamical charm, we can be certain that at temperatures at and below $T = 214$ MeV, we can rely on the $N_f = 2 + 1$ results.

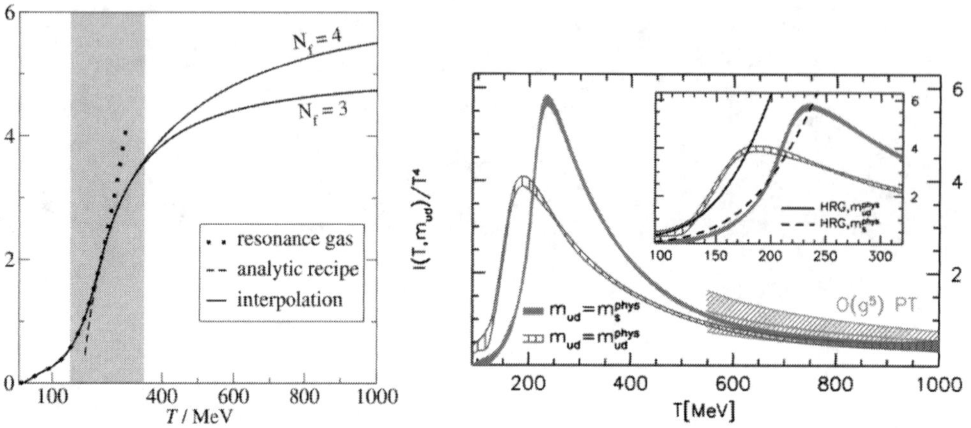

Figure 3. *Left:* Laine and Schroeder's perturbative estimate of the effect of the charm in the QCD equation of state [?]. *Right:* Wuppertal-Budapest [15] and perturbative (up to $O(g^5)$) results for the equation of state.

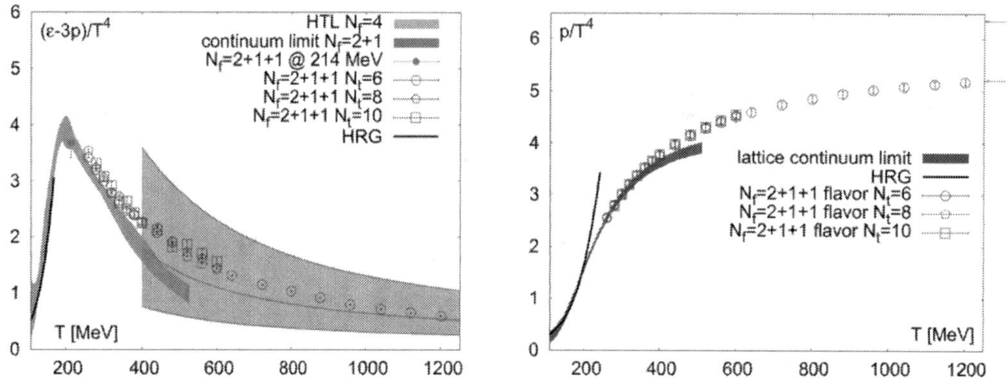

Figure 4. *Left:* Preliminary results for the charmed EoS. For comparison, we show the HRG result, the $N_f = 2+1$ band, and, at high Temperatures, the HTL result [28], where the central line marks the HTL expectation for the EoS with the band resulting from (large) variations of the renormalization scale. *Right:* Preliminary result for the pressure. All errors are statistical only.

Our preliminary results are shown in Figure 4, all errors are statistical only. Our results span a region of temperatures from $T = 214$ MeV up to $T = 1.2$ GeV. At the low end we make contact to the $N_f = 2+1$ equation of state, and at large temperatures to the HTL result. Thereby, we cover the full region of temperatures, from low temperatures, where the HRG gives reliable results, to high temperatures, where we make contact with perturbation theory. Our present set of data points will be extended to $N_t = 12$, in order to make a controlled continuum extrapolation possible.

Journal of Physics: Conference Series **535** (2014) 012016 doi:10.1088/1742-6596/535/1/012016

3. Fluctuations of conserved charges and systematic uncertainties

Correlations and fluctuations of conserved charges have been proposed long ago as observables which would unambiguously signal the QCD phase transition [?, ?]. The idea is that these quantum numbers have a very different value in a confined and deconfined system, and measuring them in the laboratory would allow to distinguish between the two phases. Furthermore, by directly comparing the results obtained on the lattice with those obtained by experiment, it is possible to calculate the freeze-out parameters from first principles [22, 30, 31].

Fluctuations of conserved charges can be obtained as linear combinations of diagonal and non-diagonal quark number susceptibilities, which can be calculated on the lattice at zero chemical potential [?, ?]. The results we discuss here, have been obtained by our collaboration using 2+1 staggered quark flavors [?] with the light and strange quark masses are set to their physical values. For further details on the action and simulation setup, we refer the reader to [?, 22, 30, 31]. For details about the simulation algorithm we refer the reader to [8]. Here, we are mostly concerned with the systematic uncertainties of our setup.

Fluctuations of conserved charges can be computed on the lattice through generalized susceptibilities. These appear in the expansion of the QCD pressure,

$$\frac{p}{T^4} = \frac{1}{VT^3} \ln Z(V,T,\mu_B,\mu_S,\mu_Q) = \sum_{lmn} \chi_{lmn}^{BSQ} \mu_B^l \mu_S^m \mu_Q^n \quad (1)$$

with

$$\chi_{lmn}^{BSQ} = \frac{1}{VT^3} \frac{\partial^{l+m+n} \ln Z(V,T,\mu_B,\mu_S,\mu_Q)}{\partial(\mu_B/T)^l \partial(\mu_S/T)^m \partial(\mu_Q/T)^n}, \quad (2)$$

being the generalized susceptibilities. These are related to the moments of the distributions of the corresponding conserved charges by

$$\begin{aligned} \text{mean}: \ M = \chi_1 \qquad & \text{variance}: \ \sigma^2 = \chi_2 \\ \text{skewness}: \ S = \chi_3/\chi_2^{3/2} \qquad & \text{kurtosis}: \ \kappa = \chi_4/\chi_2^2 \,. \end{aligned} \quad (3)$$

With these moments we can express the volume independent ratios

$$\begin{aligned} S\sigma = \chi_3/\chi_2 \quad &; \quad \kappa\sigma^2 = \chi_4/\chi_2 \\ M/\sigma^2 = \chi_1/\chi_2 \quad &; \quad S\sigma^3/M = \chi_3/\chi_1 \,. \end{aligned} \quad (4)$$

We can directly compare these quantities to experiment once the experimental setting is correctly implemented: there is no strangeness input in the colliding nuclei, and the ratio of protons and neutrons in the gold or lead atoms predeterimne the charge-to-baryon ratio in the outcoming hadrons as well. Thus [?]:

$$\langle S \rangle = 0, \qquad \langle Q \rangle = 0.4 \langle B \rangle \,. \quad (5)$$

These conditions can be respected if we introduce a strange and electric charge chemical potential in addition to the baryochemical potential. In order to extract freeze-out temperature and baryon chemical potential, the ratios $S_Q\sigma_Q^3/M_Q = R_{31}^Q(T,\mu_B) = \chi_3^Q/\chi_1^Q$ (see Figure 5) and $M_Q/\sigma_Q^2 = R_{12}^Q(T,\mu_B) = \chi_1^Q/\chi_2^Q$ can be expanded (for small chemical potentials) where $\mu_Q(\mu_B)$ and $\mu_S(\mu_B)$ are chosen to satisfy Eqs. (5), see [22].

In order to obtain reliable results from the lattice, we have to perform a continuum extrapolation, with at least 3 lattice spacings. Even with these at hand, we have to insure that we are in the scaling window, as illustrated by Figure 5. In order to control the systematic

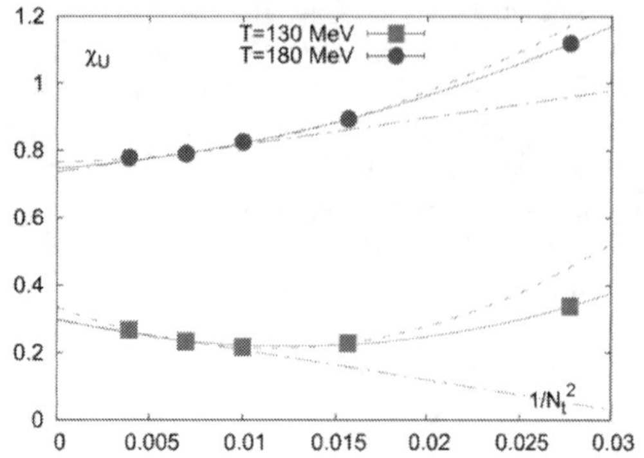

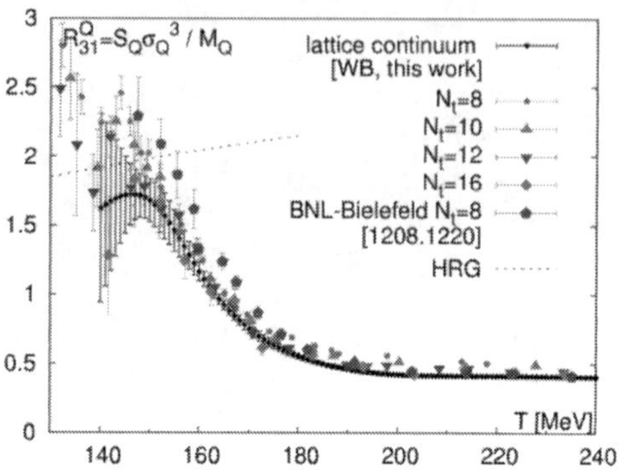

Figure 5. *Left:* χ_U, an observable severely hit both by taste breaking and by the cut-off effects in the one-link staggered dispersion relation, illustrating the need for a controlled continuum extrapolation. The data points suggest a quadratic fit in $1/N_t^2$. Here we give three possible fits both below and above the transition temperature. The solid, dashed and dash-dotted curves represent the fits on the finest five, four and three lattices, respectively. The statistical errors are much smaller than the size of the points, nevertheless the fits provide reasonable χ^2 values. *Right:* Continuum extrapolated result for $S_Q\sigma_Q^3/M_Q = R_{31}^Q$ based on $N_t = 8 \ldots 16$. For low temperatures the results agrees with the HRG expectation.

error due to the continuum extrapolation, we again follow our histogram ansatz and include several different sets of node points for our spline interpolation, extrapolate χ as well as $1/\chi$ and use different extrapolation formulae, as we did for our calculation of the EoS. We also use large N_t, as illustrated in Figure 5 and make contact with the HRG at low temperatures, where our lattice spacings are coarser.

4. Crosscheck calculations with different actions

Our results show here have all been obtained using a staggered fermion action. This and the complicated continuum limit show in Figure 5 makes crosscheck calculations using other lattice actions particularly appealing. Here, we suggest using Wilson type fermions [?], since they, while being considerably more expensive than staggered actions, do not suffer from taste breaking induced lattice artifacts. However, Wilson type fermions suffer from another problem, as they explicitly break chiral symmetry even at vanishing quark masses. We can improve on this problem by using the even more expensive overlap action, which has a unbroken lattice variant of chiral symmetry (which excludes operator mixing and additive quark mass renormalization exactly as the continuum action). However, at the moment simulations with overlap fermions have to be performed at fixed topological charge [?, 32], which will require a dedicated finite volume study in the future.

Here, we show preliminary results for two such studies, one using improved Wilson (Clover), the other using overlap type fermions. In the former case simulations include a dynamical strange quark, set to its physical value and are performed in the fixed scale approach, whereas in the latter case only two mass degenerate quark flavors are included at fixed topological charge. The results are shown in Figure 6.

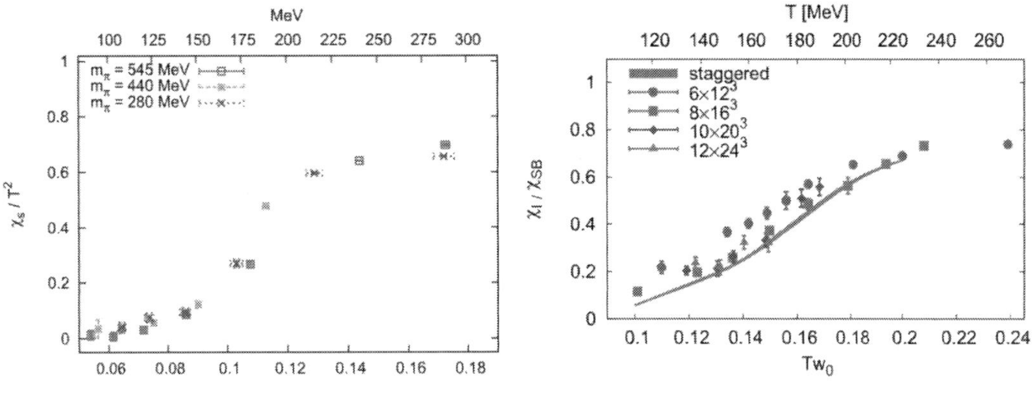

Figure 6. *Left:* Strange susceptibility at fixed lattice spacing using $N_f = 2 + 1$ flavors of Wilson (Clover) quarks, for 3 different pion masses (update of [?]). A chiral extrapolation and continuum limit are still pending, after which a direct comparison to staggered results will be possible. *Right:* $N_f = 2$ crosscheck calculation using overlap quarks [?, 32], including a dedicated $N_f = 2$ staggered calculation, all at a pion mass corresponding to about 350 MeV. The good agreement of the large N_t with the continuum extrapolated staggered result suggests that the continuum overlap and staggered results will coincide.

5. Conclusions

The precision of Lattice QCD results at finite temperature has increase significantly over the last years. We have discussed a full result (all sources of uncertainties controlled) for the $N_f = 2+1$ EoS by the Budapest-Wuppertal collaboration [1] and discussed how to include a dynamical charm quark for the $N_f = 2+1+1$ EoS and shown and discussed preliminary results. We have discussed systematic uncertainties in Lattice QCD simulations of, in particular, generalized susceptibilities and shown promising preliminary results of crosscheck calculations using other lattice actions.

Bibliography

[1] S. Borsanyi, Z. Fodor, C. Hoelbling, S. D. Katz, S. Krieg and K. K. Szabo, arXiv:1309.5258 [hep-lat].
[2] The results are available as an ancillary file to [1], where they are freely downloadable. The file contains tabulated data for the trace anomaly, pressure, entropy, energy density, and speed of sound for our temperature range.
[3] Z. Fodor and S. D. Katz, arXiv:0908.3341 [hep-ph].
[4] O. Philipsen, Prog. Part. Nucl. Phys. **70** (2013) 55 [arXiv:1207.5999 [hep-lat]].
[5] P. Petreczky, PoS ConfinementX (2012) 028 [arXiv:1301.6188 [hep-lat]].
[6] Y. Aoki, G. Endrodi, Z. Fodor, S. D. Katz and K. K. Szabo, Nature **443** (2006) 675 [hep-lat/0611014].
[7] Y. Aoki, Z. Fodor, S. D. Katz and K. K. Szabo, Phys. Lett. B **643** (2006) 46 [hep-lat/0609068].
[8] Y. Aoki, S. Borsanyi, S. Durr, Z. Fodor, S. D. Katz, S. Krieg and K. K. Szabo, JHEP **0906** (2009) 088 [arXiv:0903.4155 [hep-lat]].
[9] S. Borsanyi *et al.* [Wuppertal-Budapest Collaboration], JHEP **1009** (2010) 073 [arXiv:1005.3508 [hep-lat]].
[10] A. Bazavov, T. Bhattacharya, M. Cheng, C. DeTar, H. T. Ding, S. Gottlieb, R. Gupta and P. Hegde *et al.*, Phys. Rev. D **85** (2012) 054503 [arXiv:1111.1710 [hep-lat]].
[11] G. Endrodi, Z. Fodor, S. D. Katz and K. K. Szabo, JHEP **1104** (2011) 001 [arXiv:1102.1356 [hep-lat]].
[12] S. .Borsanyi, G. Endrodi, Z. Fodor, S. D. Katz, S. Krieg, C. Ratti and K. K. Szabo, JHEP **1208** (2012) 053 [arXiv:1204.6710 [hep-lat]].
[13] Z. Fodor and S. D. Katz, Phys. Lett. B **534** (2002) 87 [hep-lat/0104001].
[14] P. Petreczky, AIP Conf. Proc. **1520** (2013) 103.

[15] S. Borsanyi, G. Endrodi, Z. Fodor, A. Jakovac, S. D. Katz, S. Krieg, C. Ratti and K. K. Szabo, JHEP **1011** (2010) 077 [arXiv:1007.2580 [hep-lat]].
[16] Y. Aoki, Z. Fodor, S. D. Katz and K. K. Szabo, JHEP **0601** (2006) 089 [hep-lat/0510084].
[17] A. Peikert, B. Beinlich, A. Bicker, F. Karsch and E. Laermann, Nucl. Phys. Proc. Suppl. **63**, 895 (1998) [hep-lat/9709157].
[18] S. Borsanyi, S. Durr, Z. Fodor, C. Hoelbling, S. D. Katz, S. Krieg, T. Kurth and L. Lellouch *et al.*, JHEP **1209** (2012) 010 [arXiv:1203.4469 [hep-lat]].
[19] S. Durr, Z. Fodor, J. Frison, C. Hoelbling, R. Hoffmann, S. D. Katz, S. Krieg and T. Kurth *et al.*, Science **322** (2008) 1224 [arXiv:0906.3599 [hep-lat]].
[20] Hirotugu Akaike, IEEE Transactions on Automatic Control 19 (1974) 716-723
[21] C. M. Hurvich, C.-L. Tsai, Biometrika 76 (1989) 297-307
[22] S. Borsanyi, Z. Fodor, S. D. Katz, S. Krieg, C. Ratti and K. K. Szabo, Phys. Rev. Lett. **111**, 062005 (2013) [arXiv:1305.5161 [hep-lat]].
[23] S. .Borsanyi, G. Endrodi, Z. Fodor, S. D. Katz and K. K. Szabo, JHEP **1207** (2012) 056 [arXiv:1204.6184 [hep-lat]].
[24] G. Endrodi, Z. Fodor, S. D. Katz and K. K. Szabo, PoS LAT **2007** (2007) 228 [arXiv:0710.4197 [hep-lat]].
[25] P. Petreczky [for HotQCD Collaboration], PoS LATTICE **2012** (2012) 069 [arXiv:1211.1678 [hep-lat]].
[26] P. Huovinen and P. Petreczky, Nucl. Phys. A **837**, 26 (2010) [arXiv:0912.2541 [hep-ph]].
[27] S. Borsanyi, G. Endrodi, Z. Fodor, S. D. Katz, S. Krieg, C. Ratti, C. Schroeder and K. K. Szabo, PoS LATTICE **2011** (2011) 201 [arXiv:1204.0995 [hep-lat]].
[28] J. O. Andersen, L. E. Leganger, M. Strickland and N. Su, JHEP **1108** (2011) 053 [arXiv:1103.2528 [hep-ph]].
[29] G. I. Egri, Z. Fodor, C. Hoelbling, S. D. Katz, D. Nogradi and K. K. Szabo, Comput. Phys. Commun. **177** (2007) 631 [hep-lat/0611022].
[30] S. Borsanyi, Z. Fodor, S. D. Katz, S. Krieg, C. Ratti and K. K. Szabo, arXiv:1403.4576 [hep-lat].
[31] C. Ratti, these proceedings.
[32] S. Borsanyi, Y. Delgado, S. Drr, Z. Fodor, S. D. Katz, S. Krieg, T. Lippert, D. Nogradi, K. K. Szabo, B. C. Toth, PoS(LATTICE 2013) 163

Journal of Physics: Conference Series **535** (2014) 012017 doi:10.1088/1742-6596/535/1/012017

The η/s of hadrons out of chemical equilibrium

A. Wiranata[1,2,3], M. Prakash[3], P. Huovinen[4,5], V. Koch[2], and X.N. Wang[1,2]

[1]Institute of Particle Physics and Key Laboratory of Quark and Lepton Physics (MOE), Central China Normal University, Wuhan 430079, China
[2]Lawrence Berkeley National Laboratory, Nuclear Science Division; MS 70R0319, Berkeley, California 94720, USA
[3]Department of Physics and Astronomy, Ohio University, Athens, Ohio 45701, USA
[4]Frankfurt Institute for Advanced Studies, Ruth-Moufang-Straβe 1, 60438 Frankfurt am Main, Germany
[5]Institut für Theoretische Physik, Johann Wolfgang Goethe-Universität, Max-von-Laue-Strasse 1, 60438 Frankfurt am Main, Germany

E-mail: `awiranata@lbl.gov`

Abstract. We study how the shear viscosity, η, entropy density, s, and η/s depend on growing hadron chemical potentials resulting from the loss of chemical equilibrium during the evolution of a relativistic heavy-ion collision. Our calculations here are for an interacting pion gas in a system of net baryon number zero. Time evolution of the temperature and pion chemical potential are taken from ideal fluid hydrodynamic calculations of RHIC and LHC collisions. We find that the shear viscosity decreases whereas the entropy density increases with increasing pion chemical potential resulting in values of η/s that are slightly reduced from the case of chemical potentials being zero when chemical equilibrium prevails. Our results indicate that the inclusion of additional mesons and baryons will likely lead to further reduction in the value of η/s.

1. Introduction

The objective of this work is to explore the influence of nonzero hadron chemical potentials on the shear viscosity, η, entropy density s, and their ratio η/s in the hadronic phase of a relativistic heavy-ion collision. Soon after their formation, hadrons are in a dense medium in which both inelastic and elastic processes serve to keep the chemical potentials of each species zero. As evolution proceeds, the expansion of the system causes the local density to decrease which results in an imbalance of the rates of forward and backward reactions. Consequently, chemical equilibrium will be lost prior to losing thermal equilibrium so that all constituents of the system acquire sizeable chemical potentials as pointed out in Refs. [1, 2]. In an environment where only elastic collisions occur and the expansion of the fluid is adiabatic, the conservation of entropy per particle enables the determination of the chemical potentials of each species (see, e.g., Ref. [2] for a lucid and detailed account). For viscous hydrodynamic expansion in which additional entropy is generated, the ideal fluid approximation serves as a starting guess (for chemical potentials) which we will adopt here.

In our initial study here, we will focus on the situation in which the net baryon number is zero and consider an interacting pion gas in which various resonances are formed (for a list, see Table I of Ref. [3]). Examples of shear viscosity calculations with finite baryon number (in which conflicting conclusions have been drawn) can be found in Refs. [4, 5, 6]. Our calculations of η here are based on the Chapman-Enskog approach detailed in Refs. [7, 8]. Typos in Ref. [8] are corrected here and more details will be reported elsewhere. Differential cross sections required in the calculation of η are based on the K-matrix cross sections detailed in [3]. Entropy densities are calculated using the virial expansion approach as in Ref. [9]. Time evolution of the pion chemical potential is taken from ideal fluid hydrodynamic calculations for RHIC an LHC.

2. Pion chemical potentials from hydrodynamic simulations

Fig. 1 shows the time evolution of the temperature T and pion chemical potential μ from ideal fluid hydrodynamic simulations for Au+Au collisions of 200 GeV/A at RHIC and Pb+Pb collisions of 2.76 TeV/A at LHC. For the RHIC experiment, the simulation is initialized as in Ref. [10] and for the LHC experiment we use the initialization for a viscous fluid as in Ref. [11]. The calculations are for net baryon number zero and the equation of state is that of the hadron resonance gas below 183 MeV with chemical freeze-out occurring at 150 MeV. When chemical equilibrium prevails, the chemical potentials of all hadrons are, by definition, zero. With the loss of chemical equilibrium, the chemical potentials of stable (with respect to strong interactions) particles with number densities n_i are chosen such that below the chemical freeze-out temperature, n_i/s, where s is the total entropy density, is constant [2].

Results for two different centralities, 0-5% (most central collisions) and 40-50% (peripheral collisions), are shown. The hadronic life times of the chosen central and peripheral collisions are slightly different; in the most central collisions, the hadronic phase starts at a somewhat later time compared to that of the peripheral collisions. The times of onset and duration of the hadronic phase are qualitatively similar but quantitatively different at RHIC and LHC. As we can see in the top panels, temperature drops more rapidly as μ begins to grow with time compared to the case when $\mu = 0$ [12].

A caveat in our use of the above results must be mentioned. In the hadronic phase, the ideal fluid hydrodynamic calculations above employ an equation of state that includes all resonances up to 2 GeV, not just pions. The ensuing results for the time evolution of the temperature and chemical potentials of the various hadronic species are thus more realistic than would be for a single component system of interacting pions that we have chosen for our pedagogical study here. As we aim to follow up this study to include additional mesons and baryons, our results for the pionic component alone should be taken as illustrative but not quantitatively definitive.

3. Shear viscosity with finite chemical potentials

We employ the Chapman-Enskog approach, generalized to include relativistic kinematics and quantum statistics, for the calculation of shear viscosity. An interacting pion gas will be chosen to explore the effects of nonzero chemical potentials. We follow closely the formalism of Refs. [7, 8] in which the shear viscosity to first order is expressed as

$$\eta = \frac{T}{10}\frac{\gamma_0^2}{c_{00}}\,. \tag{1}$$

Above,

$$\gamma_0 = -10\,\frac{S_3^{(-2)}}{S_2^{(-1)}}\,, \quad \text{with} \quad S_n^{(\alpha)} = \sum_{k=1}^{\infty} \exp\left[\frac{km\mu}{T}\right] k^\alpha K_n(kz)\,, \tag{2}$$

where m is the mass of the particle, μ is the reduced (with the mass m) chemical potential, T is the temperature, $z = m/T$ is the relativity parameter, and $K_\nu(z)$ is the modified Bessel

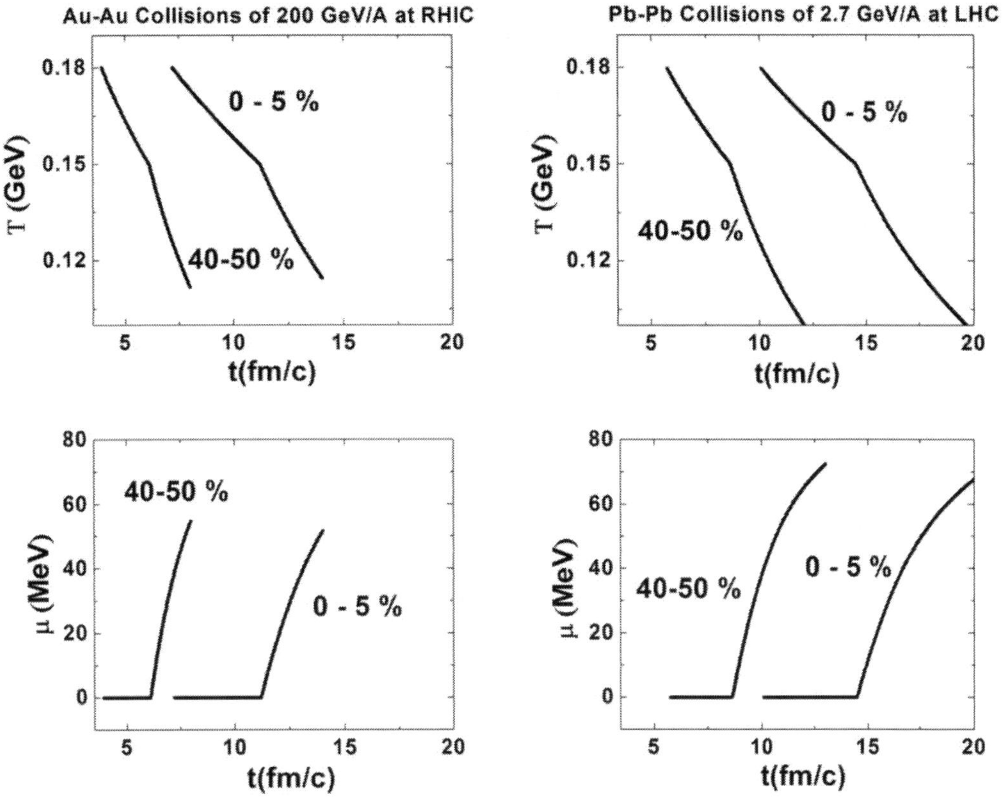

Figure 1: Temperature T and pion chemical potential μ from ideal fluid hydrodynamical simulations of Ref. [13]. The left panels are for Au+Au collisions of 200 GeV/A at the RHIC and the right panels are for Pb+Pb collisions of 2.76 TeV at the LHC. In both cases, results are shown for 40-50% and 0-5% centralities.

function of order ν. Apart from the numerical factor of -10, γ_0 represents the reduced enthalpy per particle of an ideal Boltzmann gas. The denominator in Eq. (1) is composed of terms involving the differential cross section between particles in the heat bath weighted by thermal occupation factors:

$$c_{00} = C_{00} + 2z^{-1}B_{11} + \frac{8}{3}z^{-2}A_{22}\,. \tag{3}$$

The terms C_{00}, B_{11}, and A_{22} are given by

$$\begin{aligned} C_{00} &= \frac{16z^4\,e^{(2m\mu/T)}}{\left[S_2^{-1}\right]^2}\int_0^\infty d\chi\int_0^\infty d\psi\int_0^\pi d\theta\int_0^\pi d\Theta\int_0^{2\pi} d\phi\,\sigma(\psi,\Theta)\,e^{-2z\cosh\psi\cosh\chi} \\ &\times\ \sinh^2\chi\,\cosh^3\psi\,\sinh^7\psi\,\sin\theta\left(1-\cos^2\Theta\right)\sin\Theta\left(\frac{1}{B}\right), \\ B_{11} &= \frac{8z^5\,e^{(2m\mu/T)}}{\left[S_2^{-1}\right]^2}\int_0^\infty d\chi\int_0^\infty d\psi\int_0^\pi d\theta\int_0^\pi d\Theta\int_0^{2\pi} d\phi\,\sigma(\psi,\Theta)\,e^{-2z\cosh\psi\cosh\chi} \end{aligned}$$

$$
\begin{aligned}
& \times \quad \sinh^4\chi \cosh^3\psi \sinh^7\psi \sin\theta \left(\cos^2\theta - 2\cos\theta\cos\theta'\cos\Theta + \cos^2\theta'\right)\sin\Theta\left(\frac{1}{B}\right), \\
A_{22} \quad &= \quad \frac{2z^6\, e^{(2m\mu/T)}}{\left[S_2^{-1}\right]^2}\int_0^\infty d\chi \int_0^\infty d\psi \int_0^\pi d\theta \int_0^\pi d\Theta \int_0^{2\pi} d\phi\, \sigma(\psi,\Theta)\, e^{-2z\cosh\psi\,\cosh\chi} \\
& \times \quad \sinh^6\chi \cosh^3\psi \sinh^7\psi \sin\theta\left(\cos^2\theta - \cos^2\theta'\right)\sin\Theta\left(\frac{1}{B}\right), \qquad (4)
\end{aligned}
$$

with

$$
\begin{aligned}
B = 1 \quad &+ \quad \exp\left[\frac{2m\mu}{T} - 2z\cosh\psi\,\cosh\chi\right] \\
&- \quad \exp\left[\frac{m\mu}{T} - z\left(\cosh\psi\cosh\chi - \sinh\psi\sinh\chi\cos\theta\right)\right] \\
&- \quad \exp\left[\frac{m\mu}{T} - z\left(\cosh\psi\cosh\chi + \sinh\psi\sinh\chi\cos\theta\right)\right] \qquad (5)
\end{aligned}
$$

and

$$
\cos\theta' \quad \equiv \quad \cos\theta\cos\Theta - \sin\theta\sin\Theta\cos\phi\,. \qquad (6)
$$

Hyperbolic functions of the quantity ψ characterize the relative momentum and invariant center-of-mass energy of the two scattering particles:

$$
\sinh\psi = \frac{g}{m}, \qquad \cosh\psi = \frac{P}{m}, \qquad (7)
$$

where $g = \sqrt{(p_1 - p_2)^2}/2$, $P = \sqrt{(p_1 + p_2)^2}$, and p_1 and p_2 are the initial four momenta of the colliding hadrons. Hyperbolic functions of the quantity χ are components of the fluid four-velocity $U^\alpha = (\cosh\chi, -\hat{e}\,\sinh\chi)$. The symbol $\sigma(\psi,\Theta)$ denotes the center-of-mass differential cross section. Expressions for higher orders can be found in Refs. [7, 8], but are not required for the application on hand. The result for a Boltzmann gas is obtained by retaining only the $k = 1$ term in Eq. (2) and appropriate reductions of Eqs. (4) and (5) by taking the Boltzmann limit of the Bose distribution function in the kernels. As noted in Ref. [14], η is independent of μ for a Boltzmann gas. Explicit μ dependence of η arises only when Bose or Fermi statistics is relevant.

4. Results

The scaled entropy density, s/T^3, of the pion gas is shown in Fig. 2 as a function of temperature T for the Boltzmann and Bose cases with pion chemical potentials $\mu = 0$, 25, and 50 MeV, respectively. As expected, nonzero chemical potentials yield larger values of s/T^3 in both cases, that for the Bose gas being slightly larger than the result for the Boltzmann gas at all μ/T.

In Fig. 3, the shear viscosity η is shown as a function of temperature for $\mu = 0$ and 50 MeV, respectively. With increasing T, η for the Bose case falls below that of the Boltzmann result. For the Bose gas, the magnitude of η for $\mu = 50$ MeV is slightly smaller than that for $\mu = 0$, whereas the Boltzmann result does not depend on μ in first order [7, 14].

In Fig. 4, the ratio of η/s vs. T is shown for $\mu = 0$ and 50 MeV, respectively. For both values of μ, results for the Bose gas are smaller than those for the Boltzmann gas and the differences grow with increasing T. Despite the fact that η is independent of μ for the Boltzmann case, η/s decreases with μ because of the increase in s with increasing μ. For the Bose gas, a nonzero μ increases s and decreases η relative to their values at $\mu = 0$. In conjunction, both of these

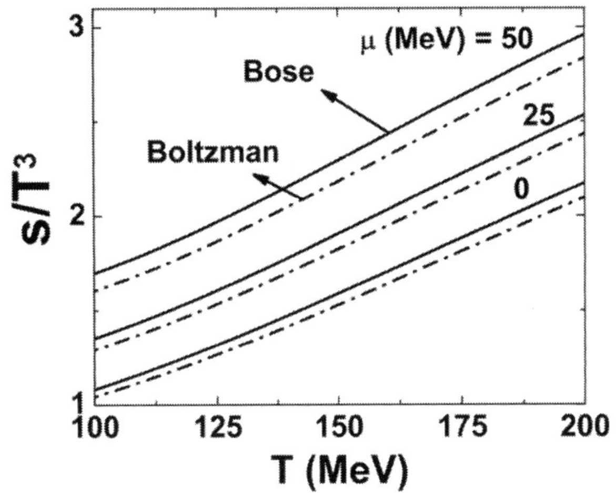

Figure 2: Scaled entropy densities s/T^3 vs. temperature T for a pion gas at the indicated values of the pion chemical potential μ. The solid (dash-dotted) curves are for a Bose (Boltzmann) gas.

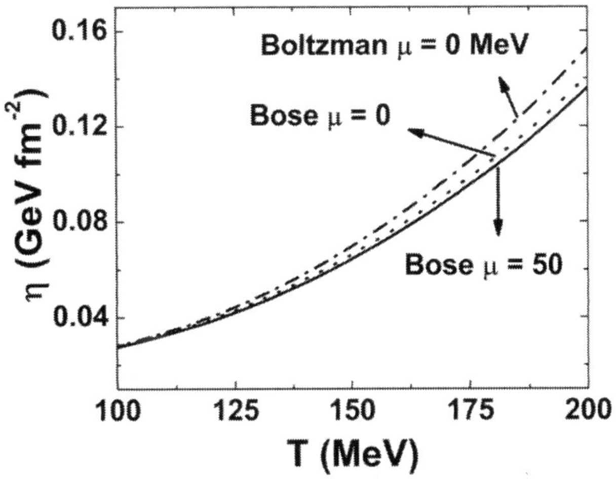

Figure 3: Shear viscosity η vs. temperature T for a pion gas. The solid curve is for a Bose gas. Other curves are for a Boltzmann gas. Pion chemical potentials μ are as indicated.

effects render η/s $(\mu \neq 0) > \eta/s$ $(\mu = 0)$ for all T. We observe, however, that for the range of temperatures of interest to hadron hydrodynamics, the decrease in η/s due to finite μ's is not significantly large.

In Fig. 5, the entropy density, s, shear viscosity, η, and η/s are shown as functions of time for RHIC and LHC experiments at centralities of 0-5% (central collisions) and 40-50% (peripheral collisions). Results here are for temperature and chemical potential profiles of the hadronic phase from hydrodynamic simulations shown in Fig. 1. Each panel of Fig. 5 contains comparisons between results when Boltzmann and Bose distribution functions are used for the calculations of η and s. For Au+Au collisions of 200 GeV/A at RHIC, η/s for RHIC lies in the range 0.36-0.5 for both centralities. In the case of Pb+Pb collisions of 7 TeV/A at LHC, η/s is in the range 0.36-0.6 (not too different from RHIC). All four panels in Fig. 5 exhibit qualitatively similar features. The nearly linear rise in η/s during the time that chemical equilibrium prevails ($T < 150$ MeV) stems from the time evolutions of η and s which in turn depend on the time

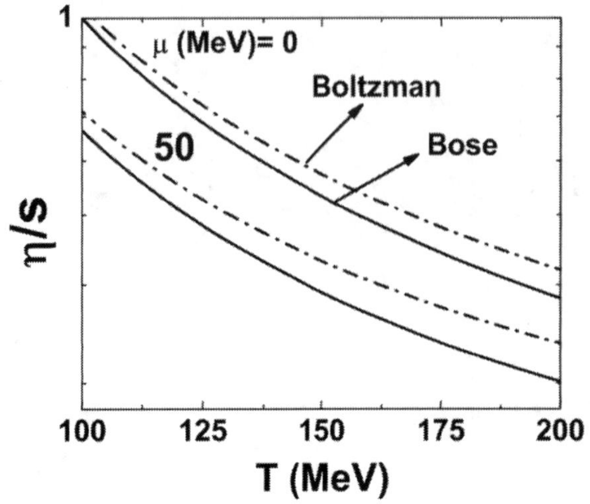

Figure 4: Ratio of shear viscosity to entropy density, η/s, of an interacting pion gas. Solid (dash-dotted) curves are for a Bose (Boltzmann) gas with pion chemical potentials $\mu = 0$ and 50 MeV, respectively.

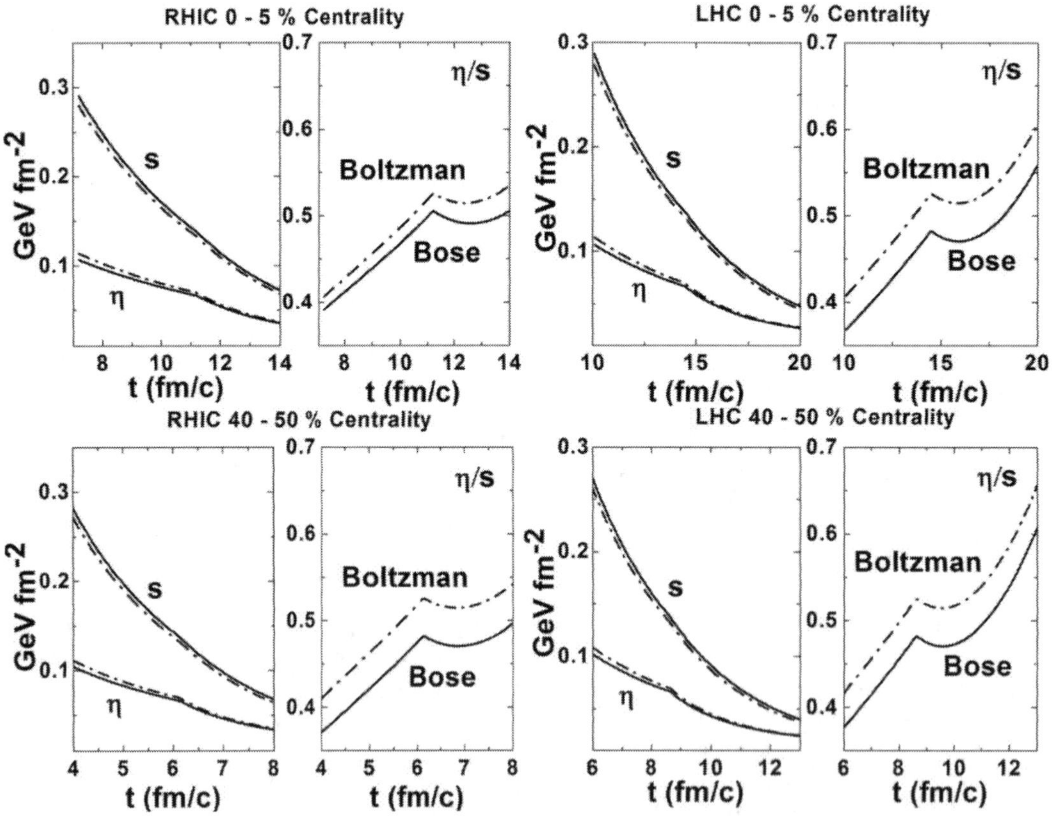

Figure 5: Entropy density, s, shear viscosity, η and η/s as functions of time for RHIC (left panels) and LHC (right panels) experiments for centralities of 0-5% (central collisions) and 40-50% (peripheral collisions). Solid (dash-dotted) curves are for Bose (Boltzmann) gases.

evolution of T. With the loss of chemical equilibrium, the time evolutions of η and s depart from their earlier behaviors as an additional time dependence enters through the time evolution of μ. Noteworthy is the feature that η/s exhibits a minimum before rising monotonically as time progresses. The minimum is a consequence of the different rates at which η and s fall with time, that of η being a combination of the rate at which T falls with t and μ grows with t (see Fig. 1).

5. Summary and conclusion

We have explored the influence of finite chemical potentials arising from the expansion and hence the dilution of the hadronic system in relativistic heavy-ion collisions of relevance to experiments at RHIC and LHC. Our study was limited to the case of an interacting pion gas in which chemical equilibrium is lost prior to thermal freeze-out. In contrast to the case when Boltzmann statistics is used and η is independent of μ, use of Bose statistics causes η to decrease with increasing μ. Additionally, Bose statistics leads to larger values of s than Boltzmann statistics. Consequently, η/s decreases as the pion chemical potential μ increases toward freeze-out. The effects of nonzero pion chemical potentials on time evolving η/s are not very large. However, our results indicate that the inclusion of additional mesons will likely lead to a further reduction in η/s. Larger reductions may be expected for nonzero net baryon and strangeness numbers. Calculations in this regard are in progress to complement those in the literature [4, 5, 6].

Acknowledgments

This work was supported by the NSFC under grant No. 11221504 & No. 1350110493, and by the US DOE under contracts #DE-AC02-05CH11231 and #DE-FG02-93ER-40756.

References

[1] S. Gavin, Nucl. Phys. **A435**, 826 (1985).
[2] H. Bebie, P. Gerber, J. L. Goity, and H. Leutwyler, Nucl. Phys. **B378**, 95 (1992).
[3] A. Wiranata, V. Koch, M. Prakash, and X. Wang, Phys. Rev. C **88**, 044917 (2013).
[4] K. Itakura, O. Morimatsu, and H. Otomo, Phys. Rev. D **77**, 014014 (2008).
[5] M. I. Gorenstein, M. Hauer, and O. N. Moroz, Phys. Rev. C **77**, 024911 (2008).
[6] G. S. Denicol, C. Gale, S. Jeon, and J. Noronha, Phys. Rev. C **88**, 064901 (2013).
[7] P. H. Polak, W. A. van Leeuwen, and S. R. de Groot, Physica **66**, 455 (1973).
[8] D. Davasne, Phys. Rev. C **53**, 3069 (1996).
[9] R. Venugopalan and M. Prakash, Nucl. Phys. **A 546**, 718 (1992).
[10] P. Huovinen and P. Petreczky, Nucl. Phys. **A 837**, 26 (2010).
[11] H. Niemi, G. S. Denicol, P. Huovinen, E. Molnar, and D. H. Rischke, Phys. Rev. C **86**, 014909 (2012).
[12] T. Hirano and K. Tsuda, Phys. Rev. C **66**, 054905 (2002).
[13] P. Huovinen, Private Communications (2014).
[14] C.-G. van Weert, P. C. de Jagher, and S. R. de Groot, Physica **66**, 474 (1973).

Journal of Physics: Conference Series **535** (2014) 012018 doi:10.1088/1742-6596/535/1/012018

Hadronic Shear Viscosity: A Comparison between the Green-Kubo and Chapmann Enskog Methods

Nasser Demir[1], **Anton Wiranata**[2,3,4]

[1] Physics Department, Faculty of Science, Kuwait University, Khaldiya, Kuwait 13060
[2] Central China Normal University, IOPP, Wuhan 430079, China
[3] Nuclear Science Division, MS 70R0319, Lawrence Berkeley National Laboratory, Berkeley, CA 94720, USA
[4] Department of Physics and Astronomy, Ohio University, Athens, OH 45701

E-mail: nasser.demir@gmail.com

Abstract. A detailed comparison between two methods to calculate shear viscosity is presented. We choose two systems in this comparison which are massless particles with current algebra cross section and a mixture comprised of pions with rho resonances. The two methods involved are the Green-Kubo method, applied using the Ultrarelativistic Quantum Molecular Dynamics (UrQMD) model to simulate the hadronic medium, and the Chapmann-Enskog method.

1. Introduction

Ultrarelativistic heavy ion collisions at the Relativistic Heavy Ion Collider (RHIC) are thought to have created a Quark Gluon Plasma (QGP) with characteristics of an ideal fluid [2, 25]. As such, the characterization of transport coefficients of hot, dense QCD matter is of significant interest. In particular, much attention has been given to the shear and bulk viscosity coefficients η and ζ, respectively, and the corresponding viscosity to entropy density ratios $\eta/s, \zeta/s$. Although an ideal fluid traditionally has been defined as having zero viscosities, semiclassical arguments have been presented that suggest a lower limit to the shear viscosity to entropy density ratio of a fluid [9], and there has been a conjecture by Kovtun, Son, and Starinets (KSS) based upon a calculation using the AdS/CFT correspondence that $\frac{\eta}{s} \geq \frac{\hbar}{4\pi k_B}$, where $\hbar$ is Planck's constant and k_B is Boltzmann's constant [21], and a modified bound of $\frac{\eta}{s} \geq \frac{4\hbar}{25\pi k_B}$ in order to avoid acausalities in gravity duals [5]. Ideal hydrodynamics calculations assuming no shear viscosity have been very successful in reproducing the elliptic flow (v_2) data from RHIC, yet the deconfined phase of the reaction cannot have zero shear viscosity. In the quest by the RHIC community to determine the viscosity of the deconfined phase, several groups have implemented viscous relativistic hydrodynamics simulations and examined what range of η/s would be permissible to reproduce the RHIC v_2 data [30, 15, 19]. Viscous hydrodynamical analyses [23, 29] finds that η/s should lie within a range of 0.08-0.24 depending upon the choice of initial conditions and equation of state. This finding is supported by lattice calculations of η/s for pure gluonic QCD, which yield values close to the KSS bound, and indirect estimates of η/s from calculations of the diffusion of heavy quarks, elliptic flow measurements, and transverse momentum correlations that arrive at roughly comparable values [24, 22, 14, 17, 1]. However, the aforementioned viscous hydrodynamics calculations assume a fixed value of η/s throughout the entire evolution of the reaction and hence neglect its temperature dependence. One should note that the shear viscosity, especially in the context of a relativistic heavy ion collision, is a *time-dependent* quantity. While the partonic phase of such a collision is expected to have a very low value of η/s, after hadronization occurs η/s is expected to rapidly increase [12]. In order to make precise quantitative statements about the viscosity of the deconfined phase of a relativistic heavy ion collision, it is necessary to separately determine the viscosity of the hadronic phase in order to quantitatively constrain the viscosity of the deconfined phase.

There have been a number of analytic calculations of η, ζ and the corresponding viscosity to entropy density ratios for a number of different hadronic systems, including pure pion gases ([16, 13, 28, 7, 8]) and binary hadronic mixtures ([28, 6, 20]). In the aforementioned analytical calculations, the linearized Boltzmann equation was solved, and the cross sections in the collision integral were treated using a variety of methods, such as chiral perturbation theory, effective NN theory, and phenomenological amplitudes [7, 6, 20]. Another analytical treatment used was the relaxation time approximation [16, 28]. There also exist numerical calculations of shear viscosity using semiclassical microscopic transport approaches of mesonic matter [26], hadronic matter at finite baryon number density [27], and for a gas of hard spheres in the van der Waals excluded volume hadron-resonance gas model [18]. However, while several calculations for the hadronic shear viscosity exist, in order to validate a method and test the limits to which a calculation can be reliable, it is necessary to perform a systematic comparison between two different methods and see where the differences exist [32] , and that is the goal of this paper.

In this paper we compare two methods of shear viscosity calculations which are the Chapmann-Enskogg approximation and the Green-Kubo Method. We chose the following two systems for the comparison: (i) massless pions with an energy-dependent current algebra cross section (ii) massive pions with both elastic scattering and inelastic scattering enabled through the rho resonance as an intermediate channel. In this work, we want to observe the origin of the difference in the result of shear viscosity calculations.

The organization of this paper is as follows. In Sec. II, the formalism and working formulae of the Chapman- Enskog and Green-Kubo methods are summarized. In Sec. III, the results of the calculations of both methods and their comparisons are given, and a summary is given in Sec. IV.

2. Formalism

In the section, we will elaborate the working expression for the shear viscosity calculations from both methods. It must be stressed that the formalisms used in this work are not new, but the application of these formalisms to test cases is new to the extent that a detailed comparison between two commonly used methods is provided. For the sake of clarity and completeness, the formalisms used in this work are summarized below, along with working formulae.

2.1. Calculating Shear Viscosity: The Chapmann Enskogg Method

In this approximation, the system, which is slightly disturbed from its equilibrium, will have a perturbed distribution function which can be expressed in term of hydrodynamic variables [11]. The perturbed distribution function contains all the information of the system including the information about the transport coefficients (e.g. shear & bulk viscosities). By means of kinetic theory, we can extract shear & bulk viscosities of the system.

For a single component gas, the first approximation to the shear viscosity takes the form of [32]

$$\eta = \frac{1}{10} k_B T \frac{\gamma_0^2}{c_{00}} , \tag{1}$$

where $\gamma_0 = -10\,\hat{h}$, $\hat{h} = K_3(z)/K_2(z)$ and $z = mc^2/k_BT$ is called the relativity parameter. The term c_{00} contains the interaction in the system which is defined as

$$c_{00} = 16\left(w_2^{(2)} - \frac{1}{z}w_1^{(2)} + \frac{1}{3\,z^2}w_0^{(2)}\right). \tag{2}$$

The quantity $w_i^{(s)}$ is so called the relativistic omega integral given by

$$w_i^{(s)} = \frac{2\pi z^3 c}{K_2(z)^2}\int d\psi\,\sinh^7\psi\,\cosh^i\psi\,K_j(2z\cosh\psi) \times \int_0^\pi d\Theta\,\sin\Theta\,\sigma(\psi,\Theta)\,(1-\cos^s\Theta), \tag{3}$$

where $j = 5/2 - 1/2(-1)^i$, $\sinh\psi = g/mc$ and $\cosh\psi = P/2mc$. The quantities g and P are the relative and center of mass momenta. The interaction in the system is hidden in the differential cross section, $\sigma(\psi,\Theta)$.

2.2. Calculating Shear Viscosity: The Green Kubo Method

The second method which we use to calculate the shear viscosity coefficient is the Green-Kubo method. In the Green-Kubo method, linear transport coefficients are cast in terms of the time integral of some sort of a correlation between fluctuations near equilibrium. For the case of the shear viscosity coefficient the formula is

$$\eta = \frac{1}{T}\int d^3r dt \langle \pi^{xy}(0)\,\pi^{xy}(t)\rangle. \tag{4}$$

The brackets refer to taking an average over the ensemble of events generated in the simulation. The

2.3. Equilibriation of Infinite Hadronic Matter

Our hadronic medium is simulated using UrQMD, which is a covariant microscopic transport model based upon the Boltzmann equation

$$\left(\frac{\partial}{\partial t} + \vec{v}\cdot\nabla_r\right) f^{(1)} = \Phi_{coll}, \tag{5}$$

where $f^{(1)}$ is the one-particle phase-space distribution function for a given species, and Φ_{coll} is the collision integral. UrQMD is composed of purely hadronic degrees of freedom, and does not contain information about any possible crossover or phase transition into partonic degrees of freedom. Interactions in our simulation are based only upon scattering, and we neglect any interparticle potential. The criterion for a collision to occur is based upon the geometric interpretation of the cross section:

$$d_{min} \leq \sqrt{\frac{\sigma_{tot}}{\pi}}, \tag{6}$$

where d_{min} is the transverse distance at closest approach between two particles. UrQMD is described in great detail in [3]. Our version of UrQMD includes 55 baryon and baryon resonance species and 32 meson and meson resonance species, (and their antiparticles). These species include strange particles, but no particles containing heavy quarks (c or b quarks).

In order to use the Kubo formalism to extract the shear viscosity coefficient, it is necessary to establish that our system reaches a state of thermal equilibrium. To force our system into equilibrium, we confine our hadronic medium to a cubic box with periodic boundary conditions in real space. This technique was also used in [4, 26, 27]. The input parameters of our system are as follows: volume of the box, initial particle species, along with their chemical potentials.

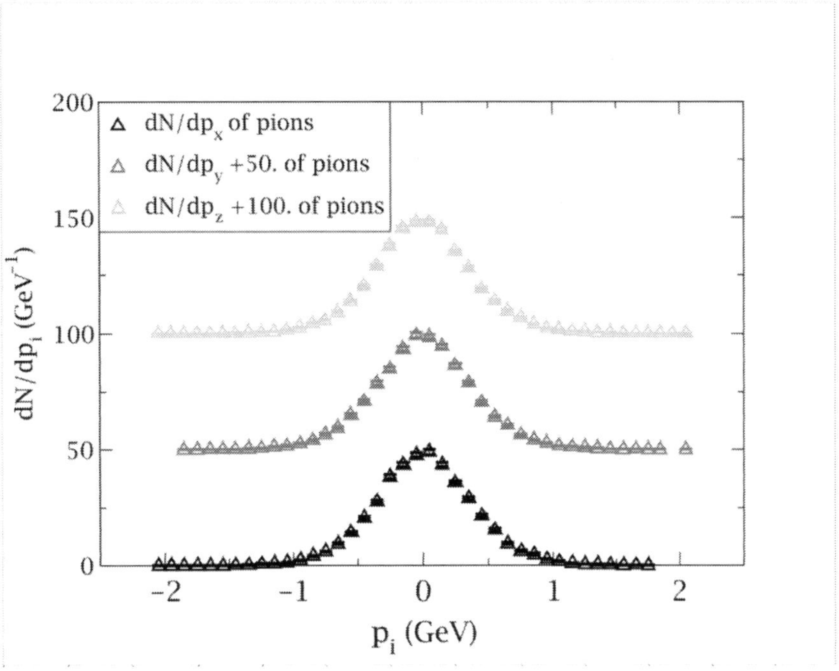

Figure 1. Momentum distributions in the x, y, and z directions, indicating the momentum distribution is indeed isotropic.

We initialize our systems to be in thermal and chemical equilibrium. For the test case involving a gas of massless pions with energy dependent cross sections, the initial yields corresponding to a given temperature and chemical potential are calculated using the Boltzmann distribution function. For the test case involving a pion rho mixture, the initial particle yields are calculated using the Statistical Hadronization with Resonances (SHARE) model [31], with the resonance and decay tables modified to represent the degrees of freedom of the system.

2.4. Thermal and Kinetic Equilibriation

As mentioned above, in order to apply the Green-Kubo method to calculate the shear viscosity coefficient, it is necessary to establish that kinetic equilibrium has been achieved. A system which has attained thermal equilibrium should have its momenta distributions following a Boltzmann distribution:

$$\frac{d^3N_i}{d^3p} \propto exp\left(-\frac{E_i}{T}\right) \tag{7}$$

for a given species N_i. We know that for a system realizing kinetic equilibrium, the momenta distributions should be isotropic, enabling us to write $d^3p = 4\pi p^2 dp$, where p is the three-momentum magnitude. Using the relativistic dispersion relation $E_i = \sqrt{p_i^2 + m_i^2}$ enables us to use the alternate observable $\frac{1}{pE}\frac{dN_i}{dE}$. Figure 1 shows the momenta distribution in the x, y, and z directions evaluated at chemical equilibriation time for the case for a chemically weighted (by the pion/rho densities) temperature of T = 126 MeV for a $\pi\rho$ mixture. Figure 2 shows the observable $\frac{1}{pE}\frac{dN}{dE}$ versus E for the same case.

Journal of Physics: Conference Series **535** (2014) 012018 doi:10.1088/1742-6596/535/1/012018

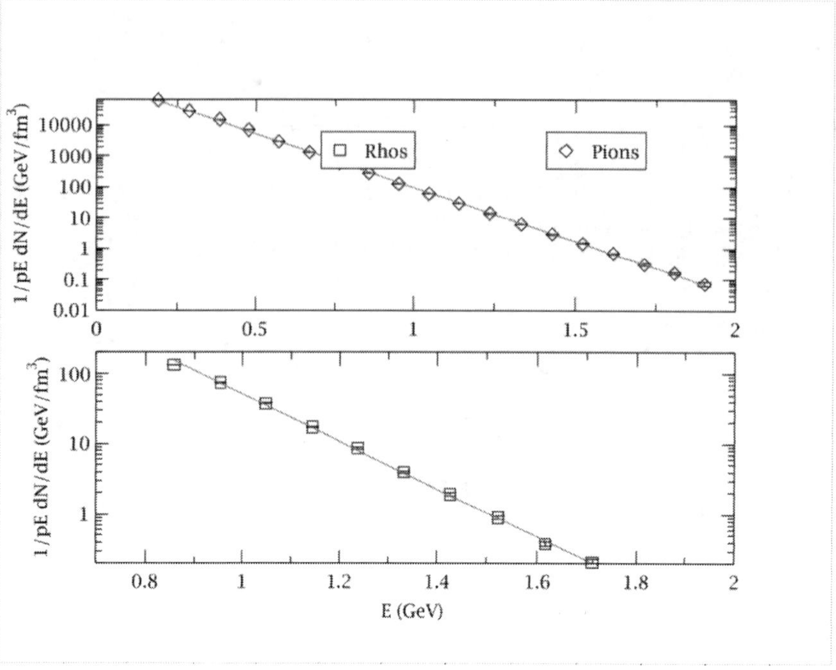

Figure 2. The observable $1/pEdN/dE$ as a function of energy. Both the pions and rhos fit to a Boltzmann distribution. The pion temperature is 125 MeV whereas the rho temperature is 129 MeV.

2.5. Shear Viscosity

Computing the shear viscosity for our system amounts to finding the time integral of the correlations of the shear component of the energy-momentum tensor about the equilibrium state. The energy-momentum tensor is related to the phase-space density of the system through the following equation

$$\pi^{\mu\nu} = \int d^3p \frac{p^\mu p^\nu}{p^0} f(x,p). \tag{8}$$

The phase space density for a system of particles uniformly distributed in phase space is given by

$$f(x,p) = \frac{1}{V} \sum_{j=1}^{N} \delta(\vec{p} - \vec{p_j}). \tag{9}$$

This enables one to calculate pi^{xy} directly

$$\pi^{xy} = \frac{1}{V} \sum_{j=1}^{N} \frac{p^x(j) p^y(j)}{p^0(j)}. \tag{10}$$

Note that since we assumed the particles were uniformly distributed in real space, the Green-Kubo formula for the shear viscosity coefficient becomes

$$\eta = \frac{V}{T} \int dt \langle \pi^{xy}(0) \pi^{xy}(t) \rangle. \tag{11}$$

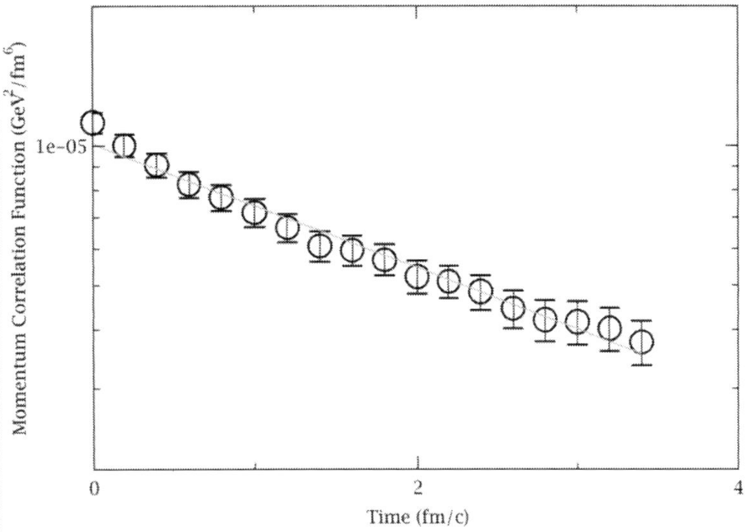

Figure 3. The momentum correlation function as a function of time t. The relaxation time for this case is $\tau_\pi = 3.1$ fm/c.

In order to simplify evaluating this integral, we assume the correlation function $\langle \pi^{xy}(0)\,\pi^{xy}(t)\rangle$ has an exponential ansatz:

$$\langle \pi^{xy}(0)\,\pi^{xy}(t)\rangle \propto exp\left(-\frac{t}{\tau_\pi}\right) \tag{12}$$

A representative sample of this correlation function is shown in Figure 3. The figure suggests that our assumption is reasonable. This ansatz has also been used in [26]. If we assume the exponential dependence of the stress tensor correlation function on post-equilibriation time t, the Kubo formula reduces to

$$\eta = \frac{V}{T}\tau_\pi \langle \pi^{xy}(0)^2\rangle. \tag{13}$$

The volume V of the system is an input parameter, the temperature T is extracted from fitting momenta distributions to a Boltzmann distribution after checking that the momenta distributions in the system are isotropic, the correlator relaxation time tau_π is obtained from fitting the momentum correlation function to an exponential, and $\langle \pi^{xy}(0)^2\rangle$ is obtained from the intercept of the momentum correlation function. Hence, the shear viscosity coefficient can be calculated in this way, and in the next section we present the results of the shear viscosity coefficient for two systems.

3. Results

In this section we present the results for the calculation of the shear viscosity via both the Chapmann-Enskog and Green-Kubo method for two systems. The first system involves a pure chiral pion gas with an energy dependent cross section given by current algebra. The

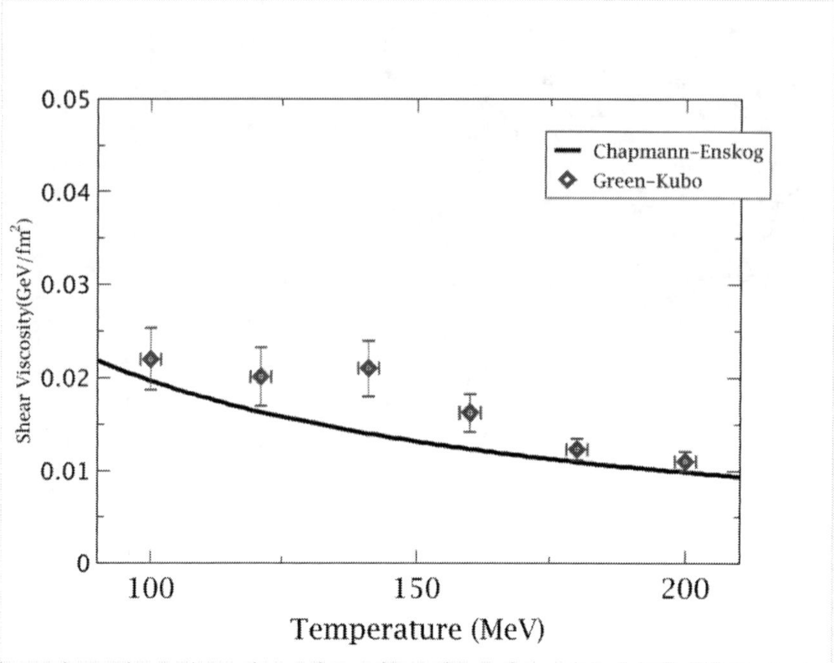

Figure 4. Shear viscosity coefficient as a function of temperature for an gas of chiral pions with energy dependent cross section.

second system is a $\pi\rho$ mixture, where massive pions scatter elastically and inelastically via an intermediate ρ resonance state. More details are given below.

3.1. Massless Pions with Energy Dependent Cross Section (with Current Algebra)
The total (energy-dependent) cross section for the first system is given by

$$\sigma\left(\sqrt{s}\right) = \frac{5s}{72\pi f_\pi^4}, \tag{14}$$

where s is the Mandelstam variable representing the total energy in the center of mass frame and $f_\pi = 93$ MeV is the pion decay constant. The results for the shear viscosity coefficient are shown as a function of temperature in Figure 4, and we find that there is very good agreement between the calculations from the Green-Kubo method and the Chapmann-Enskog method.

3.2. Pion Rho Mixture
The $\pi\rho$ mixture involves the two following processes:

- (a) $\sigma_{\pi\pi\to\pi\pi,elastic} = 17.778$ mb.
- (b) $\pi\pi \to \rho \to \pi\pi$ (inelastic)

The value of the cross section in (a) is based upon the additive quark model, which dictates that meson-meson and baryon-baryon cross sections are related via

$$\sigma_{MM} = \left(\frac{2}{3}\right)^2 \sigma_{BB}, \tag{15}$$

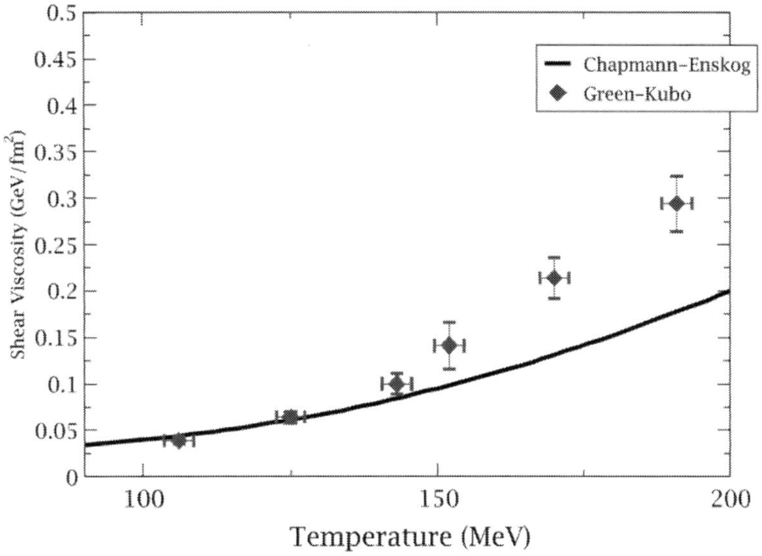

Figure 5. Shear viscosity coefficient versus temperature for a gas of massive pions scattering elastically and inelastically through an intermediate rho resonance.

where M/B stands for meson/baryon, respectively. Since the elastic nucleon-nucleon cross section is known to be approximately 40 mb, this implies that $\sigma_{\pi\pi,elastic} = 17.778$ mb.

The results for the shear viscosity coefficient are shown as a function of temperature in Figure 5. We find that, although there is good agreement between the Green-Kubo and Chapmann-Enskog methods in the temperature range $100 < T < 150)$ MeV, the results from the Green-Kubo calculation are systematically higher than those for the Chapmann-Enskog method in the temperature range $150 < T < 200$ MeV. This is a noteworthy result, particularly when one realizes that many more inelastic scattering events through the intermediate resonance channel are expected in the temperature range $T > 150$ MeV than $T < 150$ MeV. In order to fully understand why the Green-Kubo method yields a higher result for the shear viscosity than the Chapmann-Eskog method does, it is necessary to understand the effect of resonance lifetimes in UrQMD on the shear viscosity coefficient. This is a subtle effect and the effect of timescales on transport coefficients in general has been investigated in the past [10]. One of the goals of this paper is to highlight the differences in the calculation of the shear viscosity of the Chapmann-Enskog method versus the Green-Kubo method, and we find that the greatest difference is observed when one examines the results for the shear viscosity coefficient in the temperature range $150 < T < 200$ MeV, where a large number of inelastic scattering events are expected to take place. An investigation of the effect of the resonance lifetime on the shear viscosity coefficient of the system is in progress.

4. Summary

In this section, we summarize the results from both methods to calculate the shear viscosity for two different systems. The systems analyzed included a system of a pure gas of massless pions interacting via an energy-dependent cross section obtained via chiral current algebra, and a system of massive pions interacting elastically via a constant cross section given by the AQM model, with inelastic scattering enabled through an intermediate ρ resonance channel. The two methods used to calculate the shear viscosity were the Chapmann-Enskog method and the Green-Kubo method using the UrQMD model to simulate the hadronic medium in equilibrium. When the results of the calculation for the two different methods were compared, excellent agreement was found between the Chapmann Enskog and Green Kubo methods for the system of pure massless pions with an energy dependent cross section, whereas a systematic deviation occured between the Chapmann Enskog and Green Kubo methods in the temperature range $150 < T < 200$ MeV, where significant inelastic scattering is expected to occur. The Green-Kubo method systematically yielded a higher result in this temperature range. To fully understand the reason for this, a systematic investigation of the effect of resonance lifetimes on the shear viscosity needs to be performed, and this is in progress. We also plan to extend this comparison to other test systems.

Acknowledgements

A. Wiranata acknowledges NFSC grant number 11221504 and grant number 1350110493. N. Demir wishes to thank Kuwait University for its support via decree number 1906 for the 2014-2015 fiscal year.

References

[1] A. Adare et al. Energy Loss and Flow of Heavy Quarks in Au+Au Collisions at $\sqrt{s_{NN}} = 200$ GeV. *Phys. Rev. Lett.*, 98:172301, 2007.

[2] B. B. Back et al. Centrality dependence of charged hadron transverse momentum spectra in d + au collisions at s(nn)**(1/2) = 200-gev. *Phys. Rev. Lett.*, 91:072302, 2003.

[3] S. A. Bass et al. Microscopic models for ultrarelativistic heavy ion collisions. *Prog. Part. Nucl. Phys.*, 41:225–370, 1998.

[4] M. Belkacem et al. Equation of state, spectra and composition of hot and dense infinite hadronic matter in a microscopic transport model. *Phys. Rev.*, C58:1727–1733, 1998.

[5] Mauro Brigante, Hong Liu, Robert C. Myers, Stephen Shenker, and Sho Yaida. The Viscosity Bound and Causality Violation. *Phys. Rev. Lett.*, 100:191601, 2008.

[6] Jiunn-Wei Chen, Yen-Han Li, Yen-Fu Liu, and Eiji Nakano. Qcd viscosity to entropy density ratio in the hadronic phase. 2007.

[7] Jiunn-Wei Chen and Eiji Nakano. Shear viscosity to entropy density ratio of qcd below the deconfinement temperature. *Phys. Lett.*, B647:371–375, 2007.

[8] Jiunn-Wei Chen and Juven Wang. Bulk Viscosity of a Gas of Massless Pions. *Phys. Rev.*, C79:044913, 2009.

[9] P. Danielewicz and M. Gyulassy. Dissipative phenomena in quark gluon plasmas. *Phys. Rev.*, D31:53–62, 1985.

[10] Pawel Danielewicz and Scott Pratt. Delays associated with processes in nuclear reaction simulations. *Phys. Rev.*, C53:249–266, 1996.

[11] Sybren R. De Groot. *Relativistic Kinetic Theory: Principles and Applications.* North-Holland Publishing Company, third edition, 1980.

[12] Nasser Demir and Steffen A. Bass. Shear-Viscosity to Entropy-Density Ratio of a Relativistic Hadron Gas. *Phys. Rev. Lett.*, 102:172302, 2009.

[13] Antonio Dobado and Silvia N. Santalla. Pion gas viscosity at low temperature and density. *Phys. Rev.*, D65:096011, 2002.

[14] Hans-Joachim Drescher, Adrian Dumitru, Clement Gombeaud, and Jean-Yves Ollitrault. The centrality dependence of elliptic flow, the hydrodynamic limit, and the viscosity of hot QCD. *Phys. Rev.*, C76:024905, 2007.

[15] K. Dusling and D. Teaney. Simulating elliptic flow with viscous hydrodynamics. 2007.

[16] S. Gavin. Transport coefficients in ultrarelativistic heavy ion collisions. *Nucl. Phys.*, A435:826–843, 1985.

[17] Sean Gavin and Mohamed Abdel-Aziz. Measuring Shear Viscosity Using Transverse Momentum Correlations in Relativistic Nuclear Collisions. *Phys. Rev. Lett.*, 97:162302, 2006.

[18] M. I. Gorenstein, M. Hauer, and O. N. Moroz. Viscosity in the excluded volume hadron gas model. *Phys. Rev.*, C77:024911, 2008.

[19] Pasi Huovinen and Denes Molnar. The applicability of causal dissipative hydrodynamics to relativistic heavy ion collisions. *Phys. Rev.*, C79:014906, 2009.

[20] Kazunori Itakura, Osamu Morimatsu, and Hiroshi Otomo. Shear viscosity of a hadronic gas mixture. *Phys. Rev.*, D77:014014, 2008.

[21] P. Kovtun, D. T. Son, and A. O. Starinets. Viscosity in strongly interacting quantum field theories from black hole physics. *Phys. Rev. Lett.*, 94:111601, 2005.

[22] Roy A. Lacey et al. Has the QCD critical point been signaled by observations at RHIC? *Phys. Rev. Lett.*, 98:092301, 2007.

[23] Matthew Luzum and Paul Romatschke. Conformal Relativistic Viscous Hydrodynamics: Applications to RHIC results at $\sqrt{s_{NN}} = 200$ GeV. *Phys. Rev.*, C78:034915, 2008.

[24] Harvey B. Meyer. A calculation of the shear viscosity in su(3) gluodynamics. *Phys. Rev.*, D76:101701, 2007.

[25] Berndt Muller and James L. Nagle. Results from the relativistic heavy ion collider. 2006.

[26] Azwinndini Muronga. Shear viscosity coefficient from microscopic models. *Phys. Rev.*, C69:044901, 2004.

[27] Shin Muroya and Nobuo Sasaki. A calculation of the viscosity over entropy ratio of a hadronic gas. *Prog. Theor. Phys.*, 113:457–462, 2005.

[28] Madappa Prakash, Manju Prakash, R. Venugopalan, and G. Welke. Nonequilibrium properties of hadronic mixtures. *Phys. Rept.*, 227:321–366, 1993.

[29] Bjoern Schenke, Sangyong Jeon, and Charles Gale. Anisotropic flow in $\sqrt{s} = 2.76$ TeV Pb+Pb collisions at the LHC. *Phys.Lett.*, B702:59–63, 2011.

[30] Huichao Song and Ulrich W. Heinz. Causal viscous hydrodynamics in 2+1 dimensions for relativistic heavy-ion collisions. *Phys. Rev.*, C77:064901, 2008.

[31] Giorgio Torrieri, Jean Letessier, Johann Rafelski, and Steve Steinke. Statistical hadronization with resonances. *Acta Phys.Polon.*, B35:2911–2927, 2004.

[32] Anton Wiranata and Madappa Prakash. Shear Viscosities from the Chapman-Enskog and the Relaxation Time Approaches. *Phys.Rev.*, C85:054908, 2012.

Journal of Physics: Conference Series **535** (2014) 012019 doi:10.1088/1742-6596/535/1/012019

Heavy Flavor Suppression, Flow and Azimuthal Correlation: Boltzmann vs Langevin

F. Scardina [1,2], **S. K. Das**[1,2], **S. Plumari**[1,2], **D.Perricone**[1] **and V. Greco**[1,2]

[1] Department of Physics and Astronomy, University of Catania, Via S. Sofia 64, I- 95125 Catania, Italy

[2] Laboratori Nazionali del Sud, INFN-LNS, Via S. Sofia 62, I-95123 Catania, Italy

E-mail: scardinaf@lns.infn.it

Abstract. The propagation of heavy flavor through the quark gluon plasma has been treated commonly within the framework of Langevin dynamics, i.e. assuming the heavy flavor momentum transfer is much smaller than the light one. On the other hand a similar suppression factor R_{AA} has been observed experimentally for light and heavy flavors. We present a thorough comparison in terms of nuclear suppression, R_{AA}, elliptic flow, v_2, and $c\bar{c}$ back to back correlation between the Langevin equation and the full collisional Boltzmann collision integral within the framework of Boltzmann transport equation. We have shown that the Langevin dynamics overestimates the interaction and even for a fixed R_{AA} the full two-body collision integral shows that the elliptic flow is larger with respect to that predicted by a Langevin dynamics. Furthermore we have found that Boltzmann approach gives rise to a larger spreading of $c\bar{c}$ correlation in comparison with the Langevin approach.

1. Introduction

The experimental efforts at Relativistic Heavy Ion Collider (RHIC) and Large Hadron Collider (LHC) energies are aimed at creating and characterizing the properties of the quark gluon plasma (QGP). The heavy flavors, namely, charm and bottom quarks (c and b) are playing a crucial role to serve this purpose because of two-fold reason: the first, typical of particles physics, is that the mass $M_Q \gg \Lambda_{QCD}$ which makes possible the evaluation of cross section and p_T spectra within next-to-next-to-lead order (NLLO) [1, 2] in a perturbative QCD (pQCD) scheme; the second, more inherent to plasma physics is that $M_Q \gg T$ and therefore the thermal production in the QGP is expected to be negligible because it is suppressed approximately by a $\sim e^{-M/T}$ term. Hence for Heavy Quarks (HQ) one has a nearly exact flavor conservation during the evolution of the plasma in both the partonic and hadronic stages. Hence, heavy flavors are the witness of the entire phase-space evolution of the QGP and can be used as a probe to characterize the properties of the QGP. The most common tool to study heavy flavors propagation in QGP is based on the Fokker-Planck approach [3, 4, 5, 6, 7, 8, 9, 10, 11, 12, 13, 14, 15] that can be obtained from the Boltzmann equation in the limit of small angle approximation or in other words in the limit in which the momenta transfered in the collisions between the particles of the bulk and the heavy quarks are small compared to the total momentum of the heavy quarks. The Fokker Planck approximation is expected to be asymptotically valid for $m/T \to \infty$. We are presenting in this proceeding a first study on the validity of such an approximation for charm and bottom quarks. The proceeding is organized as follows. In section 2 we discuss the Boltzmann and the Fokker Planck approach which are used to describe the propagation of heavy quark through the QGP. In section 3 we discuss the evolution of the HQ spectra in a static medium. In section 4 the results for $c\bar{c}$ correlation are presented. In section 5 we compare the results we get for R_{AA} and v_2 of HQ with both the two approaches when realistic simulations of ultra relativistic heavy ion collisions are performed and finally section 6 contains the conclusions

2. Boltzmann vs Fokker-Planck dynamics

The propagation of HQ in QGP has been quite often described by the Fokker-Planck equation [3, 4, 5, 6]. The main reason is that it was believed that their motion can be encased into a Brownian motion due to their large masses that should generically lead to collisions sufficiently forward peaked and/or with small momentum transfer. Under such constraints it is known that also the Boltzmann equation reduces to a Fokker-Planck equation [4], which constitutes a significant simplification of in medium dynamics. Such a scheme has been very widely employed [5, 6, 7, 14, 8, 10, 16, 11, 12, 13, 15, 17, 18, 19] in order to calculate the nuclear suppression factor (R_{AA}) [36, 37, 38] and the elliptic flow (v_2) [36] that have been observed for the non-photonic single electron spectra that are related to the decay of charm and bottom mesons. Along with the Fokker-Planck approach in some works a description of HQ within a relativistic Boltzmann transport approach has been developed [9, 22, 34, 24, 25, 35]. The Boltzmann equation for the HQ distribution function can be written in a compact form as:

$$p^{\mu}\partial_{\mu}f_Q(x,p) = \mathcal{C}[f_Q](x,p) \tag{1}$$

where $\mathcal{C}[f_Q](x,p)$ is the relativistic Boltzmann-like collision integral in which the phase-space distribution function of the bulk medium appears as an integrated quantity, see for example Ref.s [21, 32]. This equation describes the evolution of the heavy quarks distribution function $f_Q(x,p)$. The collision integral can be written in a simplified form [3, 4] in the following way:

$$\mathcal{C}[f_Q](x,p) = \int d^3k[\omega(p+k,k)f_Q(x,p+k) - \omega(p,k)f_Q(x,p)] \tag{2}$$

where $\omega(p,k)$ express the collision rate of heavy quark per unit of momentum phase space which changes the heavy quark momentum from p to $p-k$. The first term in the integrand is the gain of probability through collisions and the second term represent the net loss out of that momentum space volume. In this work we have not considered radiative processes therefore the heavy quarks interact with the medium by mean of two-body collisions regulated by the scattering matrix of the process $g+Q \rightarrow g+Q$, therefore defining the relative velocity between the two colliding particles as v_{rel} the transition rate can be written as:

$$\omega(p,k) = \int \frac{d^3q}{(2\pi)^3} f_g(x,p) v_{rel} \frac{d\sigma_{g+Q\rightarrow g+Q}}{d\Omega} \tag{3}$$

where $\sigma_{g+Q\rightarrow g+Q}$ is related to the scattering matrix $|\mathcal{M}_{gQ}|^2$ as indicated below

$$v_{rel}\frac{d\sigma_{g+Q\rightarrow g+Q}}{d\Omega} = \frac{1}{d_c}\frac{1}{4E_pE_q}\frac{|\mathcal{M}_{gQ}|^2}{16\pi^2 E_{p-k}E_{q+k}}\delta^0(E_p + E_q - E_{p-k} - E_{q+k}) \tag{4}$$

The explicit expression of the scattering matrix is indicated in this reference [4]. We point out that the scattering matrix is the real kernel of the dynamical evolution for both the Boltzmann approach and the Fokker-Planck one. Of course all the calculations discussed in this proceeding will originate from the same scattering matrix for both cases.

The Boltzmann equation is solved numerically dividing the space into a three-dimensional lattice and using the test particle method to sample the distributions functions. The collision integral is solved by means of a stochastic algorithm in which whether a collision happen or not is sampled stochastically comparing the collision probability related to two body collisions between HQ and gluons ($P_{22} = v_{rel}\sigma_{g+Q\to g+Q} \cdot \Delta t/\Delta x$) with a random number extracted between 0 and 1 [20, 30, 21, 33] . The code has been widely tested and in particular it have been checked that the correct collision rate is reproduced and that given an initial non-equilibrium distribution the dynamical evolution brings the system towards equilibrium described by a Boltzmann-Juttner distribution. We have performed these tests as a function of cross section, temperature and mass of the particles, including non-elastic collisions [31] .
The non-linear integro-differential Boltzmann equation can be simplified employing the Landau approximation whose physical relevance can be associated to the dominance of soft scatterings with small momentum transfer $|\mathbf{k}|$ respect to the particle momentum $\mathbf{p}$. This simplification is obtained expanding $\omega(p+k,k)f(x,p+k)$ around the momentum transfer, k,

$$\omega(p+k,k)f_Q(x,p+k) \approx \omega(p,k)f(x,p) + k\frac{\partial}{\partial p}(\omega f) + \frac{1}{2}k_i k_j \frac{\partial^2}{\partial p_i \partial p_j}(\omega f) \tag{5}$$

Inserting Eq.(5) into the Boltzmann collision integral, indicated in Eq.(2), one obtains the Fokker Planck Equation:

$$\frac{\partial f}{\partial t} = \frac{\partial}{\partial p_i}\left[A_i(\mathbf{p})f + \frac{\partial}{\partial p_j}[B_{ij}(\mathbf{p})]\right] \tag{6}$$

by simply defining $A_i = \int d^3k\, w(\mathbf{p},\mathbf{k})k_i = A(\mathbf{p})p_i$ and $B_{ij} = \int d^3k\, w(\mathbf{p},\mathbf{k})k_i k_j$ that are directly related to the so called drag and diffusion coefficient. The Fokker-Planck equation can be solved by a stochastic differential equation i.e the Langevin equation that can be written as [3, 6, 14]:

$$\begin{aligned} dx_i &= \frac{p_i}{E}dt, \\ dp_i &= -Ap_i dt + (\sqrt{2B_0}P_{ij}^{\perp} + \sqrt{2B_1}P_{ij}^{\parallel})\rho_j\sqrt{dt} \end{aligned} \tag{7}$$

where dx_i and dp_i are the coordinates and momenta changes in each time step dt. A is the drag force and B the longitudinal and transverse diffusions, ρ_j is a stochastic Gaussian distributed variable and

$$P_{ij}^{\perp} = \delta_{ij} - \frac{p_i p_j}{p^2}, P_{ij}^{\parallel} = \frac{p_i p_j}{p^2}. \tag{8}$$

are the transverse and longitudinal tensor projectors. We will employ the common assumption, $B_0 = B_1 = D$ [6, 7, 14, 8, 9, 11, 15]. To achieve the equilibrium distribution $f_{eq} = e^{-E/T}$ with $E = \sqrt{p^2+m^2}$ as the final distribution one need to adjust the drag coefficient A in accordance with the Einstein relation [26] (see also [27])

$$A(p) = \frac{D(p)}{ET} - \frac{D'(p)}{p}. \tag{9}$$

3. Heavy quark diffusion in momentum space in a static medium

To investigate the similarities and the differences in between the Langevin and Boltzmann approach we have studied the heavy quark momentum evolution considering the initial charm and bottom quark distribution as a delta distribution at $p = 10$ GeV. We have solved both the Langevin and Boltzmann equation in a box where the bulk consists of only gluon at equilibrium at $T = 400 MeV$. The elastic collisions of heavy quarks with gluon has been considered within the pQCD framework. In particular we have evaluated the drag and the diffusion coefficients, which are the basic ingredients required to solve the Langevin equations, using the scattering matrix elements shown in reference [4] and with the same scattering matrix we have also evaluated the total cross section that is required in order to solve the Boltzmann collision integral. In this work we have used $\alpha_s = 0.35$. The momentum evolution of the charm and bottom quarks within the Langevin dynamics are displayed respectively in Fig. 1 (left) and Fig. 2 (left). One can notice that both the charm and bottom quarks spectra assumes a gaussian shape as expected by construction. In fact it is well known the Langevin dynamics consists of a shift of the average momenta with a fluctuation around it that includes also the possibility to gain energy for the HQ as one can observe from the tail of the momentum distribution that overshoots the initial momentum $p = 10\,\text{GeV}$ at $t = 2\,\text{fm/c}$, black solid line in Fig.1 (left).

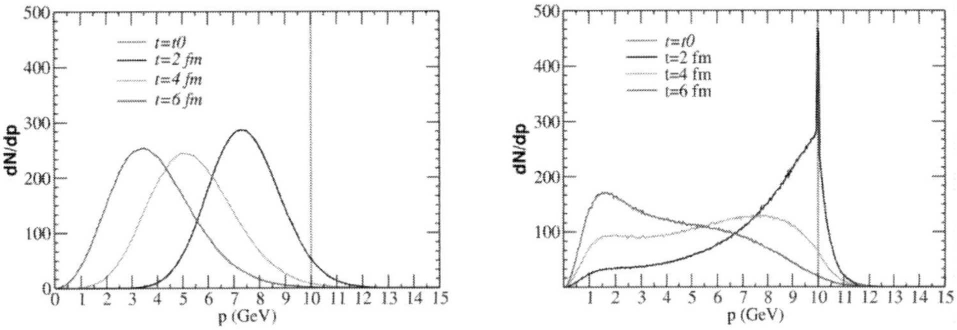

Figure 1. Evolution of the momentum distribution for charm quarks within Langevin dynamics (left) and Boltzmann equation (right) considering the initial momentum distribution of the charm quarks as a delta distribution at p=10 GeV.

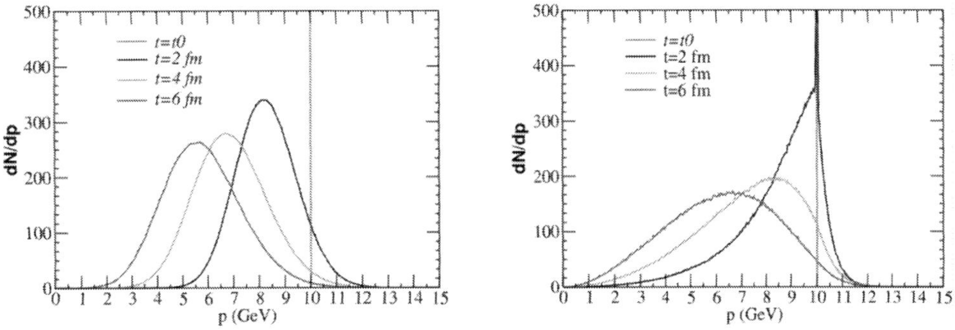

Figure 2. Evolution of the momentum distribution for bottom quarks within Langevin dynamics (left panel) and the Boltzmann (right panel) considering the initial momentum distribution of the bottom quark as a delta distribution at p=10 GeV.

In Fig. 1 (right) we depict the momentum distribution for charm quark within the Boltzmann equation. From this figures it appears a very different evolution of the charm quark momentum distribution which does not have a Gaussian shape and already at $t = 2\,\text{fm/c}$ has a very different spread in momentum with a larger contribution from processes where the charm quark can gain energy and a long tail at low momenta. Such a tail indicates that charms during their propagation thorough the medium can lose a quite large amount of energy. The fact that charm spectra at different time does not show a gaussian shape essentially indicates that for a particle with $M \sim \langle p \rangle \sim 3T$, as it is for the charm at a temperature $T = 0.4\,\mathrm{GeV}$, the momentum evolution is not of Brownian type. For the bottom quarks, shown in Fig. 2 , the momentum evolution gives a much better agreement between the Boltzmann and the Langevin evolution because $M_{bottom}/T \simeq 10$.

4. Azimuthal correlations of heavy quarks

In the initial hard scatterings heavy quark pairs are produced back-to-back at leading order due to momentum conservation and this lead to a back-to-back correlation in azimuthal angle. These produced heavy quarks pairs move in opposite direction and hence suffer different interactions depending if they propagate towards the inner part of the bulk or towards the outer. This would drastically alter their initial back-to-back correlation [28, 29]. We have studied the azimuthal correlation of the $c\bar{c}$ pairs within both the Boltzmann and Langevin dynamics in a box. We initialize the $c\bar{c}$ pairs at the boundary of the box with position x=z=0 and y=-2.5fm, and having momentum $p_x = p_z$=0 and p_y=10 GeV. Hence, the initial azimuthal correlation will be a delta around $\Delta\phi = 0$. As a consequence of the interaction with the thermal bath this initial azimuthal correlation will broaden around $\Delta\phi = 0$. We have studied the final azimuthal correlation for charm quarks within the momentum cut $3 < p < 4GeV$.

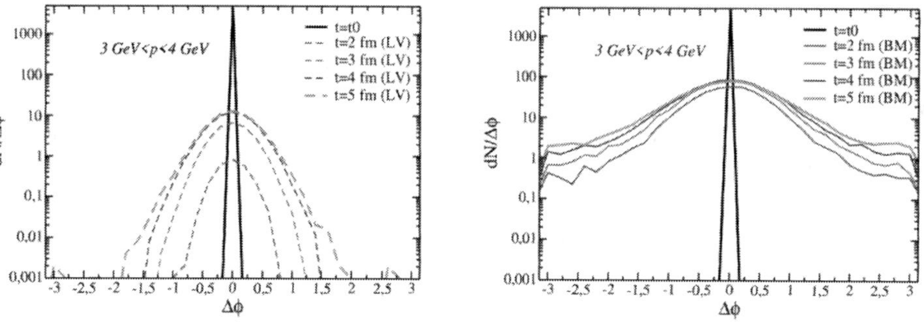

Figure 3. Left: Azimuthal correlation of $c\bar{c}$ within Langevin dynamics at different times. ; Right: Azimuthal correlation of $c\bar{c}$ within Boltzmann equation at different times.

In Fig. 3 (left) the azimuthal correlation of $c\bar{c}$ within Langevin dynamics has been displayed at different times while in Fig. 3 (right) the azimuthal correlation of $c\bar{c}$ within Boltzmann equation has been shown. A larger spreading of momentum distribution in case of Boltzmann equation give rise a larger spreading of $c\bar{c}$ correlation in comparison with the Langevin dynamics.

5. Heavy flavor suppression and flow

We have compared the Boltzmann and Langevin approach also for realistics simulations of heavy ion collisions. We have carried out simulations of $Au + Au$ collisions at $\sqrt{s} = 200GeV$ for $20 - 30\%$ centrality. The initial conditions for the bulk in coordinate space are given by the standard Glauber model while in momenta space we employ a Boltzmann-Juttner distributions up to a transverse momentum $p_T = 2GeV$ and a minijet distributions for higher momenta as calculated by pQCD at Next to leading order. For the initial distributions of charm quarks in coordinate space we use $f^c_{t=0} = 4.1 * 10^2/(0.7 + 0.09p)^{15.44}$ while for bottoms quarks we use $f^b_{t=0} = 1/(57.74 + p^2)^{5.04}$. In coordinate space the HQ are distributed according to the number

of binary collisions N_{coll}. Moreover the bulk that we obtained using the transport approach is then used as background for the Langevin evolution of HQ. This is a novelty of our work since usually the background for the Langevin is given by hydrodynamic simulations.
Our purpose is to perform the comparison between the Langevin and Boltzmann transport equations for different momentum transfer scenario that can be directly related to the angular distribution of scattering matrix or cross section. This can be achieved by using three different values of the Debye screening masses (m_D) needed to shield the divergence associated with the t-channel of the scattering matrix. We have chosen three values for m_D: $m_D = 0.4$GeV which implies a very forward peaked cross section and thus a small average momentum transferred; $m_D = 0.83$GeV and $m_D = 1.6\,$GeV which implies an almost isotropic cross section and thus a quite large average momentum transfered.

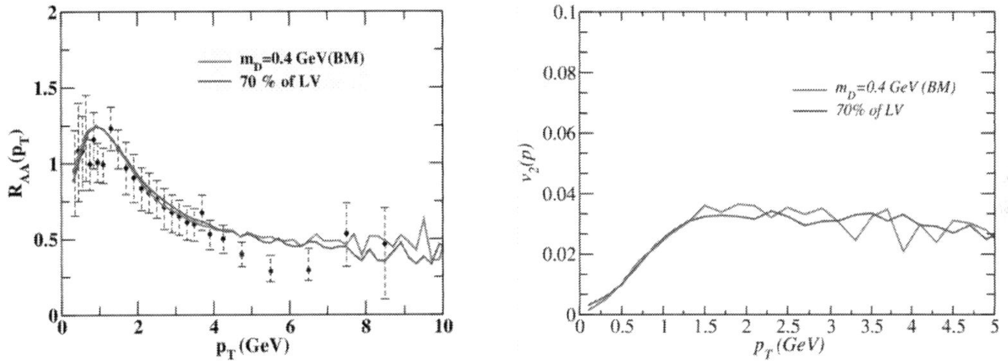

Figure 4. R_{AA} (left) and v_2 (right) as a function of p_T at m_D=0.4 GeV within both Langevin (LV) and Boltzmann (BM) approach.

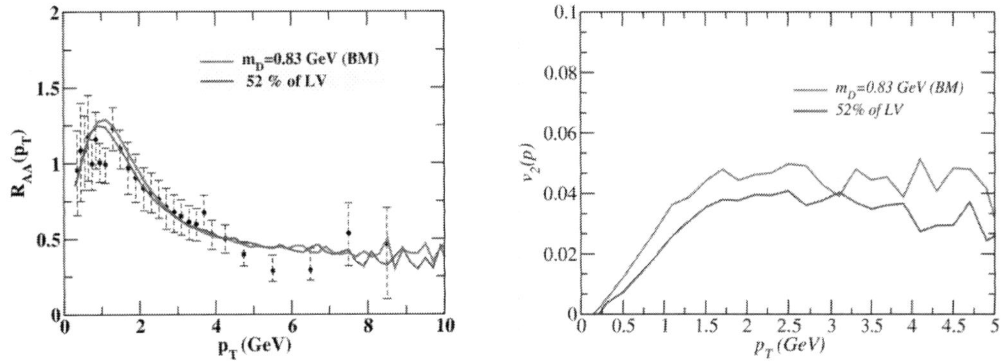

Figure 5. R_{AA} (left) and v_2 (right) as a function of p_T at m_D=0.83 GeV within both Langevin (LV) and Boltzmann (BM) approach.

In the present work we have tried to reproduce the same R_{AA} in both Langevin and Boltzmann approach and we have studied the relation with the corresponding v_2 for all the three m_D case.

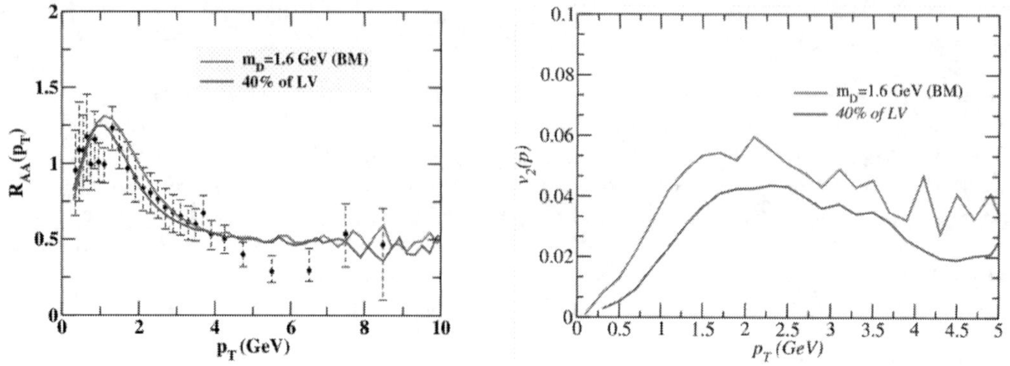

Figure 6. R_{AA} (left) and v_2 (right) as a function of p_T at m_D=1.6 GeV within both Langevin (LV) and Boltzmann (BM) approach.

In Fig. 4 we have displayed the R_{AA} (left) and v_2 (right) at m_D=0.4 GeV within both Langevin and Boltzmann approach. As expected, in this near forward peaked angular distribution of the cross section case the Langevin dynamics is quite close to the Boltzmann one (for details we refer to our earlier work [24]). In this case to reproduce the same R_{AA} we need to reduce the interaction from the Langevin side by 30% (or equivalently we need only the 70% of LV interaction). In Fig. 5 we have displayed the results same as Fig 4 with m_D=0.83 GeV. In this case to have the same R_{AA} (within LV and BM), we need to reduce the interaction from the Langevin side by almost 50%. In this case Boltzmann dynamics generate a lager v_2 in comparison with the Langevin dynamics. In Fig. 6 we have displayed the R_{AA} (left) and v_2 (right) at m_D=1.6 GeV within both Langevin and Boltzmann approach. In this near isotropic angular distribution of the cross section case, we need to reduce the interaction by 60% from the Langevin side to get the same R_{AA} as of Boltzmann . In this case ever reducing the Langevin interaction to have the same R_{AA} obtained with the Boltzmann we get in the first case a smaller elliptic flow and this indicates that the Boltzmann approach is more efficient in the generation of the v_2.

6. Conclusions

In this proceeding we have presented a study of the approximations involved by Langevin equation making a direct comparison with the full collisional integral within the framework of Boltzmann transport equation in a box where the bulk consists of only gluon at T=0.4 GeV as well as for a realistic simulations of heavy ion collisions. We have found that the Langevin approach is a good approximation for bottom quark whereas for charm quark Langevin approach deviates from Boltzmann approach and the difference is larger for larger m_D. It has also been found that to get a similar suppression factor for both the approach we need to reduce the interaction of the Langevin approach. For the same R_{AA} Boltzmann dynamics produced larger v_2, which may be helpful for a simultaneous reproduction of R_{AA} and v_2 for the same set of inputs. Moreover we have found that the Boltzmann approach gives rise to a larger spreading of $c\bar{c}$ correlation in comparison with the Langevin approach.

Acknowledgements
We acknowledge the support by the ERC StG under the QGPDyn Grant n. 259684

References

[1] M. Cacciari, P. Nason and R. Vogt, Phys. Rev. Lett. **95** (2005) 122001
[2] M. Cacciari, S. Frixione, N. Houdeau, M. L. Mangano, P. Nason and G. Ridolfi, JHEP **1210** (2012) 137
[3] Rapp R and Hees H van 2010 Hwa R C, Wang X N (Ed.) Quark Gluon Plasma 4, *World Scientific*, **111** [arXiv:0903.1096 [hep-ph]]
[4] Svetitsky B 1988 *Phys. Rev.* D **37**, 2484
[5] Mustafa M G, Pal D and Srivastava D K 1998 *Phys. Rev.* C **57**, 889
[6] Moore G D, Teaney D 2005 *Phys. Rev.* C **71** 064904
[7] Hees H van, Greco V and Rapp R 2006 *Phys. Rev.* C **73**, 034913
[8] Hees H van, Mannearelli M, Greco V and Rapp R 2008 *Phys. Rev. Lett.* **100** 192301
[9] Gossiaux P B, Aichelin J 2008 *Phys. Rev.* C **78** 014904 ; Gossiaux P B, Aichelin J, Gousset T and Guiho V 2010 *J. Phys.* G **37** 094019
[10] Akamatsu Y, Hatsuda T, Hirano T 2009 *Phys. Rev.* C **79** 054907
[11] Das S K, Alam J, Mohanty 2009 *Phys. Rev.* C **80** 054916; 2010 *Phys. Rev.* C **82** 014908; Majumdar S, Bhattacharyya T, Alam J, Das S K 2012 *Phys. Rev.* C **84** 044901
[12] Alberico W M *et al.* 2011 *Eur. Phys. J.* C **71** 1666; Alberico W M *et al.* 2013 *Eur. Phys. J.* C **73** 2481
[13] Young C, Schenke B, Jeon S and Gale C 2012 *Phys. Rev.* C **86**, 034905
[14] Cao S, Bass S A Bass 2011 *Phys. Rev.* C **84** 064902
[15] Lang T, Hees H, Steinheimer J and Bleicher M, arXiv:1208.1643 [hep-ph]
[16] P. B. Gossiaux *et al.*, arXiv:1102.1114 [hep-ph]
[17] M. He, R. J. Fries and R. Rapp, Phys. Rev. Lett. **110**, 112301 (2013)
[18] S. K. Das and A. Davody, Phys. Rev. C **89**, 054912 (2014)
[19] Hao-jie Xu, Xin Dong, Li-juan Ruan, Qun Wang, Zhang-bu Xu, and Yi-fei Zhang, arXiv:1305:7302
[20] Ferini G, Colonna M, Toro M Di and Greco V, 2009 *Phys. Lett.* **B670**,325; Greco V, Colonna M, Toro M Di and Ferini G, 2009 *Progr. Part. Nucl. Phys.* **62**, 562
[21] Xu Z, Greiner C 2005 *Phys. Rev.* C **71**, 064901
[22] Uphoff J, Fochler O, Xu Z and Greiner C 2011 *Phys. Rev.* C **84** 024908
[23] Younus M, Coleman-Smith C E,Bass S A and Srivastava D K, arXiv:1309.1276 [nucl-th]
[24] S.K. Das, F. Scardina and V. Greco, arXiv:1312.6857 [nucl-th].
[25] S. K. Das, F. Scardina, S. Plumari and V. Greco, arXiv:1309.7930 [nucl-th]
[26] D. B. Walton and J. Rafelski, Phys. Rev. Lett. **84**, 31 (2002)
[27] S. Majumdar, T Bhattacharyya and J. Alam, arXiv:1305.6445 [nucl-th]
[28] Zhu X, Xu N, Zhuang P, 2008 *Phys. Rev. Lett.* **100**, 152301 (2008)
[29] Nahrgang M, Aichelin J, Gossiaux P B and Werner K, arXiv:1305.3823 [hep-ph]
[30] M. Ruggieri, F. Scardina, S. Plumari and V. Greco, Phys. Lett. B **727** (2013) 177
[31] F. Scardina, M. Colonna, S. Plumari and V. Greco, Phys. Lett. B **724** (2013) 296
[32] A. Lang, W. Cassing, U. Mosel, H.H. Reusch, and K. Werner, Journal of Computational Physics 106, 391 (1993)
[33] A. Lang *et al.*, Jour. of Comp. Phys. **106**, 391 (1993)
[34] J. Uphoff, O. Fochler, Z. Xu and C. Greiner, Phys. Lett. B **717** (2012) 430
[35] B. Zhang, L. -W. Chen and C. -M. Ko, "Charm elliptic flow at RHIC," Phys. Rev. C **72** (2005) 024906
[36] A. Adare *et al.* (PHENIX Collaboration), Phys. Rev. Lett. **98**, 172301 (2007).
[37] B. I. Abeleb *et al.* (STAR Collaboration), Phys. Rev. Lett. **98**, 192301 (2007).
[38] B. Abelev *et al.*,(ALICE Collaboration) JHEP **1209** (2012) 112

Journal of Physics: Conference Series **535** (2014) 012020 doi:10.1088/1742-6596/535/1/012020

Self-consistent Cooper-Frye freeze-out of a viscous fluid to particles

Zack Wolff and Denes Molnar

Department of Physics and Astronomy, Purdue University, 525 Northwestern Ave, West Lafayette, IN 47901, USA

E-mail: zwolff@purdue.edu

Abstract. Comparing hydrodynamic simulations to heavy-ion data inevitably requires the conversion of the fluid to particles. This conversion, typically done in the Cooper-Frye formalism, is ambiguous for viscous fluids. We compute self-consistent phase space corrections by solving the linearized Boltzmann equation and contrast the solutions to those obtained using the ad-hoc "democratic Grad" ansatz typically employed in the literature where coefficients are independent of particle dynamics. Solutions are calculated analytically for a massless gas and numerically for both a pion-nucleon gas and for the general case of a hadron resonance gas. We find that the momentum dependence of the corrections in all systems investigated is best fit by a power close to 3/2 rather than the typically used quadratic ansatz. The effects on harmonic flow coefficients v_2 and v_4 are substantial, and should be taken into account when extracting medium properties from experimental data. PACS: 12.38.Mh, 24.10.Lx 24.10.Nz, 25.75.Ld, 51.10.+y

1. Introduction

The most widely used framework for describing the early stages of a heavy-ion collision is relativistic hydrodynamics[1]. Using hydrodynamics requires one to convert the evolving fluid into a particle description in order to calculate experimental observables such as particle spectra and flow coefficients. This "particlization"[2] is usually done using the Cooper-Frye[3] formalism to calculate the particles emitted from a drop of the fluid on a constant temperature hypersurface. This requires knowledge of the distribution function of each particle species. If the fluid is in local thermal equilibrium, i.e. an ideal fluid, then one can uniquely calculate these distributions from the hydrodynamic variables. If the fluid contains dissipative effects such as those arising in a viscous fluid, infinitely many of these particle distributions will match the hydrodynamic fields. Other theory input is required to uniquely and self-consistently determine the particle distributions.

This ambiguity in the viscous particle distributions is typically ignored and in practice, corrections to the ideal thermal distributions are assumed to be quadratic in momentum. In addition to this quadratic "Grad ansatz", the distributions are also usually taken to be independent of the particle collision rates which keep the hadron gas near equilibrium. In this work we resolve this ambiguity by calculating the distributions self-consistently by taking reaction rates of the hadrons into account using the Boltzmann equation at particlization. We first calculate the distributions under the "Grad ansatz" which assumes the corrections are quadratic in momentum and then later relax this constraint. Many of the calculational details absent here can be found in [4].

2. Phase space corrections from linearized transport

The Cooper-Frye particlization prescription requires as input distribution functions for each particle species i which are calculable from the hydrodynamic fields by inverting $T^{\mu\nu}(x) \equiv \sum_i \int \frac{d^3p}{E} p^\mu p^\nu f_i(x, \mathbf{p})$. For ideal fluids in local equilibrium with known chemical potentials, the conversion is well defined since knowledge of the energy density in the energy-momentum tensor is enough to calculate the temperature needed for the thermal particle distributions assuming the equation of state is a known input of the hydrodynamic simulation.

However, if the fluid has nonzero viscosity, it is not in perfect local equilibrium and the energy-momentum tensor acquires non-ideal corrections. In addition to the previously known ideal component, one must now solve for the viscous corrections to the distributions by inverting

$$\delta T^{\mu\nu} = \pi^{\mu\nu} + \Pi(u^\mu u^\nu - g^{\mu\nu}) \equiv \sum_i \int \frac{d^3p}{E} p^\mu p^\nu \delta f_i(x, \mathbf{p}) \tag{1}$$

Without additional theory constraint on the functional form of δf_i, infinitely many functions are possible and one is left to "guess" at the correct distributions. If one takes corrections from shear only and ignores bulk viscosity, one common choice that satisfies the above constraint is the "democratic Grad"[5] ansatz which assumes corrections quadratic in momentum

$$\phi_i^{\text{dem}}(x, \mathbf{p}) = \frac{\pi^{\mu\nu}(x) p_\mu p_\nu}{2[e(x) + p(x)]T^2(x)}, \qquad (\delta f_i \equiv \phi_i f_i^{eq}) \tag{2}$$

This simple choice has no dependence on the microscopic dynamics of the particles and ignores the fact that species which interact more should be closer to equilibrium then those that rarely scatter off other particles. In contrast to this democratic ansatz, we calculate the viscous corrections self-consistently from linearized covariant transport theory where evolution is governed by the Boltzmann equation

$$p^\mu \partial_\mu f_i(x, \mathbf{p}) = S_i(x, \mathbf{p}) + \sum_{jk\ell} C^{ij\to k\ell}[f_i, f_j, f_k, f_\ell](x, \mathbf{p}) \tag{3}$$

with S the source term specifying the initial conditions and collision terms here taken as

$$C^{ij\to k\ell}[f_i, f_j, f_k, f_\ell](x, \mathbf{p}_1) \equiv \iiint_{234} \left(\frac{g_i g_j}{g_k g_\ell} f_{3k} f_{4\ell} - f_{1i} f_{2j} \right) \bar{W}_{12\to 34}^{ij\to k\ell} \delta^4(12 - 34) \tag{4}$$

with shorthands $\int_a \equiv \int d^3p_a/(2E_a)$, $f_{ai} \equiv f_i(x, \mathbf{p}_a)$, and $\delta^4(ab - cd) \equiv \delta^4(p_a + p_b - p_c - p_d)$. and g_i the internal degrees of freedom.

If the viscous corrections are small, one can linearize the Boltzmann equation in these corrections δf_i. Most systems quickly relax to a solution of these equations which is governed by gradients of the equilibrium distribution, i.e. the Navier-Stokes regime where derivatives of δf_i are higher order corrections. The tensor structure of this equation imposes the form of the solutions[4]

$$\phi_i(x, \mathbf{p}) = \chi_i(|\tilde{\mathbf{p}}|) P^{\mu\nu} X_{\mu\nu} \tag{5}$$

with $\tilde{\mathbf{p}}$ being momentum over temperature in the fluid rest frame and tensor shorthands

$$P_a^{\mu\nu} \equiv \frac{1}{T^2}\left[\Delta_\alpha^\mu \Delta_\beta^\nu p_a^\alpha p_a^\beta - \frac{1}{3}\Delta^{\mu\nu}(\Delta_{\alpha\beta} p_a^\alpha p_a^\beta)\right], \quad X^{\mu\nu} \equiv \frac{\sigma^{\mu\nu}}{T} = \frac{\pi_{NS}^{\mu\nu}}{\eta_s T}, \quad \Delta^{\mu\nu} \equiv g^{\mu\nu} - u^\mu u^\nu \tag{6}$$

with $\sigma\mu\nu \equiv \nabla^\mu u^\nu + \nabla^\nu u^\mu - \frac{2}{3}\Delta^{\mu\nu}(\partial u)$ The only freedom left in the distribution function corrections is a function of dimensionless momentum $\chi_i(|\tilde{\mathbf{p}}|)$ for each particle species. One can show that solving the original transport equation is equivalent to extremizing the functional[6]

$$Q[\chi] = \frac{1}{2T^2}\sum_i \int_1 P_1 \cdot P_1 f_{1i}^{\rm eq}\chi_{1i} \; + \; \frac{1}{2T^4}\sum_{ijk\ell} \iiiint_{1234} f_{1i}^{\rm eq} f_{2j}^{\rm eq} \bar{W}_{12\to 34}^{ij\to k\ell}\, \delta^4(12-34)\cdot$$
$$(\chi_{3k}P_3 \cdot P_1 + \chi_{4\ell}P_4 \cdot P_1 - \chi_{1i}P_1 \cdot P_1 - \chi_{2j}P_2 \cdot P_1)\chi_{1i} \quad (7)$$

and the extremal value of Q gives the shear viscosity of the mixture

$$\eta_s = \frac{2}{15T^3}\sum_i \int \frac{d^3p}{E} p^4 f_i^{\rm eq}\chi_i \Rightarrow \eta_s = \frac{8}{5}Q_{max}T^3 \quad (8)$$

In contrast, the Grad ansatz corresponds to giving all particles the same coefficient $\chi_i^{\rm dem} = \frac{\eta_s T}{2(e+p)} = \frac{\eta_s}{2s}$ independent of momentum and particle species.

3. Results for massless two-component system

In this section we consider the so-called Grad ansatz where $\chi_i(|\tilde{\mathbf{p}}|) = \chi_i^{Grad} = const$ such that the viscous corrections ϕ_i are quadratic in momentum. For a two-component system of massless particles evolving with elastic two-body interactions with simple isotropic, energy-independent cross sections σ_{AA}, σ_{BB}, and σ_{AB} in the Grad approximation we can solve the functional method analytically[4] and get

$$\chi_A^{Grad} = \frac{3LT}{20}\frac{5K_{B(B)} + 7K_{B(A)} + 2K_{A(B)}}{K_{A(A)}[5K_{B(B)} + 7K_{B(A)}] + K_{A(B)}[9K_{B(A)} + 7K_{B(B)}]} \quad (9)$$

$$\chi_B^{Grad} = \frac{3LT}{20}\frac{5K_{A(A)} + 7K_{A(B)} + 2K_{B(A)}}{K_{B(B)}[5K_{A(A)} + 7K_{A(B)}] + K_{B(A)}[9K_{A(B)} + 7K_{A(A)}]} . \quad (10)$$

where $K_{i(j)} = Ln_j\sigma_{ij}$ are partial inverse Knudsen numbers for species i scattering off species j and L is the characteristic length scale of gradients in the system.

These results were derived using the linearized transport equation assuming small deviations from local equilibrium which corresponds to the Navier-Stokes limit of hydrodynamics where the system has relaxed to a solution governed by gradients of the hydrodynamic variables. In heavy-ion collisions, this relaxation to equilibrium has to compete with expansion diluting the system. We now test how well this Navier-Stokes solution works for expanding systems with a two-component massless system undergoing boost-invariant 0+1D Bjorken expansion as in [7]. The system starts out in equilibrium at τ_0 but expansion causes viscous corrections to quickly develop. These corrections can be expressed simply in terms of the partial sheer stresses of the two species. If we assume dissipative corrections quadratic in momentum then our χ_i's are constant $\equiv c_i$. In the late-time Navier Stokes regime, linearized kinetic theory predicts

$$\frac{c_B}{c_A} = \frac{5K_A + 2(K_{A(B)} + K_{B(A)})}{5K_B + 2(K_{A(B)} + K_{B(A)})} \qquad (K_i \equiv \sum_j K_{i(j)}) . \quad (11)$$

While the "democratic Grad" approach gives $c_i = 1$ for all species, so $c_B/c_A = 1$.

Fig. 1 compares these two extreme cases to fully nonlinear transport solutions from Molnar's Parton Cascade (MPC) [8]. The five different scenarios listed in the table of Fig. 1, all keeping species A closer to equilibrium than species B, were simulated with MPC. All scenarios start in equilibrium where $c_B/c_A = 1$, but then expansion forces them out of equilibrium. At late times the dissipative corrections predicted by the linear transport equation are correct within 10% while the commonly used "democratic Grad" ansatz fails to account for the species dependence of corrections. This gives confidence that the Navier-Stokes regime described by the linearized transport equation can be appropriate to use for heavy-ion expanding systems.

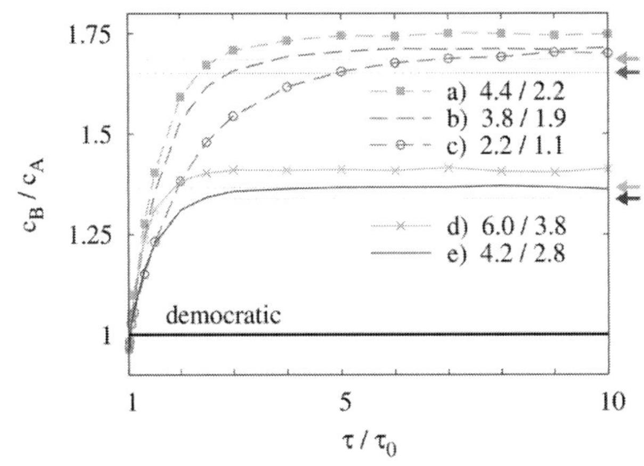

Scenario	K_A	K_B	$n_A : n_B$	$\sigma_{AA} : \sigma_{AB} : \sigma_{BB}$
a)	4.4	2.2	3 : 1	20 : 10 : 5
b)	3.8	1.9	2 : 2	20 : 10 : 5
c)	2.2	1.1	2 : 2	12 : 6 : 3
d)	6	3.8	1 : 3	24 : 24 : 12
e)	4.2	2.8	2 : 2	20 : 13.3 : 8.89

Figure 1. Ratio of viscous corrections as a function of dimensionless proper time for a massless two-component system undergoing 0+1D Bjorken expansion. Thin dotted lines and arrows refer to the expectation based on the linear transport equation and the numbers give the ratio of inverse Knudsen numbers, K_A/K_B as shown in the table.

4. Results for massive two-component system: pion-nucleon gas

In [4] it was shown that a nonrelativistic particle in the dynamic Grad approach described in the previous section receives a correction inversely proportional to the square root of its mass which reproduces the canonical viscosity expression in this limit:

$$\chi^{Grad} = \frac{5\sqrt{\pi}}{32}\sqrt{\frac{T}{m}}\frac{T}{n\sigma_{TOT}} \quad \Rightarrow \quad \eta_s^{Grad} = \frac{5\sqrt{\pi}}{16}\frac{\sqrt{mT}}{\sigma_{TOT}}\,, \tag{12}$$

In this section, we now consider a more realistic pion-nucleon system with relativistic kinematics. Lumping isospin states and antiparticles into a single species, this is a two-component system with $m_\pi = 0.14$ GeV, $g_\pi = 3$, $m_N = 0.94$ GeV, $g_N = 4$. For temperatures 120 MeV $\lesssim T \lesssim$ 165 MeV of interest, we approximate the two-body cross sections with constant, energy-independent, effective values $\sigma_{\pi\pi}^{eff} = 30$ mb, $\sigma_{\pi N}^{eff} = 50$ mb, and $\sigma_{NN}^{eff} = 20$ mb which reproduce the pion and proton relaxation times in [9]. Note that pions are much more prevalent at these temperatures so the nucleon-nucleon scattering is mostly irrelevant. The ratio of pion to proton coefficients in this temperature range $c_\pi/c_N \sim 2$ turns out to be near the nonrelativistic calculation predictions, even though pions are relativistic. This means nucleons are about twice as close to equilibrium as pions of the same momentum. This pion-proton difference will be manifest in identified particle observables if the self-consistent, species dependent viscous corrections are used in Cooper-Frye freeze-out.

To find the effect on pion and proton elliptic flow in heavy-ion collisions we run a hydrodynamic simulation of a RHIC Au+Au collision at $\sqrt{s_{NN}} = 200$ GeV with impact parameter $b = 7$ fm. The calculation was done with AZHYDRO[10] version 0.2p2 which is a 2+1D hydro code with longitudinal boost invariance and the fairly up-to-date s95-p1 equation of state parametrization[11] that matches lattice QCD results to a hadron resonance gas. Since AZHYDRO does not contain dissipation, the shear stress is estimated on the conversion hypersurface from gradients of the flow using the Navier-Stokes estimate: $\pi^{\mu\nu} = \eta_s\sigma^{\mu\nu}$.

Fig. 2 shows differential elliptic flow for pions and protons for freeze-out at $T_{conv} = 140$ MeV (left) and $T_{conv} = 165$ MeV (right). In the ideal case (filled circles) one can already see the mass ordering of the pion and proton v_2 curves characteristic of hydrodynamic flow. When dissipative effects are calculated with the commonly used democratic Grad ansatz (open boxes), the mass ordering persists but elliptic flow is strongly suppressed by viscous effects at moderate momentum. In contrast, self-consistent species-dependent freeze-out (crosses) causes a clear pion-proton elliptic flow splitting at the higher values of p_T with proton v_2 being $\sim 30\%$ higher than that of the pion. The qualitative picture is similar at the two temperatures, with the

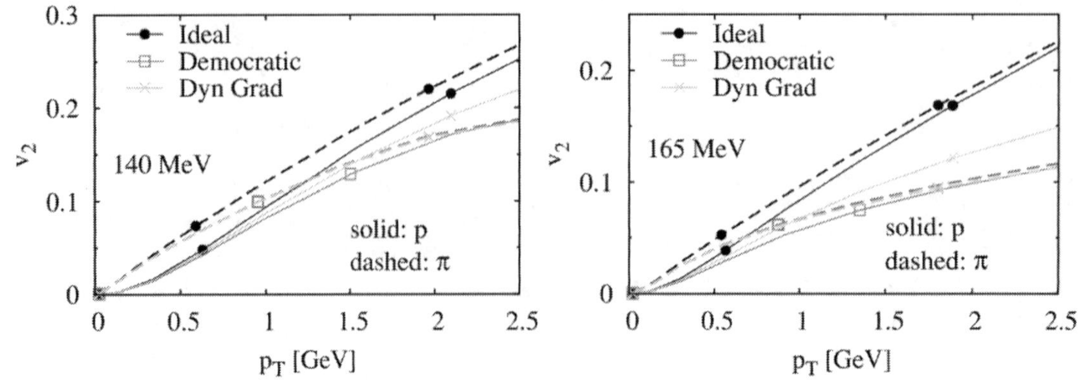

Figure 2. Differential elliptic flow $v_2(p_T)$ of pions and protons from the AZHYDRO simulation described in the text for T_{conv} = 140 MeV (left) and T_{conv} = 165 (right). Pions are shown in dashed lines and protons in solid. The commonly used "democratic Grad" approach (open boxes) is compared to self-consistent shear corrections (crosses) computed for a pion-nucleon gas from linearized kinetic theory. Results with no viscous corrections ($\delta f = 0$) (filled circles) are also shown for reference.

noticeable difference of reduced viscous effects at 140 MeV due to smaller flow gradients at later "times".

5. Results for a multicomponent hadron gas

In the previous section, self-consistent viscous corrections were calculated for a pion-nucleon gas. While these are the main stable particles whose flow is measured in experiments (along with kaons), this is a rather large simplification of the system expected to be present after hadronization in a heavy-ion collision. The hadron gas is made up of many more species and resonances whose scattering rates with all other species should be taken into account in the calculation. Here we use two different simplifying scenarios to categorize the cross sections present in a full multicomponent hadron gas. In the first scenario we give all species the same constant, energy-independent scattering cross section with every other species, namely 30mb similar to [12]. In the second scenario we use the additive quark model (AQM)[13] to motivate cross sections which scale as the product of the number of quarks in the two hadrons, namely the meson-baryon cross sections will be in ratios of $\sigma_{MM} : \sigma_{MB} : \sigma_{BB} = 4 : 6 : 9 = 30 : 45 : 67.5$ mb. In both cases we consider only elastic $ij \to ij$ channels (allowing for $i = j$). As before, we group members of the same isospin multiplet and antiparticles into a single species with degeneracy factors adjusted accordingly. The calculation was done including all hadrons up to $m = 1.672$ GeV, i.e. the $\Omega(1672)$, which translates into 49 effective species.

Fig. 3 shows pion and proton elliptic flow $v_2(p_T)$ in Au+Au at RHIC at b=7 fm from a calculation analogous to the pion-nucelon system in the previous section with particlization at $T_{conv} = 165$ MeV, except now the self-consistent calculation has been done for a multicomponent hadron gas. The left plot shows that with the same constant cross section between all species, the result is similar to the democratic case, as found previously in [12].

In contrast, the right plot shows that if the more realistic hadron gas with AQM motivated cross sections is used, the results on pion and proton identified flows are noticeable. Again they cross at $p_T \sim 1$ GeV and behavior is roughly similar to the pion-nucleon gas. Though not shown, results are similar at a decoupling temperature $T_{conv} = 140$ MeV.

As mentioned previously, the functional extremization approach used here automatically gives

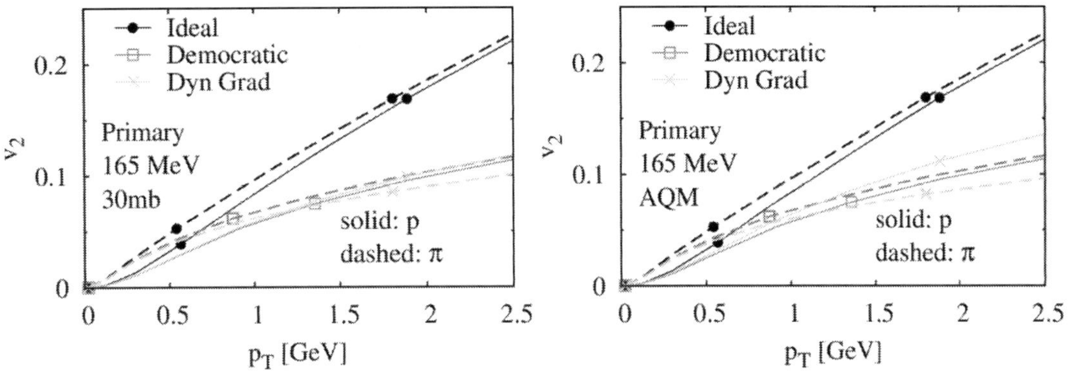

Figure 3. Same as Fig. 2 but the self-consistent species-dependent coefficients were calculated for the multicomponent hadron gas with 49 effective species with decoupling temperature T_{conv} = 165 MeV. The left plot shows the first scenario described in the text where all species have a constant 30 mb cross section with each other species. The right plot shows the second scenario motivated by the additive quark model (AQM) with $\sigma_{MM} : \sigma_{MB} : \sigma_{BB} = 30 : 45 : 67.5$ mb.

the shear viscosity of the gas after particlization and results for 4 systems are given in Fig. 4. As one can see, the value of the shear viscosity to entropy density $\frac{\eta}{s}$ of the hadron gas falls close to the proposed limit of $1/4\pi$ at 165 MeV due to the rapid rising of the entropy at high temperature.

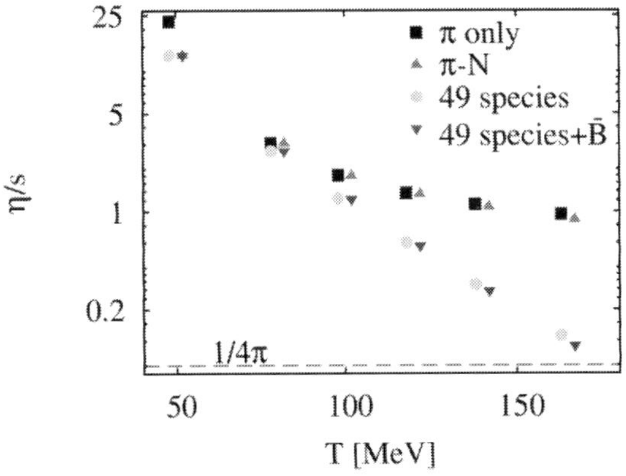

Figure 4. Shear viscosity to entropy density ratio calculated from the functional method described in the text for pions (squares), a pion-nucleon gas (triangles), the hadron gas without antiparticles (circles), and the full hadron gas (upside-down triangles).

The Cooper-Frye prescription gives the momentum distribution of particles coming directly from the fluid. Whether one couples the hydrodynamic simulation to a hadron transport code, the so-called "afterburner" approach, or the particles are allowed to free stream to the detectors, one should allow all of the unstable resonances to decay to stable pions and protons and include these secondary particles in the spectra and elliptic flow calculations of these stable particles. This resonance feed down has not been calculated in Fig. 3, but is included in Fig. 5 below with the help of the AZHYDRO companion code RESO[14]. For ideal freeze-out (δf =0), the democratic freeze-out, and the constant cross section scenario, the main effect of the resonance decays is reduction of the pion-proton splitting. At T_{conv} = 165 MeV, there is almost no difference in pion and proton elliptic flow (although for lower conversion temperatures, some difference does remain). Again in contrast to the other freeze-out scenarios, the self-consistent freeze-out based on the linear transport equation gives a measurable increase in proton elliptic flow, $\sim 30\%$ compared to pion flow around the higher momentum of around 2 GeV. This amount of splitting should be detectable in experiment and can be used to distinguish between different freeze-out distributions.

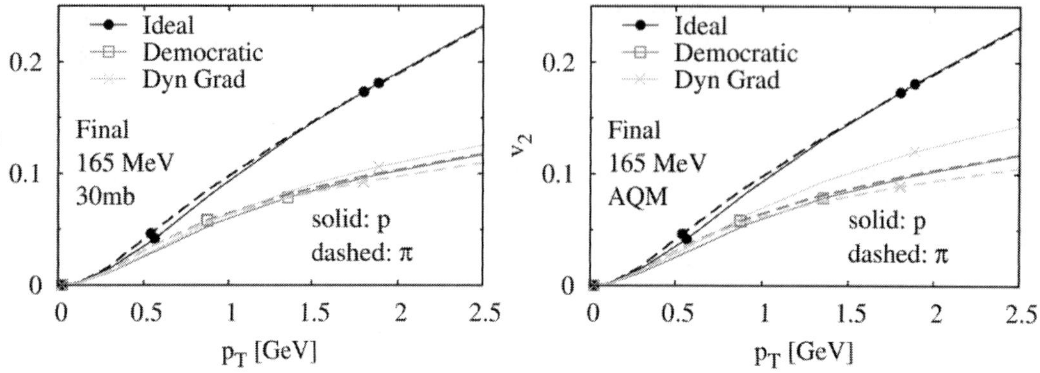

Figure 5. Same as Fig. 3 except after feed down from resonance decays using the RESO code in the AZHYDRO package[10].

6. Results for hadron gas with $\delta f_i \propto p^{3/2}$

Up to this point, we have stayed with the commonly used assumption that dissipative corrections to phase space distributions from shear viscosity are quadratic in momentum. The quadratic dependence used so far was essentially a simple guess at the real solution, while our integral equations for χ_i still contain an unknown function of the magnitude of the momentum. A better method then "guessing" at the correct form of the solution would be to either calculate which simple functional form better extremizes the functional and/or compare the results for observables obtained from different choices of the functional form of the corrections. We chose to do the former and using our analytic solution to the massless problem calculated the single power that best maximizes the functional. The answer we came up with was near $\frac{3}{2}$, a value which has been found in other theoretical and phenomenological studies[15][16]. We then chose this momentum to the $\frac{3}{2}$ power as our functional form of χ_i and again calculated what the results for pion and proton elliptic flow would be assuming hydrodynamic freeze-out to a hadron gas. The result after resonance decays is shown in the left plot of Fig. 6.

When the viscous corrections are calculated using a momentum to the $\frac{3}{2}$ power instead of quadratic, the elliptic flow of pions and protons comes closer together at higher momentum. Both pion and proton v_2 go above the democratic curves at high momentum do to the smaller viscous corrections there. This is compensated by the self-consistent curves being below the democratic ones at low momentum. The effects of this new momentum dependence on higher harmonics were also studied and the results for v_4 are shown in the right plot of Fig. 6.

The effect on identified particle v_4 is quite pronounced. The self-consistent calculation brings the proton up to near zero and actually positive in the quadratic case not pictured here. The effect on the sixth coefficient is less pronounced as both the quadratic and $p^{1.5}$ corrections have small deviations from the democratic case in this simulation and are thus not shown here.

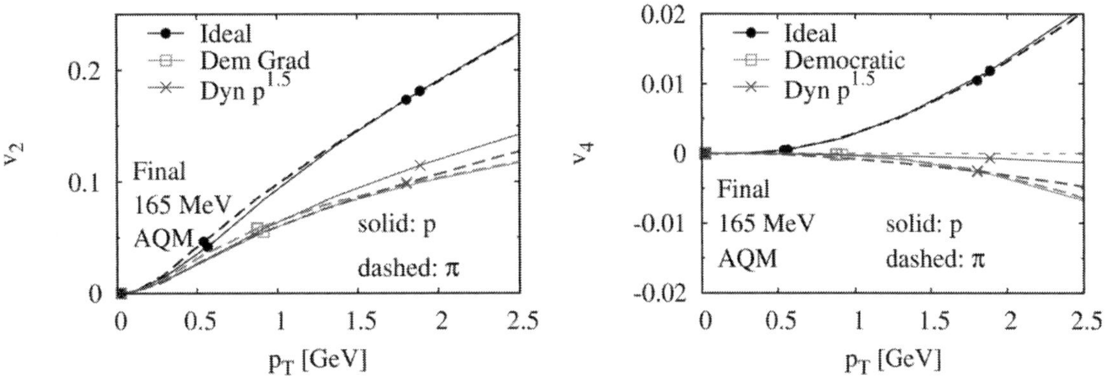

Figure 6. Same as Fig. 5 (right) except the self-consistent calculation here was done for corrections proportional to momentum to the 1.5 instead of quadratic for v_2 (left) and v_4 (right).

7. Summary

Extracting accurate values for medium properties from heavy-ion data requires proper treatment of the viscous corrections to hadron distribution functions in the Cooper-Frye freeze-out whether one uses hydrodynamics or a hybrid hydrodynamics+transport simulation. Here we have calculated these corrections self-consistently using the linearized Boltzmann equation in contrast to the commonly used democratic Grad ansatz which ignores differences in particle scattering rates. The effect on harmonic flow coefficients should be detectable in identified particle v_2 and v_4 in heavy-ion experiments. While this study suggests that the method used to calculate the corrections will impact particle observables, it is by no means the final word on the effects. Improvements are being made to use a realistic viscous hydrodynamics code along with a full list of energy-dependent cross sections of hadrons for future work.

Acknowledgments

This work was supported by the U.S. Department of Energy, Office of Science, under grants DE-AC02-98CH10886 [RIKEN BNL] and DE-PS02-09ER41665.

References

[1] P. Huovinen and P. V. Ruuskanen, Ann. Rev. Nucl. Part. Sci. **56**, 163 (2006) [nucl-th/0605008];
[2] P. Huovinen and H. Petersen, Eur. Phys. J. A **48**, 171 (2012) [arXiv:1206.3371 [nucl-th]].
[3] F. Cooper and G. Frye, Phys. Rev. D **10**, 186 (1974).
[4] D. Molnar and Z. Wolff arXiv:1404.7850v2 [nucl-th] (2014)-
[5] D. Molnar, J. Phys. G **38**, 124173 (2011) [arXiv:1107.5860 [nucl-th]].
[6] P. B. Arnold, G. D. Moore and L. G. Yaffe, JHEP **0011**, 001 (2000) [hep-ph/0010177]
[7] P. Huovinen and D. Molnar, Phys. Rev. C **79**, 014906 (2009) [arXiv:0808.0953 [nucl-th]]
[8] D. Molnar, MPC 1.8.13. This transport code is available at http://karman.physics.purdue.edu/OSCAR
[9] M. Prakash, M. Prakash, R. Venugopalan and G. Welke, Phys. Rept. **227**, 321 (1993)
[10] P. F. Kolb, J. Sollfrank and U. W. Heinz, Phys. Rev. C **62**, 054909 (2000) [hep-ph/0006129] available at http://karman.physics.purdue.edu/OSCAR
[11] P. Huovinen and P. Petreczky, Nucl. Phys. A **837**, 26 (2010) [arXiv:0912.2541 [hep-ph]]
[12] G. S. Denicol and H. Niemi, Nucl. Phys. A **904-905**, 369c (2013) [arXiv:1212.1473 [nucl-th]]
[13] E. M. Levin and L. L. Frankfurt, JETP Lett. **2**, 65 (1965); V. V. Anisovich, M. N. Kobrinsky, J. Nyiri and Y. .M. Shabelski, Sov. Phys. Usp. **27**, 901 (1984) [Usp. Fiz. Nauk **144**, 553 (1984)]
[14] P. F. Kolb and U. W. Heinz, In *Hwa, R.C. (ed.) et al.: Quark gluon plasma* 634-714 [nucl-th/0305084]
[15] K. Dusling, G. D. Moore and D. Teaney, Phys. Rev. C **81**, 034907 (2010) [arXiv:0909.0754 [nucl-th]]
[16] M. Luzum, J.Y. Ollitrault, Phys. Rev. C **82**, 014906 (2010)

Journal of Physics: Conference Series **535** (2014) 012021 doi:10.1088/1742-6596/535/1/012021

Heavy flavor production in the STAR experiment

Barbara Trzeciak[1] **for the STAR Collaboration**

[1]Faculty of Nuclear Sciences and Physical Engineering, Czech Technical University in Prague, Brehova 7, 115 19 Praha 1, Czech Republic

E-mail: `trzecbar@fjfi.cvut.cz`

Abstract. In this paper, recent STAR heavy flavor measurements in proton-proton and heavy-ion collisions are highlighted. We report studies of open charm mesons, reconstructed directly from hadronic decay products, and studies of electrons from semi-leptonic decays of heavy flavor hadrons. We also present J/ψ measurements via the di-electron decay channel at various collision systems and energies. In Au+Au collisions the energy dependence of J/ψ production measured at $\sqrt{s_{NN}}$ = 39, 62.4 and 200 GeV is shown. Finally, prospects of heavy flavor measurements with the STAR detector upgrades are discussed.

1. Introduction

Heavy quarks are produced at the initial hard-scattering stage of the ultra-relativistic heavy-ion collisions at Relativistic Heavy Ion Collider (RHIC). Heavy quarks are expected to interact with the hot and dense medium created in RHIC high energy collisions differently from light quarks therefore they serve as unique probes of the properties of the Quark-Gluon Plasma (QGP). Heavy quarks may also help to understand the parton energy loss mechanism and the degree of the medium thermalization since their production and elliptic flow are sensitive to the dynamics of the medium.

Moreover, studies of production of various quarkonium states in heavy-ion collisions can provide insight into the thermodynamic properties of the hot and dense medium. It was proposed that due to a Debye screening of the quark-antiquark potential in the hot medium quarkonia are dissociated and thus this "melting" can be a signature of the QGP formation. Also, since different quarkonium states have different binding energies they are expected to dissociate at different temperatures and can be treated as a QGP thermometer. But there are other mechanisms that can alter quarkonium yields in heavy-ion collisions relative to $p+p$ collisions, for example cold nuclear matter effects (shadowing/anti-shadowing, final state nuclear absorption) or statistical coalescence of heavy quark-antiquark pairs. Systematic measurements of the quarkonium production for different colliding systems, centralities and collision energies may help to understand the quarkonium production mechanisms in heavy-ion collisions as well as the medium properties.

2. Open Heavy Flavor

2.1. D mesons

Open charm mesons can be directly reconstructed via their hadronic decays products (e.g. $D^0 \rightarrow K^- + \pi^+$). In this way, a direct access to the heavy flavor hadron kinematics is obtained. But it is difficult to trigger and the reconstruction is challenging without a good secondary vertex resolution. Since STAR did not have a good secondary-vertex detector when the data that are presented in this paper were recorded, the large combinatorial background contribution is estimated and subtracted on statistical basis using particle combinations with wrong charge signs or from different events.

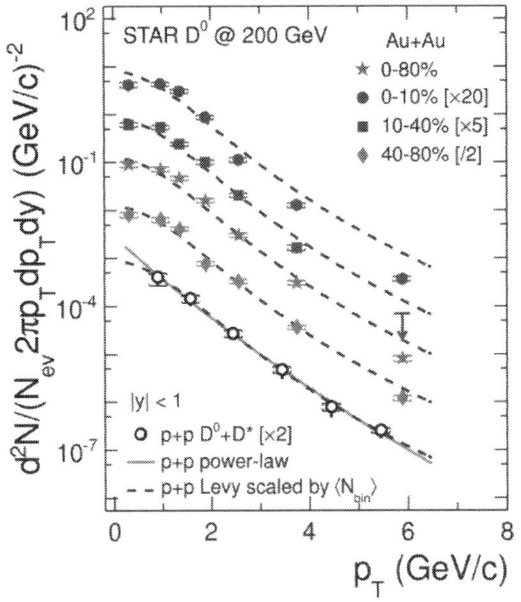

Figure 1. D^0 p_T spectra in Au+Au collisions (full symbols) and $p + p$ collisions (open symbols) at $\sqrt{s_{NN}} = 200$ GeV at mid-rapidity in different centrality bins [1]. Dashed lines represent Levy function fit to $p + p$ scaled by the number of binary collisions.

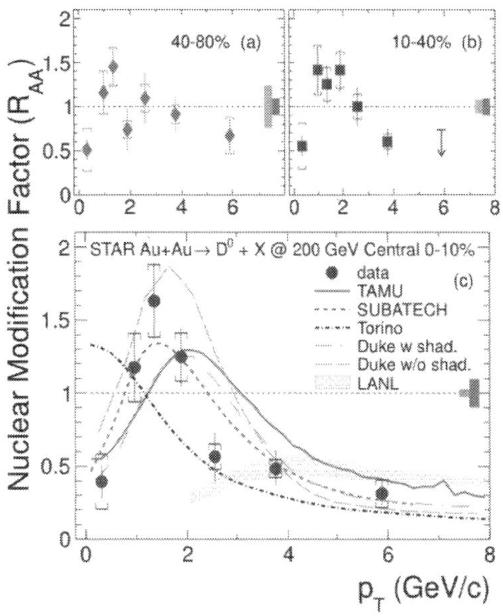

Figure 2. D^0 R_{AA} as a function of p_T for Au+Au collisions at $\sqrt{s_{NN}} = 200$ GeV at mid-rapidity with different model calculations [1].

Figure 1 presents the D^0 p_T spectra in Au+Au collisions at $\sqrt{s_{NN}} = 200$ GeV at mid-rapidity for different centralities [1]. The $D^0 + D^*$ measurement in p+p collisions at $\sqrt{s} = 200$ GeV [2] is also shown as open circles. The dashed curves represent Levy functions fitted to the $p + p$ data and scaled up by the number of binary collisions (N_{bin}) for each centrality bin. The charm production cross-section at mid-rapidity in the 0-10% central collisions is found to be $d\sigma^{NN}_{c\bar{c}}/dy = 148 \pm 15$ (stat.) $\pm$ 19 (sys.) μb and the cross-sections per nucleon-nucleon collision show a number-of-binary-collision scaling from $p+p$ to Au+Au that indicates that charm quarks are mostly produced in the initial hard processes [1].

In the most central Au+Au collisions a strong suppression at high p_T compared with the corresponding Levy function is observed. To quantify the suppression, the nuclear modification factor (R_{AA}) is calculated and shown in Fig. 2 as a function of p_T for different centrality bins [1]. No suppression is seen for peripheral collisions, while for the most central collisions the R_{AA} is about 0.5 for $p_T > 3$ GeV/c. This suppression at high p_T is also consistent with those of non-photonic electrons (see Sec. 2.2, shown in Fig. 6) and light hadrons, $\pi^{\pm}$, (shown in Fig. 3(d)). The 0-10% result is compared with different model predictions [3, 4, 5, 6, 7, 8, 9, 10]. The mid-p_T enhancement seen in the data can be described by models that include strong charm-medium interactions and charm quark hadronization via coalescence with light quarks.

Figure 3 shows D^0 p_T spectra and R_{AA} as a function of p_T measured in U+U collisions at $\sqrt{s_{NN}}$ = 193 GeV at mid-rapidity (full symbols) together with the mid-rapidity Au+Au result (open symbols). In the same centrality bin, energy density in U+U collisions could be 20% higher than in Au+Au collisions [11]. Data show similar R_{AA} trend in U+U and Au+Au collisions.

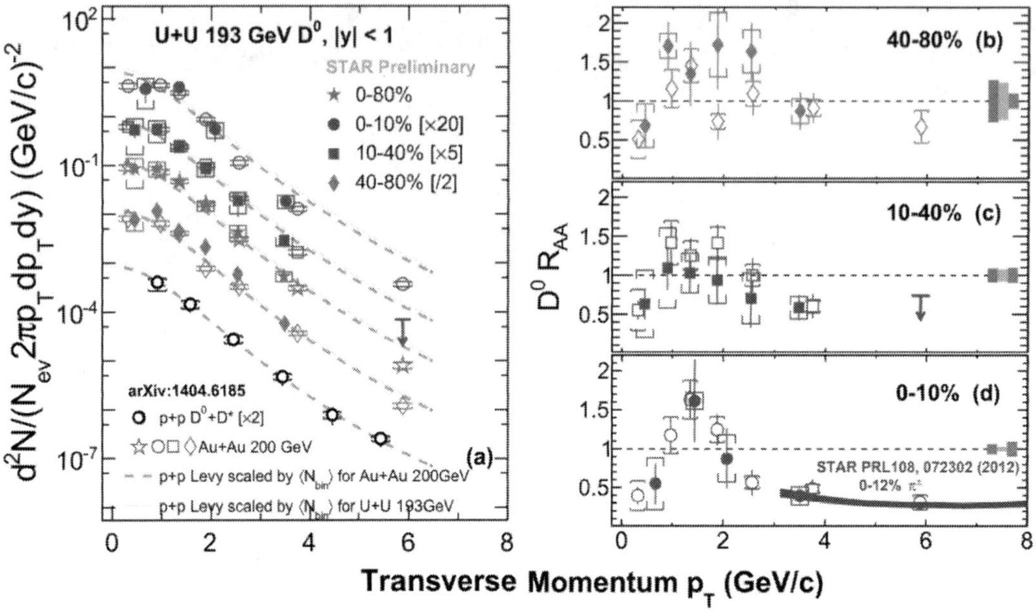

Figure 3. D^0 p_T spectra (left panel) and R_{AA} (right panel) in U+U collisions (full symbols) at $\sqrt{s_{NN}}$ = 193 GeV and Au+Au collisions (open symbols) at $\sqrt{s_{NN}}$ = 200 GeV at mid-rapidity. Solid and dashed lines represent Levy function scaled by the number of binary collisions for U+U and Au+Au, respectively.

2.2. Non-photonic electrons

Open heavy flavor mesons can also be studied using electrons from their semi-leptonic decays (non-photonic electrons, NPE). The branching ratio for this process is higher and it is relatively easy to trigger on high-p_T electrons. The disadvantage of this method is that it gives an indirect access to the heavy flavor hadron kinematics and it contains contributions from both charm and beauty hadron decays.

NPE p_T spectra for Au+Au collisions at $\sqrt{s_{NN}}$ = 200 GeV at mid-rapidity together with the FONLL upper limit calculations [12] scaled up by N_{bin} are shown in Fig. 4 for different centralities. The NPE production in the most central and mid-central collisions is suppressed compared to the FONLL calculations. STAR has also obtained the first result of the NPE production in Au+Au collisions at $\sqrt{s_{NN}}$ = 62.4 GeV, shown in Fig. 5. Solid and dashed lines represent the upper-limit and the central value of FONLL calculations scaled by N_{bin}, respectively. In contrast to the result in 200 GeV, no suppression at $p_T < 5.5$ GeV/c is observed compared to the FONLL calculations. The result is not corrected yet for the J/ψ and Drell-Yan contributions. The J/ψ contribution is about 20% at high p_T at $\sqrt{s_{NN}}$ = 200 GeV. Even taking this into account, no significant suppression would be seen at 62.4 GeV.

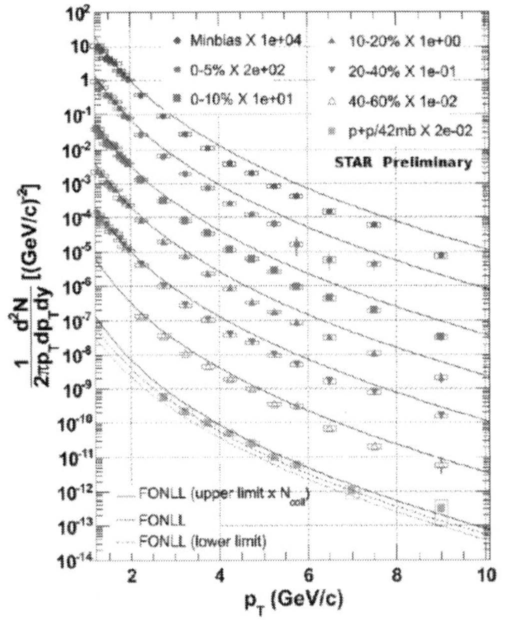

Figure 4. NPE p_T spectra in Au+Au collisions for different centrality bins and $p+p$ collisions (green squares) at $\sqrt{s_{NN}} = 200$ GeV at mid-rapidity. Solid lines represent the upper limit of FONLL calculations scaled by a number of binary collisions in a given centrality.

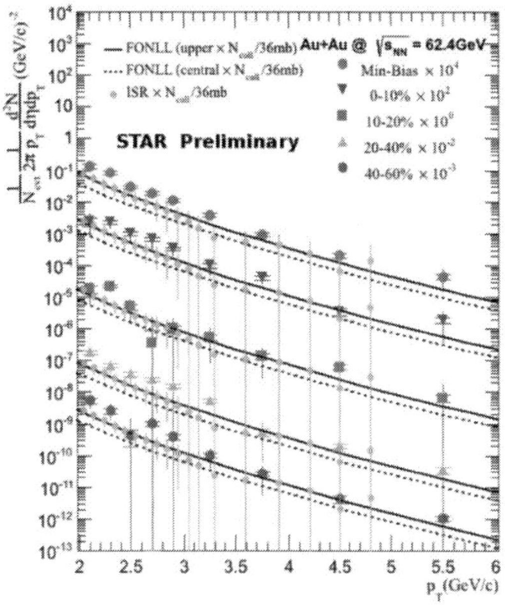

Figure 5. NPE p_T spectra in Au+Au collisions at $\sqrt{s_{NN}} = 62.4$ GeV at mid-rapidity. Solid lines represent the upper limit of FONLL calculations scaled by a number of binary collisions in a given centrality. ISR results [13] are also shown.

NPE R_{AA} in the most central Au+Au collisions (0-10%) at $\sqrt{s_{NN}} = 200$ GeV at mid-rapidity is shown in Fig. 6 as a function of p_T. A strong suppression, similar to above-mentioned D^0 result, is seen at high p_T. Figure 7 shows the measured elliptic flow of NPE in Au+Au collisions at $\sqrt{s_{NN}} = 200$ GeV at mid-rapidity [14]. The measurement is done using the two- ($v_2\{2\}$) and four-particle ($v_2\{4\}$) correlations. $v_2\{2\}$ and $v_2\{4\}$ are finite at $p_T > 0.5$ GeV/c and results obtained from both methods are consistent with each other. Also, at $p_T > 4$ GeV/c an increase of v_2 is observed that is probably due to jet-like correlations [14].

The measured R_{AA} and v_2 are compared with different model predictions. The gluon radiation scenario alone (DGLV) [15] fails to describe the large NPE suppression while adding collisional energy loss improves the agreement with the data. The R_{AA} can be described by the collisional dissociation model [10] and AdS/CFT calculations [16]. The partonic transport model, BAMPS [17, 18] describes the v_2 data well but it underpredicts the NPE R_{AA} result. The He *et al.* [3, 19] and the Gossiaux *et al.* [5, 20, 21] models underpredict $v_2\{2\}$ but they are in a good agreement with the measured NPE R_{AA}. The Gossiaux *et al.* model includes radiative and collisional energy loss and the He *et al.* model is a TMatrix interactions model. In general, it is challenging to describe well NPE R_{AA} and v_2 simultaneously.

3. Quarkonia

Quarkonia at STAR are measured via the di-electron decay channel at mid-rapidity ($|y| < 1$). STAR has observed significant suppressions of Υ production in central Au+Au collisions at

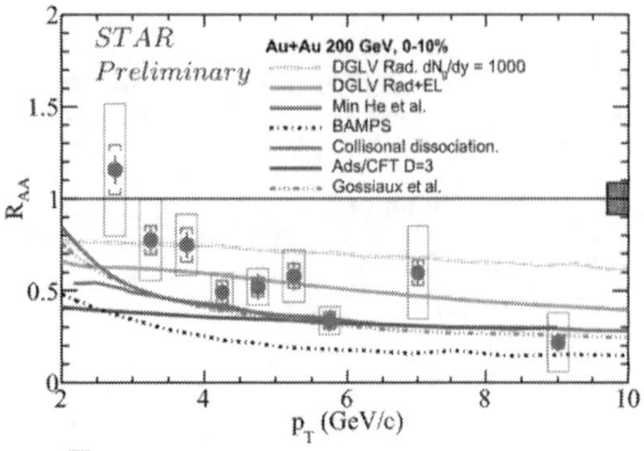

Figure 6. NPE R_{AA} as a function of p_T for Au+Au collisions at $\sqrt{s_{NN}} = 200$ GeV at mid-rapidity in 0-10% most central events with different model predictions.

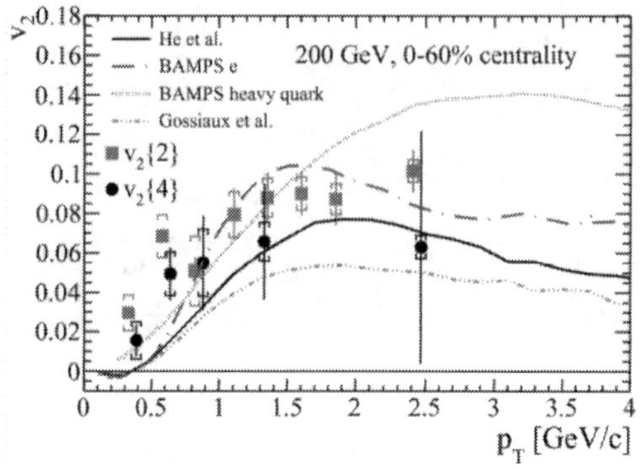

Figure 7. NPE v_2 for Au+Au collisions at $\sqrt{s_{NN}} = 200$ GeV at mid-rapidity in 0-60% central events with different model predictions [14].

$\sqrt{s_{NN}} = 200$ GeV compared to $p + p$ collisions, more details can be found here [22]. Here, we will focus on low and high-p_T inclusive J/ψ measurements at different collision energies.

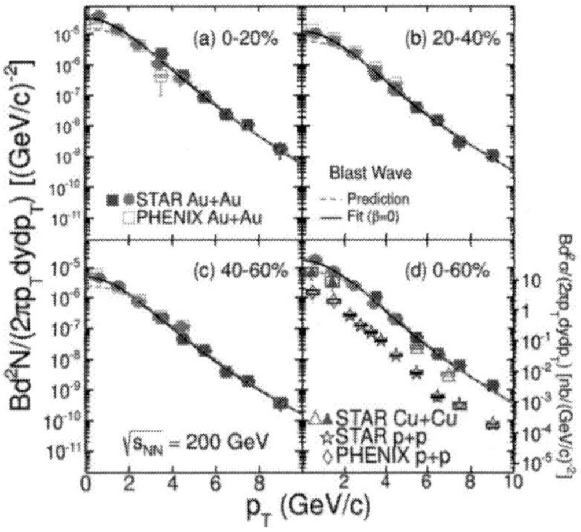

Figure 8. J/ψ p_T spectra in Au+Au collisions at $\sqrt{s_{NN}} = 200$ GeV at mid-rapidity, for different centrality bins, with Tsallis Blast-Wave predictions. On panel (d) $p+p$ results as stars and Cu+Cu results as triangles are also presented.

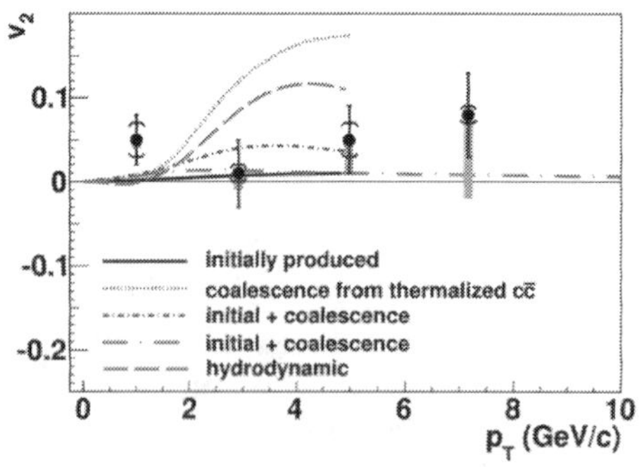

Figure 9. J/ψ v_2 in Au+Au collisions at $\sqrt{s_{NN}} = 200$ GeV at mid-rapidity in 0-80% central events with different model predictions [27].

Figure 8 shows STAR J/ψ p_T spectra in Au+Au collisions at $\sqrt{s_{NN}} = 200$ GeV for different centrality bins [23, 24] as full symbols. The spectra are compared to the Tsallis Blast-Wave (TBW) prediction assuming that J/ψ flows like lighter hadrons [25, 26], shown as dashed lines. The J/ψ p_T spectra are softer than the TBW prediction p_T and agree with TBW fit with radial flow $\beta = 0$ (solid lines). This may indicate small J/ψ radial flow or contribution from the recombination to low-p_T J/ψ production.

Measurement of J/ψ v_2 may provide additional information about the J/ψ production mechanisms. Figure 9 shows J/ψ v_2 measured in STAR in Au+Au collisions at $\sqrt{s_{NN}} = 200$ GeV [27]. At $p_T > 2$ GeV/c v_2 is consistent with zero. Comparing to different model predictions [28, 29, 30, 31], data disfavor the scenario that J/ψ with $p_T > 2$ GeV/c are dominantly produced by coalescence from thermalized (anti-)charm quarks.

Figure 10 shows J/ψ R_{AA} as a function of the number of participant nucleons (N_{part}) in Au+Au collisions at $\sqrt{s_{NN}} = 200$ GeV. The low-p_T (< 5 GeV/c) result [23] is shown as black full circles and the high-p_T (> 5 GeV/c) measurement [24] as red full circles. Suppression increases with collision centrality and R_{AA} for high-p_T J/ψ is systematic higher than the low-p_T J/ψ R_{AA}. At high p_T no significant suppression in peripheral and mid-central collisions is seen while in central collisions (0-30%) a strong suppression is observed which may indicate color screening features - at $p_T > 5$ GeV/c the recombination and initial parton scattering effects are expected to be negligible [24].

Figure 11 shows low-p_T J/ψ R_{AA} in Au+Au collisions at lower collision energies, $\sqrt{s_{NN}}$ = 62.4 and 39 GeV together with the 200 GeV result. For lower initial energy densities the interplay between direct J/ψ production and recombination from c and $\bar{c}$ quarks is expected to be different. Similar level of suppression is seen in the data for all three energies, $\sqrt{s_{NN}} = 200$, 62.4 and 39 GeV. However, one should note that due to lack of $p + p$ measurements at 62.4 and 39 GeV Color Evaporation Model calculations [32] are used as baselines, which introduce large uncertainties shown as boxes in Fig. 11.

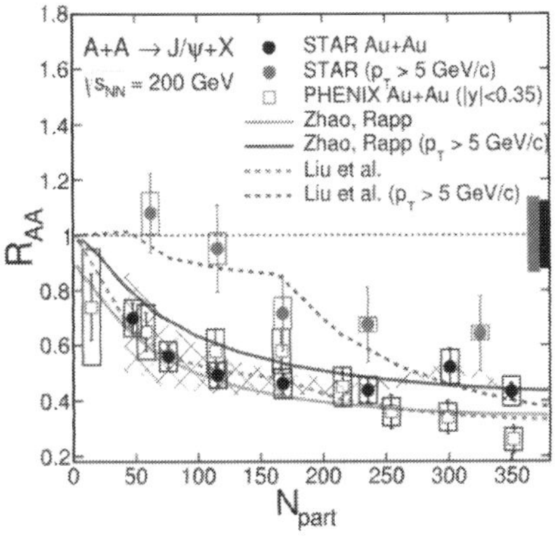

Figure 10. J/ψ R_{AA} as a function of N_{part} in Au+Au collisions at $\sqrt{s_{NN}} = 200$ GeV at mid-rapidity with two model predictions. The low-p_T (< 5 GeV/c) result is shown as black full circles and the high-p_T (> 5 GeV/c) measurement as red full circles.

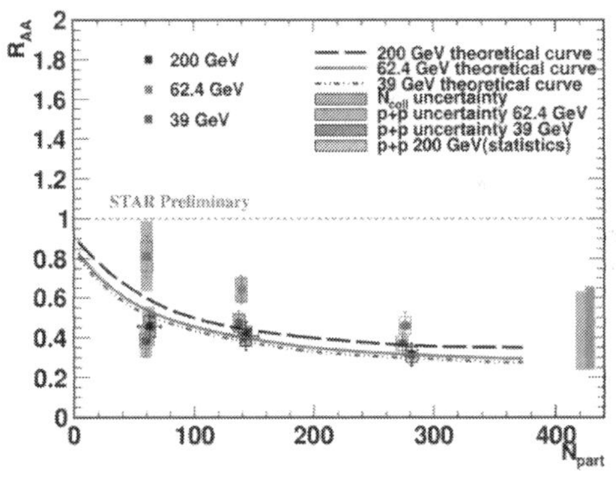

Figure 11. J/ψ R_{AA} as a function of N_{part} in Au+Au collisions at $\sqrt{s_{NN}} = 200$ (black), 62.4 (red) and 39 (blue) GeV at mid-rapidity with model predictions.

The R_{AA} results at $\sqrt{s_{NN}} = 200$ GeV in Fig. 10 are compared with two models, Zhao and Rapp [33] and Liu *et al.* [34]. Both models take into account direct J/ψ production with the color screening effect and J/ψ produced via the recombination of c and $\bar{c}$ quarks. Green lines represent low-p_T predictions and both agree well with the measurement. High-p_T predictions are shown as blue lines. At high p_T the Liu *et al.* model describes the data well while the Zhao and Rapp model underpredicts the measured R_{AA}. For $\sqrt{s_{NN}} = 62.4$ and 39 GeV the Zhao and Rapp model predicts similar suppression levels, and within the uncertainties the model agrees with the data (Fig. 11).

4. Outlook

STAR has two new detectors, fully installed and taking data in 2014: Heavy Flavor Tracker (HFT) and Muon Telescope Detector (MTD) [35]. The HFT is a silicon vertex detector. It allows precise open heavy flavor measurements, v_2 or R_{AA}, and to distinguish between charm and bottom. Projected D^0 elliptic flow v_2 measurement using the HFT detector is shown in Fig. 12. With the HFT it will also be possible to study non-prompt J/ψ production from B meson decays using the displaced vertex. The MTD is situated outside the STAR magnet which is used as an absorber. The detector enables muon identification in STAR and so quarkonia measurements in the di-muon decay channel which is cleaner comparing with the di-electron decay channel. It has also excellent mass resolution that is particularly important for measurement of different Υ states. Figure 13 shows precision projection for R_{AA} of different Υ states with the MTD detector. With the MTD open heavy flavor studies can be also performed, using electron-muon correlations.

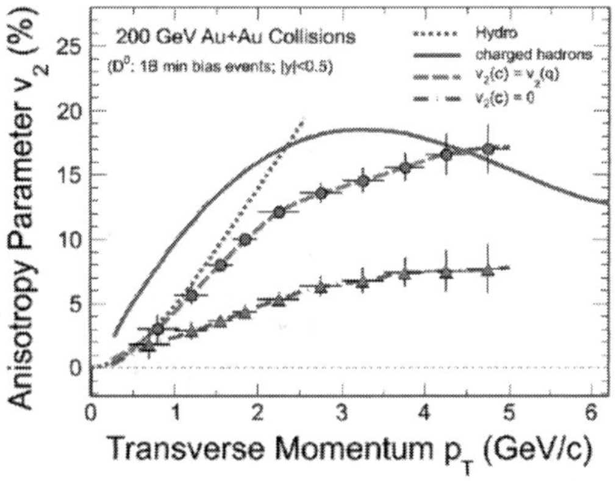

Figure 12. Projected D^0 v_2 measurement with the HFT detector, dashed lines represent different model assumptions.

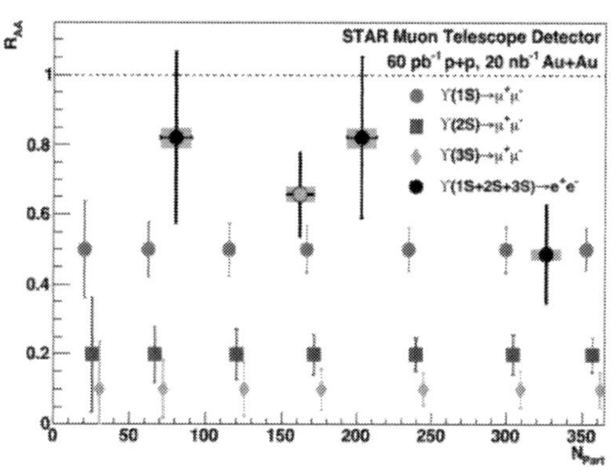

Figure 13. Statistical projection of Υ R_{AA} measurement for different Υ states with the MTD detector. The figure also shows in black measured STAR Υ R_{AA} [22], and has been modified from what was shown originally in order to include the published data.

5. Summary

In summary, STAR observed strong NPE and D^0 suppression at high p_T in Au+Au collisions at $\sqrt{s_{NN}} = 200$ GeV, but no NPE suppression is seen at $\sqrt{s_{NN}} = 62.4$ GeV at $p_T < 5.5$ GeV/c compare to pQCD calculations. Also, similar sizes of D^0 nuclear modification factors, R_{AA}, are seen in Au+Au and U+U collisions. D^0 R_{AA} at $\sqrt{s_{NN}} = 200$ GeV can be described by calculations that include strong charm-medium interactions and coalescence hadronization. Significant suppression of low p_T J/ψ is seen at $\sqrt{s_{NN}} = 200$, 62.4 and 39 GeV Au+Au collisions, but no strong energy dependence is observed. At $\sqrt{s_{NN}} = 200$ GeV high-p_T J/ψ are strongly suppressed in central Au+Au collisions, that may indicate color screening features. New STAR upgrades will allow to perform more precise open heavy flavor and quarkonia measurements in next years.

Acknowledgements
This publication was supported by the European social fund within the framework of realizing the project „Support of inter-sectoral mobility and quality enhancement of research teams at Czech Technical University in Prague", CZ.1.07/2.3.00/30.0034.

References

[1] Adamczyk L *et al.* (STAR Collaboration) 2014 (*Preprint* 1404.6185)
[2] Adamczyk L *et al.* (STAR Collaboration) 2012 *Phys.Rev.* **D86** 072013
[3] He M, Fries R J and Rapp R 2012 *Phys.Rev.* **C86** 014903
[4] He M, Fries R J and Rapp R 2013 *Phys.Rev.Lett.* **110** 112301
[5] Gossiaux P, Aichelin J, Gousset T and Guiho V 2010 *J.Phys.* **G37** 094019
[6] Gossiaux P, Nahrgang M, Bluhm M, Gousset T and Aichelin J 2013 *Nucl.Phys.* **A904-905** 992c–995c
[7] Alberico W, Beraudo A, De Pace A, Molinari A, Monteno M *et al.* 2011 *Eur.Phys.J.* **C71** 1666
[8] Alberico W, Beraudo A, De Pace A, Molinari A, Monteno M *et al.* 2013 *Eur.Phys.J.* **C73** 2481
[9] Cao S, Qin G Y and Bass S A 2013 *Phys.Rev.* **C88** 044907
[10] Sharma R, Vitev I and Zhang B W 2009 *Phys.Rev.* **C80** 054902
[11] Kikola D, Odyniec G and Vogt R 2011 *Phys.Rev.* **C84** 054907
[12] Cacciari M, Nason P and Vogt R 2005 *Phys.Rev.Lett.* **95** 122001
[13] Basile M, Cara Romeo G, Cifarelli L, Contin A, D'Ali G *et al.* 1981 *Nuovo Cim.* **A65** 421
[14] Adamczyk L *et al.* (STAR Collaboration) 2014 (*Preprint* 1405.6348)
[15] Djordjevic M, Gyulassy M, Vogt R and Wicks S 2006 *Phys.Lett.* **B632** 81–86
[16] Horowitz W 2008 *J.Phys.* **G35** 044025
[17] Uphoff J, Fochler O, Xu Z and Greiner C 2011 *Phys.Rev.* **C84** 024908
[18] Uphoff J, Fochler O, Xu Z and Greiner C 2012 *Phys.Lett.* **B717** 430–435
[19] van Hees H, Mannarelli M, Greco V and Rapp R 2008 *Phys.Rev.Lett.* **100** 192301
[20] Gossiaux P and Aichelin J 2008 *Phys.Rev.* **C78** 014904
[21] Aichelin J, Gossiaux P and Gousset T 2012 *Acta Phys.Polon.* **B43** 655–662
[22] Adamczyk L *et al.* (STAR Collaboration) 2014 *Phys.Lett.* **B735** 127
[23] Adamczyk L *et al.* (STAR Collaboration) 2013 (*Preprint* 1310.3563)
[24] Adamczyk L *et al.* (STAR Collaboration) 2013 *Phys. Lett.* **B 722** 55–62
[25] Tang Z, Xu Y, Ruan L, van Buren G, Wang F *et al.* 2009 *Phys.Rev.* **C79** 051901
[26] Tang Z, Yi L, Ruan L, Shao M, Chen H *et al.* 2013 *Chin.Phys.Lett.* **30** 031201
[27] Adamczyk L *et al.* (STAR Collaboration) 2013 *Phys.Rev.Lett.* **111** 052301
[28] Yan L, Zhuang P and Xu N 2006 *Phys.Rev.Lett.* **97** 232301
[29] Greco V, Ko C and Rapp R 2004 *Phys.Lett.* **B595** 202–208
[30] Zhao X and Rapp R 2008 (*Preprint* 0806.1239)
[31] Liu Y, Xu N and Zhuang P 2010 *Nucl.Phys.* **A834** 317C–319C
[32] Nelson R, Vogt R and Frawley A 2013 *Phys.Rev.* **C87** 014908
[33] Zhao X and Rapp R 2010 *Phys.Rev.* **C82** 064905
[34] Liu Y p, Qu Z, Xu N and Zhuang P f 2009 *Phys.Lett.* **B678** 72–76
[35] Wang Y (STAR Collaboration) *These proceedings* STAR upgrade program and future physics

STAR upgrade program and future physics

Yaping Wang[1,2] for the STAR Collaboration
[1]Department of Physics, University of Illinois at Chicago, Chicago, IL 60607, USA
[2]Institute of Particle Physics, Central China Normal University, Wuhan 430079, China

E-mail: ypwang@rcf.rhic.bnl.gov

Abstract. In this paper, we will present STAR's future plan in terms of both the detector upgrade and physics measurement to study matter with colour degrees of freedom. We will first discuss the status of the newly installed Heavy Flavor Tracker and Muon Telescope Detector, and their physics prospect in 2014-2016. We will then describe the proposed detector upgrades for the second phase of Beam Energy Scan program in 2018-2019 to study the QCD phase diagram. Finally we will present STAR's plan with detector upgrades in the forward directions for the anticipated pp/pA physics program in 2021-2022 and ep/eA in 2025+. The upgraded STAR experiment will be in an excellent position to perform precision measurements of the partonic structures of the nucleon and nuclei.

1. Introduction

The STAR Collaboration has identified eight key questions for understanding the QCD matter and strong force [1]. With the large acceptance tracking, calorimetry and particle identification detectors, STAR was able to make important measurements helping to answer these questions, such as the discovery of the strongly coupled Quark-Gluon Plasma (sQGP) at RHIC. In this proceedings, we present STAR's future plan to further improve our knowledge in order to address these fundamental questions, including the phase diagram of QCD and partonic structures of the nucleon and nuclei.

In section 2, we will discuss the most recent detector upgrades, including the Heavy Flavor Tracker (HFT) [2] and Muon Telescope Detector (MTD) [3] completed in the beginning of 2014. They will play key roles in studying interactions of heavy flavor particles with the sQGP and properties of the sQGP through measurements of heavy flavor hadrons. In section 3, we will describe upgrade proposals for the second phase of Beam Energy Scan (BES) program (BES-II) in 2018-2019. These upgrades, including the inner sectors of the Time Projection Chamber (iTPC) [4] and a new Event Plane and centrality Detector (EPD), will greatly enhance the STAR capability of studying the QCD phase diagram. In sections 4 and 5, we will present upgrade plans in the forward directions for the anticipated p+p and p+A physics program in 2021-2022 [5] and e+p and e+A physics program in 2025+ [6], respectively. These upgrades will allow high precision measurements to study the unexplored partonic structures of the nucleon and nuclei in low Bjorken x domain.

2. Heavy flavor physics program

Heavy flavor quarks (c, b) are dominantly created from initial hard scatterings at RHIC. They are regarded as an ideal probe to study the properties of the medium. The HFT and MTD were designed for measurements of open heavy flavor hadrons and quarkonia, respectively.

2.1. Heavy Flavor Tracker

The HFT [2] is a silicon vertex detector utilizing the state-of-the-art Monolithic Active Pixel Sensor (MAPS) and conventional silicon pad/strip detector technologies. It consists of 3 sub-detector systems, the Silicon Pixel (PXL) detector, Intermediate Silicon Tracker (IST) and Silicon Strip Detector (SSD). The PXL detector is made of two layers of thin MAPS installed close to the beam pipe at the radii of 2.8 and 8 cm, respectively. The excellent position resolution of the PXL detector (~12 μm) combined with its low material budget (0.4-0.6% radiation length per layer) allows precise measurements of track impact parameters even at low transverse momenta. The expected track impact parameter resolution is 60 (30) μm for Kaons with p_T=0.75 (>1.5) GeV/c. The IST (SSD) is made of one layer of single-sided Silicon Pad (double-sided Silicon Strip) sensors at the radius of 14 (22) cm. The IST and SSD not only provide the bridge between the PXL detector and TPC with a track pointing resolution ~1 mm, but also suppress contributions from pile-up tracks in offline reconstruction and data analyses, thanks to their fast readout electronics.

The HFT was installed and fully commissioned during cosmic rays and Au+Au collisions at $\sqrt{s_{NN}}$=14.5 GeV in the beginning of 2014. Subsequently, it was included in the physics data taking. About 1.2 billion Minimum Bias (MB) events for Au+Au collisions at $\sqrt{s_{NN}}$=200 GeV have been recorded with the HFT in 2014. In 2015 and 2016, we expect to collect p+p, p+Au and additional Au+Au data, allowing precise measurements of the nuclear modification factors R_{CP}, R_{pA} and R_{AA}, elliptic flow and particle correlations, for fully reconstructed open charm hadrons, as well as electrons or muons from semi-leptonic decays of open heavy flavor hadrons. Separate studies of charm and bottom contributions to the latter can be done using track impact parameters measured by the HFT.

Figure 1 illustrates the expected measurement precision of open heavy flavor hadrons by the HFT. Left panel shows D^0 elliptic flow extracted from 1 billion min-bias (MB) events in Au+Au collisions at $\sqrt{s_{NN}}$=200 GeV. A transport model is used for the projection in which charm quarks either flow as light quarks (red circles) or do not flow at all (green triangles). Right panel shows the ratio of the nuclear modification factors, R_{CP}, between Λ_c baryons and D^0 mesons extracted from 250 million central and 2 billion minimum-bias events in Au+Au collisions at $\sqrt{s_{NN}}$=200 GeV. The black (red) points assume no baryon-over-meson enhancement (the same enhancement as that of Λ/K_s^0).

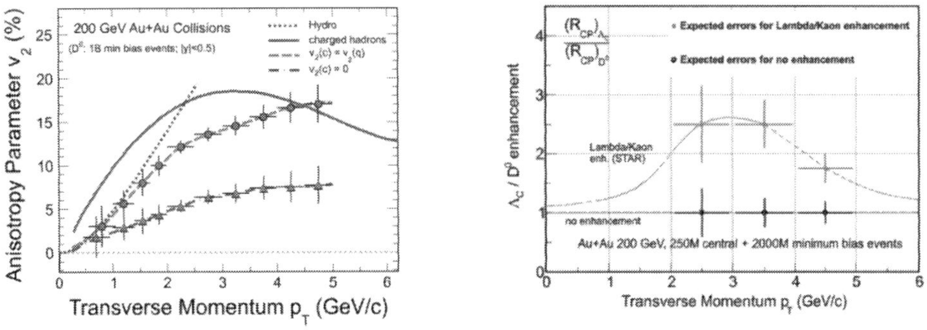

Figure 1. Projected open charm measurement precision with the Heavy Flavor Tracker in Au+Au collisions at $\sqrt{s_{NN}}$=200 GeV [2]. Left panel: D^0 elliptic flow extracted from 1 billion min-bias events. Right panel: ratio between Λ_c/D^0 R_{CP} extracted from 250 million central and 2 billion min-bias events. See text for details.

2.2. Muon Telescope Detector

The MTD is made of Multi-gap Resistive Plate Chambers (MRPC) operating in the avalanche mode [3]. It is installed outside the STAR Solenoid magnet, covering 45% of the 2π azimuthal angle in the pseudo-rapidity range $-0.5<\eta<0.5$. While 10% modules were installed in 2012 and 63% in 2013, the full MTD was commissioned in the beginning of 2014. Three MTD-related triggers, single muon, di-muon and electron-muon coincidence, were included in the physics data taking. 10 nb^{-1} of di-muon data have been recorded for Au+Au collisions at $\sqrt{s_{NN}}$=200 GeV in 2014. More data will be taken for p+p, p+Au and Au+Au collisions in 2015 and 2016.

The MTD enables identification and online triggering of high-p_T muons for the first time at STAR. It opens a door to many interesting physics studies. For example, quarkonium measurements have been traditionally performed in the di-electron channel at STAR. As muons are less affected by bremsstrahlung than electrons, separated measurements of different Υ states become possible with the MTD. Figure 2 illustrates the projected quarkonium measurements by the MTD. Measurements of $c\bar{c}$ correlations and its contribution to dilepton production can also be done through electron-muon pairs with small background contribution.

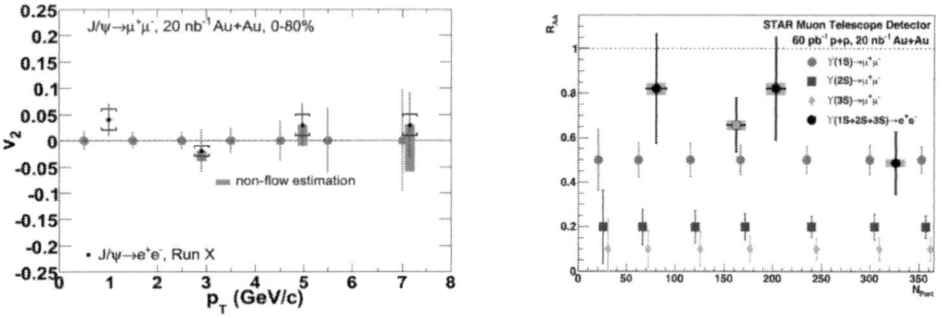

Figure 2. Projected quarkonium measurement precision with the Muon Telescope Detector in Au+Au collisions at $\sqrt{s_{NN}}$=200 GeV [3]. Recently STAR results are also shown as grey points [7]. Left panel: J/ψ elliptic flow from 20 nb^{-1} data; Right panel: nuclear modification factors for different Υ states from 20 nb^{-1} data. Black points with error bars are from [8].

3. RHIC BES-II program

Understanding the QCD phase diagram is one of the major scientific goals for heavy-ion physics [4]. RHIC has completed the first phase of BES (BES-I) program to search for the critical point and first order phase transition boundary with Au+Au collisions at $\sqrt{s_{NN}}$=200, 62, 39, 27, 19.6, 14.5, 11.5 and 7.7 GeV. The results have narrowed the region of interest to collision energies between 7.7 and 20 GeV. However, better measurement precisions and theoretical interpretations will be needed for a definite conclusion. The former could be achieved in the second phase of the BES program (BES-II) in 2018-2019, with the following STAR detector upgrades:

(1) iTPC upgrade includes filling the inner sectors with active readout pads and renewing the wires. The iTPC upgrade will improve the track reconstruction efficiency, momentum and dE/dx resolutions for the mid-rapidity $-1<\eta<1$. But more importantly it can extend the TPC coverage from $-1<\eta<1$ to $-1.7<\eta<1.7$. Such improvements will enable STAR to better measure the longitudinal extent of "the ridge", which provides the best means to disentangle contributions to the correlations from various stages of the collisions, such as the initial state, plasma phase and freeze-out.

(2) EPD is a new detector proposed for the forward rapidity region $1.5<|\eta|<5$. The EPD will provide the precise measurements of both the collision centrality and event plane.

(3) End-cap Time-Of-Flight (ETOF) is proposed to provide particle identification capabilities at large rapidity, covering $1<|\eta|<2$. The ETOF, using similar hardware techniques of the current barrel STAR TOF detector, will be placed at the east and west surfaces of the TPC.

With these upgrades, STAR will make precise measurements to explore the QCD phase diagram (see Figure 3). Note that these upgrades will not only be essential to the BES Phase-II program, but also benefit the future pp/pA and ep/eA programs, as described in sections 4 and 5.

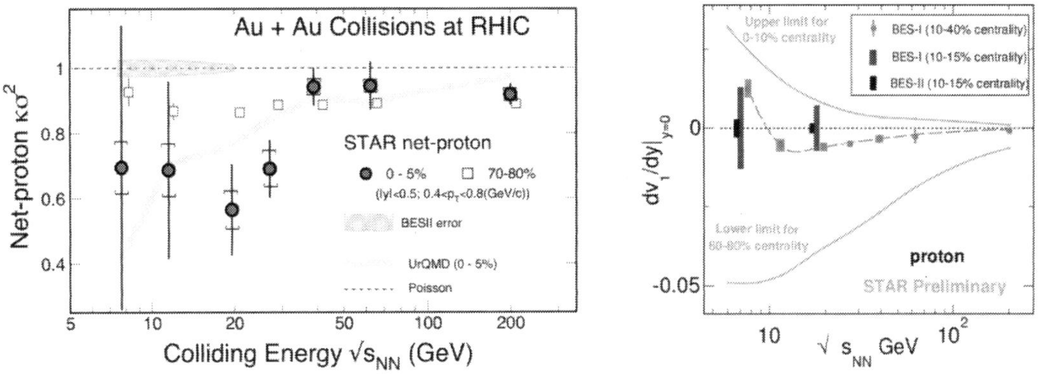

Figure 3. Left panel: Collision energy dependence of net-proton in 0-5% (filled-circles) and 70-80% (open-squares) Au+Au collisions from the BES-I [9]. The vertical bars (parentheses) are statistical (systematic) uncertainties. The green band shows the expected statistical uncertainties from the BES-II. Right panel: Statistical uncertainties with narrow centrality bins for proton dv_1/dy near midrapidity at two sample beam energies for BES-I data sets (blue error bars) and with the proposed BES-II data sets (black error bars). The upper (lower) green curve shows the approximate upper (lower) bound of the measured dv_1/dy range for protons in central collisions (peripheral collisions). The red points show the BES-I measurement.

4. Polarized p+p and p+A program

For the anticipated p+p and p+A physics program in 2021-2022, STAR is considering to build a Forward electromagnetic and hadron Calorimetric System (FCS) and Forward Tracking System (FTS), as shown in Figure 4. FCS prototypes have been built and tested in test beams and the performance meets the expectation. The FTS design is still under study, using either Silicon or GEM technology. Note that the FTS and FCS will also be crucial for ep/eA physics program as described in section 5.

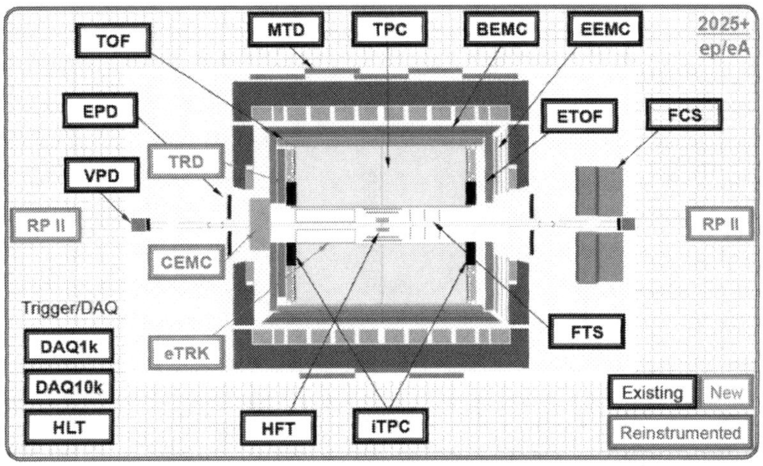

Figure 4. The planned STAR forward instrumentation upgrade, which is essential for detailed studies of polarized p+p/p+A and eSTAR programs. In polarized p+p/p+A program, the ion beams are from right to left (eastward) while proton beams from left to right (westward). In eSTAR program, the electron beams are from right to left (eastward) while hadron beams from left to right (westward).

Journal of Physics: Conference Series **535** (2014) 012022 doi:10.1088/1742-6596/535/1/012022

The physics opportunities for polarized pp/pA collisions have been discussed in a Letter of Intent (LoI) by the STAR collaboration [5]. Through measurements of forward photons, J/ψ, Drell-Yan (DY), inclusive jets and di-jets, and forward-forward gamma/hadron/jet correlation, STAR will be able to study in great detail the partonic structure of the nucleon and nuclei. For example, by measuring transverse spin asymmetries of prompt photons, $W^{\pm}/Z$ bosons (see Figure 5) and inclusive jets, one can understand the origin of the large transverse spin asymmetries at high x_F, and also constrain the transverse momentum dependent partron distribution and fragmentation functions. Through measuring double spin asymmetries in forward di-jet/hadron-jet/direct photon-jet productions, we can access the gluon helicity distribution at small x as shown in Figure 6.

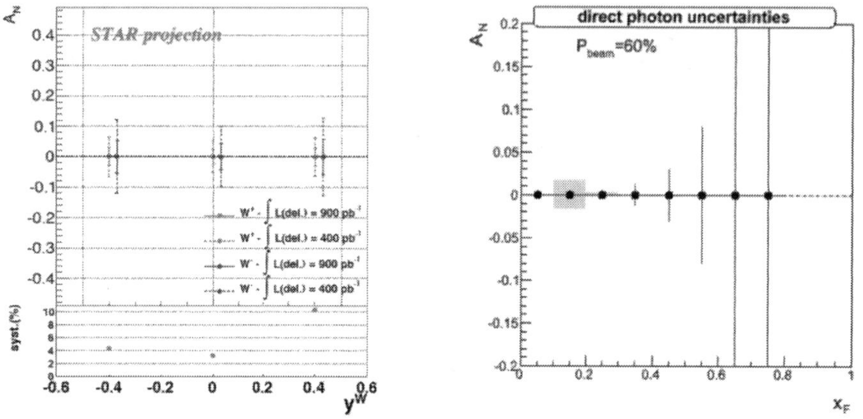

Figure 5. The projected uncertainties for transverse single spin asymmetries of $W^{\pm}$ (left panel) and direct photons after background subtraction (right panel).

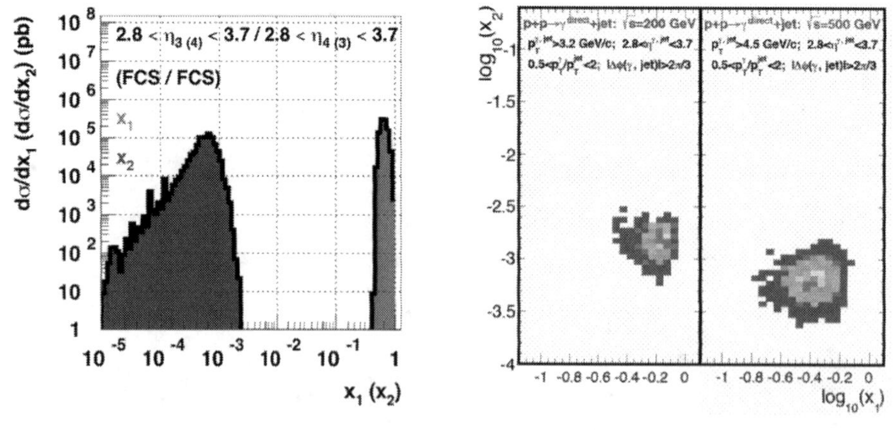

Figure 6. Left panel: The x_1/x_2 range for the forward STAR acceptance region in $2.8 < \eta < 3.7$ for di-jet measurement with FCS-FCS combinations. Right panel: Bjorken x distribution of hard scattering partons at small x via direct γ-jet coincidence channel.

5. eSTAR program

The nuclear physics community in the US is discussing the possibility of building a high-energy Electron Ion Collider (EIC) in the longer-term future. Such a machine will allow precise measurements of the gluon polarization contribution to the nucleon spin, 3D partonic structure of the nucleon and nuclei, gluon saturation at small x, and parton energy loss in cold nuclear matter. One possibility of realizing such a machine is to convert RHIC into an EIC machine (eRHIC) by adding a new electron ring.

By adding a few more detectors in the electron going direction, the new STAR detector on eRHIC (eSTAR) will be well suited with the physics program [6]. The anticipated detector upgrades as shown in Figure 4 include a Crystal ElectroMagnetic Calorimeter (CEMC) with a pre-shower detector in front covering $-4<\eta<-2$, a Transition Radiation Detector (TRD) placed between the TPC and ETOF, covering $-2<\eta<-1$, and an upgraded Roman Pot detector (RP-II) with an additional silicon plane added to the existing RP detector. With the generic EIC R&D framework supported by BNL, the STAR collaboration is looking into the possibility of using BSO crystals for the CEMC and using a GEM detector in combination with a Xe+CO2 active volume for the TRD.

The physics potential of the eSTAR detector has been described in great details in [6]. Figure 7 shows two such examples, the projected measurement precisions for the polarized proton structure function $g_1(x,Q^2)$ and Deeply Virtual Compton Scattering (DVCS) cross-section at eSTAR. The former is extracted from inclusive electron measurements in longitudinally polarized electron-proton Deep Inelastic Scattering (DIS), from which one can extract the quark and gluon polarization contributions to the nucleon spin. The latter is sensitive to Generalized Parton Distributions (GPDs). The study of GPD evolution will allow an access to gluon GPDs.

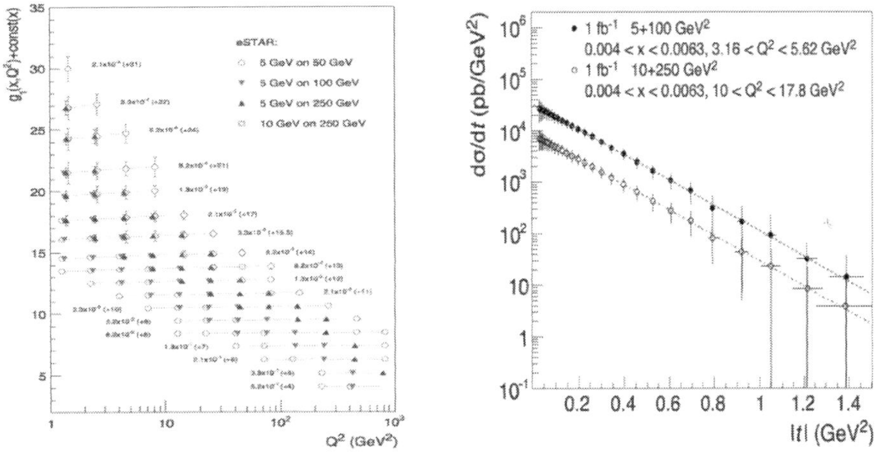

Figure 7. Left panel: projected $g_1(x,Q^2)$ measurement results at eSTAR. Right panel: projected measurement results for differential DVCS cross-sections at eSTAR.

6. Conclusion

The STAR experiment is a dedicated facility for studying matter with color degrees of freedom. Two recently completed detector upgrades, the Heavy Flavor Tracker and Muon Telescope Detector, will greatly help heavy flavor measurements. For the future, STAR has developed a compelling physics and detector upgrade plan to understand the dynamical evolution from the cold nuclear matter to the hot quark-gluon plasma.

References

[1] STAR Decadal Plan, 2010, *http://www.bnl.gov/npp/docs/STAR_Decadal_Plan_Final[1].pdf*

[2] Technical Design Report: The STAR Heavy Flavor Tracker, 2011, *https://drupal.star.bnl.gov/STAR/files/HFT_TDR_Final.doc*

[3] L. Ruan et al., J. Phys. G: Nucl. Part. Phys. **36** (2009) 095001

[4] Beam Energy Scan White Paper: Studying the Phase Diagram of QCD Matter at RHIC, 2014, *https://drupal.star.bnl.gov/STAR/files/BES_WPII_ver6.9_Cover.pdf*
[5] Letter of Intent: A polarized p+p and p+A program for the next years, 2014, *https://drupal.star.bnl.gov/STAR/files/pp.pA_.LoI_.pp_.pA_.v7.pdf*
[6] eSTAR: A Letter of Intent, 2014, *https://drupal.star.bnl.gov/STAR/files/eSTAR-LoI_v30_0.pdf*
[7] L. Adamczyk et al., Phys. Rev. Lett. **111** (2013) 052301
[8] L. Adamczyk et al., Phys. Lett. B **735** (2014) 127-137
[9] L. Adamczyk et al., Phys. Rev. Lett. **112** (2014) 032302

Latest results on heavy flavor di-lepton (p-Pb and Pb-Pb)

Dong Ho Moon on behalf of the CMS collaboration
University of Illinois at Chicago
E-mail: dmoon@cern.ch

Abstract. The Compact Muon Solenoid (CMS) has measured various quarkonium states via their decays into muon pairs in pp, PbPb and pPb collisions at $\sqrt{s_{NN}} = 2.76$ and 5.02 TeV. Quarkonia are especially relevant for studying the quark-gluon plasma since they are produced at early times of the collision and propagate through the medium, mapping its evolution. The most recent results on the production of prompt J/ψ in PbPb and the three Υ states in pp and pPb collisions will be presented.

1. Introduction

One strong signature of the existence for a Quark-Gluon-Plasma (QGP) is quarkonium suppression. Recent results from quarkonia measurements performed in CMS have been found to support theoretical predictions called the sequential melting scenario [1, 2, 3]. Quarkonimum production in heavy-ion collisions is affected by initial effects such as nuclear parton distribution functions (nPDFs) [4], parton energy loss, and the Cronin effect [5, 6], and final state effects such as Debye color screening and statistical recombination [1].

The result of study of the prompt J/ψ azimuthal anisotropy will be reported in this paper. Expected effects contributing to an observed non-zero elliptic anisotropy include recombination of thermalized charm quarks and (or) a path-length difference for absorption of quarkonia traversing the different direction of the almond-shaped hot and dense medium created in mid-central PbPb collisions.

This paper also reports the results of $\mathrm{b\bar{b}}$ states produced in proton-lead (pPb) collisions. This system is a good reference to understand initial state effects and may allow insight into cold nuclear effects that differ from the suppression effects observed in previous measurements [7].

2. Environmental setup and analysis procedure

The central feature of the CMS apparatus is a superconducting solenoid, providing an axial magnetic field of 3.8 T. Immersed in the magnetic field are the silicon pixel and strip tracker, the lead-tungstate crystal electromagnetic calorimeter, and the brass/scintillator hadron calorimeter. Muons are measured in gas ionization detectors embedded in the steel return yoke and in the pseudorapidity window $|\eta| < 2.4$, with detection planes made of three technologies: Drift Tubes, Cathode Strip Chambers, and Resistive Plate Chambers. Matching the muons to the tracks measured in the silicon tracker results in a transverse momentum resolution better than 1.5% for p_T smaller than 100 GeV/c. A more detailed description of the CMS detector can be found in Ref. [8].

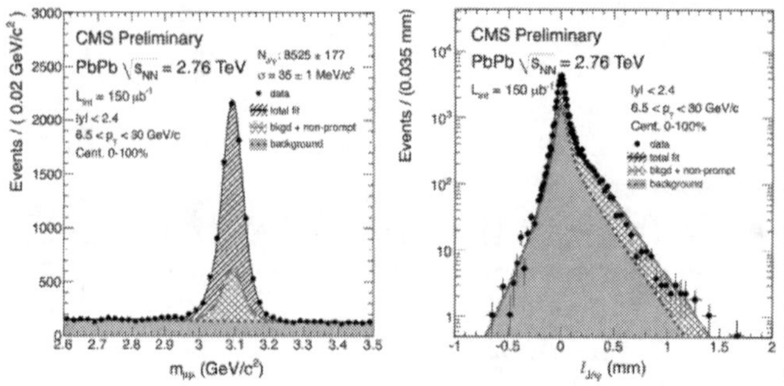

Figure 1. Invariant-mass spectra (left) and pseudo-proper decay length distribution (right) of $\mu^+\mu^-$ pairs integrated over centrality. The spectra are integrated over the rapidity range 0 < $|y|$ < 2.4 and the p_T range 6.5 < p_T < 30 GeV/c. The projections of the two-dimensional fit onto the respective axes are overlaid as solid black lines. The dashed red lines show the fitted contribution of non-prompt J/ψ. The fitted background contributions are shown as dotted blue lines [9].

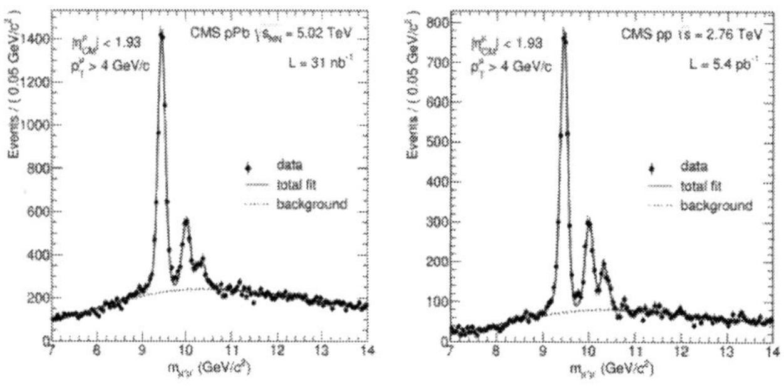

Figure 2. Invariant mass spectrum in pPb (left) and pp collisions (right) of $\mu^+\mu^-$ pairs with single muons with $p_T^\mu > 4$ GeV/c and $|\eta_{CM}^\mu| < 1.93$. The data (black circles) overlaid with the fit (solid blue line). The background component of the fit is represented by the dashed blue line [10].

The pp. data used in the study corresponded to an integrated luminosity of 5.4 pb^{-1} at a center-of-mass energy $\sqrt{s_{NN}}$ = 2.76 TeV. The corresponding integrated luminosity of pPb data is 31 nb^{-1} with a center-of-mass energy $\sqrt{s_{NN}}$ = 5.02 TeV. The direction of the higher-energy proton beam was initially set up to be clockwise, and was reversed after an integrated luminosity of 18 nb^{-1} of data was recorded. As a result of the energy difference of the colliding beams, the nucleon-nucleon center-of-mass in the pPb collisions are not at rest with respect to the laboratory frame. Massless particles emitted at $|\eta_{CM}| = 0$ in the nucleon-nucleon center-of-mass frame are detected at η = -0.465 (clockwise proton beam) or +0.465 (counterclockwise proton beam) in the laboratory frame. The PbPb data used for prompt J/ψ azimuthal anisotropy measurement were collected with a corresponding integrated luminosity of 150 μb^{-1}.

The invariant mass spectrum of all $\mu^+\mu^-$ pairs used in the PbPb analysis is shown in the left of Figure 1. The black curve is an extended unbinned maximum likelihood fit to the spectrum, using the sum of a Crystal Ball (CB) and a Gaussian function for signal, and an exponential for the background. To measure the fraction of non-prompt J/ψ (the so called b-fraction), the pseudo-proper decay length $l_{J/\psi}$, as shown in the right of Figure 1, is computed as an estimate of the b-fraction decay length by $L_{xy} m_{J/\psi}/p_T$. Here, the L_{xy} is defined as,

$$L_{xy} = \frac{\hat{u}^T S^{-1} \vec{r}}{\hat{u}^T S^{-1} \hat{u}}, \tag{1}$$

where $\hat{u}$ is the unit vector in the direction of the J/ψ p_T and S is the sum of the primary and secondary vertex covariance matrices. In the final step, the invariant mass spectrum of $\mu^+\mu^-$ pairs and their $l_{J/\psi}$ distribution are fitted simultaneously in a two-dimensional (2D), extended unbinned maximum likelihood fit.

Figure 2 shows the mass spectra for Υ(nS) states in pp (left) and pPb (right). The yields are extracted from an unbinned maximum likelihood fit to the invariant dimuon mass spectra. The reconstructed mass line shape of each Υ(nS) state is modeled by a CB function, i.e. a Gaussian function with the low-side tail replaced by a power law function describing final-state radiation. Reasonable variations with multiplicity are considered in the systematic uncertainties. The Υ(nS) mass ratios are fixed to the their world average values [11], with ground shape is modeled by an exponential function multiplied by an error function and all its parameters are left free in the fit, as in Ref. [12].

3. Results

3.1. Charmonia in PbPb collisions

After extracting the prompt J/ψ yields in each rapidity, p_T, centrality, and $\Delta\phi$ bin, the v_2 is calculated with a fit of the $\frac{1}{N_{total}^{J/\psi}} \frac{dN_{J/\psi}}{d\phi}$ vs $\Delta\phi$ distributions with the function

$$1 + 2v_2 \cos(2\Delta\phi). \tag{2}$$

The $N_{total}^{J/\psi}$ is the yield in $\Delta\phi = 2/\pi$ for each kinematic bin. An example of such fits is given in Figure 3, for the centrality integrated case. The measured prompt J/ψ v_2, for 10-60% event centrality, integrated over $6.5 < p_T < 30$ GeV/c and $|y| < 2.4$ is

$$0.054 \pm 0.013(\text{stat}) \pm 0.006(\text{syst}), \tag{3}$$

with a significance for a nonzero v_2 of 3.8σ. The v_2 results versus centrality, p_T, and rapidity are shown in Figure 4. For each of these results, the dependence on one variable is studied by averaging over the other two. A non-zero v_2 is measured in all the kinematic bins studied. The observed anisotropy shows no strong centrality or rapidity dependence when integrated over rapidity and centrality, respectively. In 10-60% centrality events, the result are compatible with a p_T-independent anisotropy, whether measured at low-p_T ($3 < p_T < 6.5$ GeV/c) in the forward rapidity interval $1.6 < |y| < 2.4$, or at high-p_T ($6.5 < p_T < 30$ GeV/c) in rapidity interval $|y| < 2.4$.

3.2. Bottomonia in pp and pPb collisions

The pp and pPb data are further analyzed separately as a function of event activity variables measured in two different rapidity regions. Specifically, the single ratios, Υ(2S)/Υ(1S) and Υ(3S)/Υ(1S) are measured in bins of: (1) $E_T^{|\eta|>4}$, the raw transverse energy deposited in the most forward part of the HF calorimeters at $4.0 < |\eta| < 5.2$, and (2) $N_{tracks}^{|\eta|<2.4}$, the number of charged particles, not including the two muons, with $p_T > 400$MeV/c reconstructed in the tracker at $|\eta| <2.4$ and originating from the same vertex as the Υ. In Figure 5, for both pp and pPb, the results are shown as a function of forward transverse energy ($E_T^{|\eta|>4}$, left panel), and as a function of mid-rapidity track multiplicity ($N_{tracks}^{|\eta|<2.4}$, right panel). The difference observed

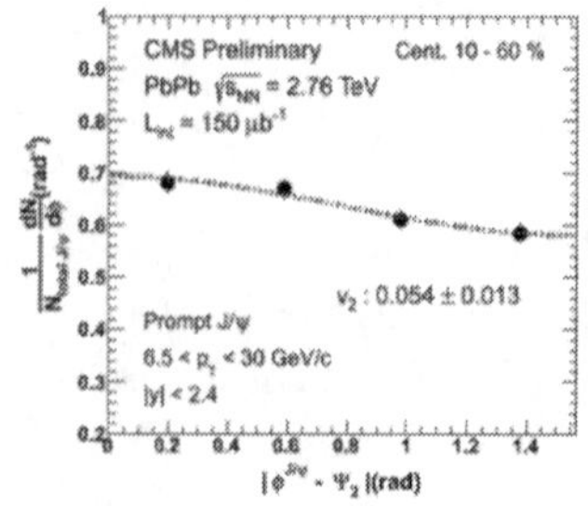

Figure 3. The $\Delta\phi$ distribution of high-p_T ($6.5 < p_T < 30$ GeV/c) prompt J/ψ raw yields, measured in the rapidity range $|y| < 2.4$ and event centrality 10-50%, and normalized by the bin width and the sum of the yields in all four $\Delta\phi$ bins. The dashed red line in all figures represents the function $1+2v_2\cos(2\Delta\phi)$ used to extract v_2 [9].

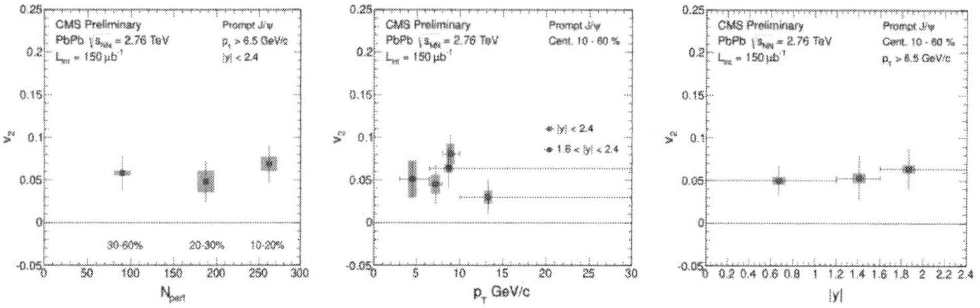

Figure 4. Prompt J/ψ centrality (left), p_T (middle) and rapidity (right) dependence of v_2. The boxes represented point-by-point systematic uncertainties. Horizontal bars indicate the bin width [9].

between the Υ states when binning in $N_{tracks}^{|\eta|<2.4}$ can be explained in two opposite ways. If, on one hand, the Υ(1S) is systematically produced with more particles than the excited states, it would influence the underlying distribution of charged particles and create an artificial effect when selected in small multiplicity bins. This effect should be sensitive to the underlying multiplicity distribution and would result in a larger correlation if one reduces the size of the multiplicity bins. On the other hand, if the Υ are interacting with the surrounding environment, the Υ(1S) is expected, as the most tightly bound state and the one of smallest size, to be less affected than Υ(2S) and Υ(3S), leading to a decrease of the Υ(nS)/Υ(1S) ratios with increasing multiplicity. In either cases, the ratios will continuously decrease from the pp to pPb to PbPb systems, as a function of event multiplicity.

In addition, self-normalized to their activity-integrated values, the individual Υ(nS) yields are calculated. The results are shown in Figure 6 in bins of $E_T^{|\eta|>4}/\langle E_T^{|\eta|>4}\rangle$ (top) and $N_{tracks}^{|\eta|<2.4}/\langle N_{tracks}^{|\eta|<2.4}\rangle$ (bottom), for pp and pPb collisions, where the denominator is average of yields in all events. All the self-normalized cross section ratios increase with increasing forward transverse energy and mid-rapidity particle multiplicity in the event. In the cases where Pb ions are involved, the increase observed in both variables can arise from the increase in the number of nucleon-nucleon collisions. The pp results are reminiscent of a similar J/ψ measurement made in pp collisions at 7 TeV [13].

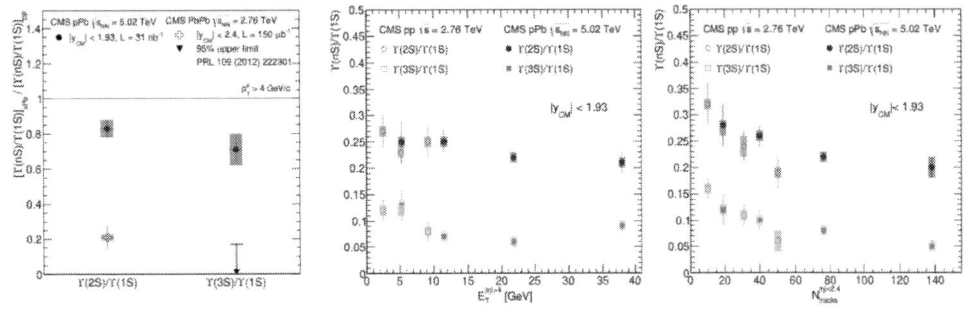

Figure 5. (Left) Event activity integrated double ratios of the excited states, $\Upsilon(2S)$ and $\Upsilon(3S)$, to the ground state, $\Upsilon(1S)$ in pPb collisions at $\sqrt{s_{NN}} = 5.02$ TeV with respect to pp collisions at $\sqrt{s} = 2.76$ TeV (circles), compared to the corresponding ratios for PbPb (cross) collisions at $\sqrt{s_{NN}} = 2.76$ TeV from Ref. [12], which used a different dataset for the pp normalization. (Middle, Right) Single cross section ratios $\Upsilon(2S)/\Upsilon(1S)$ and $\Upsilon(3S)/\Upsilon(1S)$ for $|y_{CM}| < 1.93$ versus transverse energy measured in $4.0 < |\eta| < 5.2$ (left) and number of charged tracks measured in $|\eta| < 2.4$ (right), for pp collisions at $\sqrt{s} = 2.76$ TeV (open symbols) and pPb collisions at $\sqrt{s_{NN}} = 5.02$ TeV (closed symbols). In both figures, the error bars indicate the statistical uncertainties, and the boxes represent the point-to-point systematic uncertainties. The global uncertainties on the pp results are 7% and 8% for $\Upsilon(2S)/\Upsilon(1S)$ and $\Upsilon(3S)/\Upsilon(1S)$, respectively, while in the pPb results they amount to 8% and 9%, respectively [10].

4. Summary

We presented the azimuthal anisotropy of prompt J/ψ that shows a 3.8σ significance for a non-zero v_2 in the centrality 10-60% of high p_T (>6.5 GeV/c) over rapidity range $|y| < 2.4$. No strong centrality, p_T and rapidity dependence is observed within the measured uncertainties for results that integrate the other variables. This result could give a hint of the path-length dependence of partonic energy loss in a deconfined medium.

The results of double ratios $[\Upsilon(nS)]/[\Upsilon(1S)]_{pPb}/[\Upsilon(nS)]/[\Upsilon(1S)]_{pp}$ suggests the presence of final-state suppression effects in the pPb collisions compared to pp collisions which affect more strongly the excited states ($\Upsilon(2S)$ and $\Upsilon(3S)$) compared to the ground state ($\Upsilon(1S)$). The excited-to-ground-states cross section ratios, $\Upsilon(nS)/\Upsilon(1S)$, are found to decrease with increasing charged-particle multiplicity as measured in the $|\eta| < 2.4$ interval that contains the region in which the Υ are measured. The self-normalized cross section ratios, $\Upsilon(nS)/\langle\Upsilon(nS)\rangle$ increase with event activity.

References

[1] T. Matsui and H. Satz, *Phys. Lett. B* **178** (1986) 416
[2] S. Digal, P. Petreczky, and H. Satz, *Phys. Rev. C* **64** (2001) 904015
[3] A. Mocsy and P. Petreczky, *Phys. Rev. Lett* **99** (2007) 211602
[4] R. Vogt, *Phys. Rev. C* **81** (2010) 044903
[5] R. Sharma and I. Vitev, *Phys. Rev. C* **87** (2013) 044905
[6] F. Arleo and S. Peigne, *JHEP* **03** (2013) 122
[7] CMS Collaboration, *JHEP* **1205** (2012) 063
[8] CMS Collaboration, *JINST* **3** (2008) S08004
[9] CMS Collaboration, CMS Physics Analysis Summary CMS-HIN-12-001 (2012)
[10] CMS Collaboration, *JHEP* **04** (2014) 103
[11] Particle Data Group Collaboration *Phys. Rev. D* **86** (2012) 010001
[12] CMS Collaboration, *Phys. Rev. Lett.* **109** (2012) 222301
[13] ALICE Collaboration, *Phys. Rev. Lett. B* **712** (2012) 165

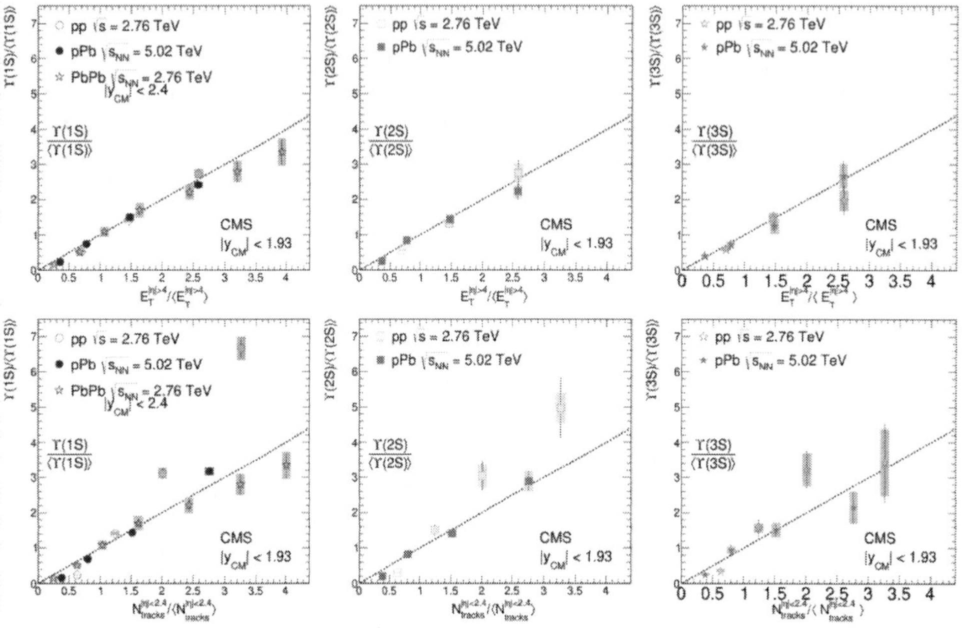

Figure 6. The Υ(nS) cross section versus transverse energy measured at $4 < |\eta| < 5.2$ (top row) and versus charged-track multiplicity measured in $|\eta| < 2.4$ (bottom row), measured in $|y_{CM}| < 1.93$ in pp collisions at $\sqrt{s} = 2.76$ TeV and pPb collisions at $\sqrt{s_{NN}} = 5.02$ TeV. For Υ(1S), the PbPb data at $\sqrt{s_{NN}} = 2.76$ TeV (open stars) are overlaid. Cross sections and x-axis variables are normalized by their corresponding activity-integrated values. For all points, the abscissae are at the mean value in each bin. The dotted line is a linear function with a slope equal to unity. The error bars indicate the statistical uncertainties, and the boxes represent the point-to-point systematics uncertainties [10].

Journal of Physics: Conference Series **535** (2014) 012024 doi:10.1088/1742-6596/535/1/012024

Probing chiral restoration with ρ-meson melting through sum rules

Paul M. Hohler

E-mail: pmhohler@comp.tamu.edu
Cyclotron Institute and Department of Physics and Astronomy, Texas A&M University, College Station, TX 77843-3366, USA

Ralf Rapp

E-mail: rapp@comp.tamu.edu
Cyclotron Institute and Department of Physics and Astronomy, Texas A&M University, College Station, TX 77843-3366, USA

Abstract. We investigate whether the rho-meson melting scenario is compatible with chiral symmetry restoration utilizing a comprehensive evaluation of the QCD and Weinberg sum rules at nite temperature. As input to this analysis, in-medium vector spectral functions which describe dilepton data from ultra-relativistic heavy-ion collisions are used along with temperature dependent condensates from lattice calculations, when available, or approximated by a hadron resonance gas. The combined deployment of QCD and Weinberg sum rules turns out to be rather stringent in constructing axialvector spectral functions consistent with (partial) chiral restoration.

1. Introduction

In the CD vacuum, chiral symmetry is spontaneously broken by the presence of non-zero quark condensates. Lattice- CD calculations reveal that this symmetry becomes restored at higher temperatures as the condensates progress through a pseudo-critical region around $T_{\rm pc} \simeq 160\,\mathrm{MeV}$ [1, 2]. A long standing problem in hadronic physics is identifying an experimental signal for this transition. Ideally, this could be achieved by simultaneously measuring the medium modifications of the spectral functions of chiral partners, *e.g.*, vector (ρ) and axialvector (a_1) mesons chiral restoration is then characterized by a degeneracy of the two channel's spectral functions. The vector channel has been extensively explored through dilepton spectra in ultra-relativistic heavy-ion collisions [3, 4, 5]. Theoretical calculations from hadronic effective theory using microscopic interactions [6] are consistent with the experimental data across a wide range of collision energies. These studies reveal that the ρ-meson resonance melts without an appreciable mass shift as the fireball cools through the pseudo-critical region [7]. Experimental access to the axialvector channel ($a_1 \to \pi\gamma$) is difficult due to its small branching and large width. Thus establishing this melting as a signal for chiral restoration remains outstanding. A theoretical determination of the axialvector spectral function is therefore needed to provide the necessary connection between the ρ melting and chiral symmetry.

The direct approach would be to calculate the axialvector spectral function from effective field theory paralleling the previous work for the ρ. However, this has proven challenging [8] and has lead some to give up the natural implementation of the mesons as local gauge bosons of the chiral symmetry [9, 10]. A recent study has been able to overcome these initial problems while maintaining a local gauging procedure in vacuum [11] work is ongoing to extend these calculations to finite temperatures.

In this paper, based on Ref. [12], we take the more modest approach by performing a sum rule analysis to ask whether the melting scenario is compatible with chiral restoration. By using the experimentally tested in-medium ρ spectral functions and constraints from lattice data, we study whether an axialvector spectral function can be found which satisfies the sum rules. For our analysis, we use both the Weinberg sum rules (WSRs) [13, 14, 15], which are responsive to chiral symmetry restoration, and CD sum rules (CDSRs) [16, 17], which involve additional chirally invariant condensates, to provide maximal constraints on the in-medium axial-/vector spectral functions. As a by-product, we test the CDSRs for the vector channel. Similar works have been performed, e.g., in the low-temperature limit [18, 19], for heavy-quark channels [23], or by focusing on chirally odd condensates in the vector channel only [24]. A recent study [20] parallels some of the philosophies of the current work by investigating whether finite energy WSRs can be satisfied with axial-/vector spectral functions constrained by finite energy CDSRs [21, 22].

The results presented here build upon our earlier study [25], where we quantitatively evaluated the vacuum CDSRs and WSRs using spectral functions that accurately describe hadronic τ-decay data [26, 27]. There we found that the WSRs (predominately WSR-1 and -2) were particularly selective in constraining the spectral functions: by requiring that they be satisfied dictated the presence of an excited a_1' state beyond the kinematic limit of the τ-data. Such selectivity makes the sum rules a promising tool in determining in-medium axialvector spectral functions, so new insights into the mechanism of chiral symmetry restoration can be expected.

The proceeding is organized as follows. In Sec. 2, the in-medium CDSRs and WSRs are introduced. Section 3 describes the temperature dependence of the pertinent condensates, while in Sec. 4, the axial-/vector spectral function are specified. The quantitative sum rule analysis is presented in Sec. 5. We conclude in Sec. 6.

2. Finite Temperature Sum Rules

Sum rules equate dispersion integrals of spectral functions to a space-like expansion where the non-perturbative coefficients are given by low energy condensates. For our purposes, the relevant spectral functions are the isovector current-current correlator in the vector (V) and axialvector (A) channels,

$$\Pi^{\mu\nu}_{V,A}(q^2) = -i\int d^4x\; e^{ixq}\left\langle T\vec{J}^{\mu}_{V,A}(x)\vec{J}^{\nu}_{V,A}(0)\right\rangle \;, \tag{1}$$

where the currents, in the basis of two light flavored quarks, are given by $\vec{J}^{\mu}_{V} = \bar{q}\vec{\tau}\gamma^{\mu}q$ and $\vec{J}^{\mu}_{A} = \bar{q}\vec{\tau}\gamma^{\mu}\gamma_5 q$, ($\vec{\tau}$: isospin Pauli matrices). Here, only the charge-neutral states (isospin I_3=0) are considered, thus isospin indices will be dropped. In vacuum, the correlator can be decomposed as

$$\Pi^{\mu\nu}_{V,A}(q^2) = \Pi^{T}_{V,A}(q^2)\left(-g^{\mu\nu} + \frac{q^{\mu}q^{\nu}}{q^2}\right) + \Pi^{L}_{V,A}(q^2)\frac{q^{\mu}q^{\nu}}{q^2} \tag{2}$$

into the 4D-transverse and longitudinal polarizations. In the vector channel, $\Pi^{L}_{V}(q^2)$=0, due to vector current conservation, while the partial conservation of the axial-current involves the pion pole,

$$\mathrm{Im}\Pi^{L}_{A}(q^2) = -\pi f_{\pi}^2 q^2 \delta(q^2 - m_{\pi}^2) \;. \tag{3}$$

At finite temperature, Lorentz symmetry is broken such that the 4D-transverse polarization is further decomposed into 3D-transverse and -longitudinal parts. In this paper, we concentrate on the case of vanishing 3-momentum ($\vec{q}$=0) for which the 3D modes are again degenerate. We define the pertinent spectral functions as

$$\rho_{V,A} = -\frac{\mathrm{Im}\Pi^T_{V,A}}{\pi} \ , \ \rho_{\bar{A}} = \rho_A - \frac{\mathrm{Im}\Pi^L_A}{\pi} \ . \tag{4}$$

For the finite temperature CDSRs, the dispersion integral of these spectral functions constitutes the left-hand side (LHS) of the sum rule, while the right-hand side (RHS) is constructed from an operator product expansion (OPE). There is a unique sum rule for the vector and axialvector channels given by [28, 29, 30]

$$\begin{aligned}\frac{1}{M^2}\int_0^\infty ds \frac{\rho_{V,\bar{A}}(s)}{s} e^{-s/M^2} &= \frac{1}{8\pi^2}\left(1+\frac{\alpha_s}{\pi}\right) + \frac{m_q\langle\bar{q}q\rangle}{M^4} + \frac{1}{24M^4}\langle\frac{\alpha_s}{\pi}G^2_{\mu\nu}\rangle \\ &- \frac{\pi\alpha_s}{M^6}\frac{(56,-88)}{81}\langle\mathcal{O}_4^{V,A}\rangle + \sum_h \frac{\langle\mathcal{O}_h^{d=4,\tau=2}\rangle_T}{M^4} + \frac{\langle\mathcal{O}_h^{d=6,\tau=2}\rangle_T}{M^6} + \frac{\langle\mathcal{O}_h^{d=6,\tau=4}\rangle_T}{M^6}\dots\ ,\end{aligned} \tag{5}$$

where the space-like q^2 is traded for the Borel mass M^2 by the standard Borel transform. The OPE is comprised of all operators up to dimension-6, both chirally invariant and chiral order parameters. This includes the conventional scalar operators (quark, gluon, and 4-quark condensates, $\langle\bar{q}q\rangle$, $\langle\frac{\alpha_s}{\pi}G^2_{\mu\nu}\rangle$, and $\langle\mathcal{O}_4^{V,A}\rangle$, respectively), as well as non-scalar operators induced by thermal hadrons (h), organized by dimension (d) and twist (τ).

On the other hand, the dispersion integrals of the WSRs are associated with the difference between the vector and axialvector spectral functions. As such, the corresponding OPEs are expressed only in terms of chiral order parameters, while the different WSRs correspond to different moments of the dispersion integrals. The finite temperature sum rules, first formulated in Ref. [15], read

$$(\mathrm{WSR\,1}) \quad \int_0^\infty ds \frac{\rho(s)}{s} = f_\pi^2 \ , \tag{6}$$

$$(\mathrm{WSR\,2}) \quad \int_0^\infty ds \quad \rho(s) = f_\pi^2 m_\pi^2 = -2m_q\langle\bar{q}q\rangle \ , \tag{7}$$

$$(\mathrm{WSR\,3}) \quad \int_0^\infty dss \quad \rho(s) = -2\pi\alpha_s\langle\mathcal{O}_4^{SB}\rangle \ , \tag{8}$$

where $\rho = \rho_V - \rho_A$. The chiral breaking 4-quark condensate can be expressed in terms of the axial-/vector ones as

$$\langle\mathcal{O}_4^{SB}\rangle = \frac{16}{9}\left(\frac{7}{18}\langle\mathcal{O}_4^V\rangle + \frac{11}{18}\langle\mathcal{O}_4^A\rangle\right) \ . \tag{9}$$

The presence of only chiral order parameters renders the WSRs particularly sensitive to chiral restoration, whereas the channel-specific CDSRs provide independent constraints.

3. In-Medium Condensates

The in-medium behavior of each condensate can be estimated from a hadron resonance gas (HRG) calculation. Here, we consider all resonances with mass $m_h \leq 2\,\mathrm{GeV}$ [31]. Including only the effects from single-hadron matrix elements for each operator, $\mathcal{O}$, leads to

$$\langle\mathcal{O}\rangle_T \simeq \langle\mathcal{O}\rangle_0 + \sum_h d_h \int \frac{d^3k}{(2\pi)^3\,2E_h}\langle h(\vec{k})|\mathcal{O}|h(\vec{k})\rangle n_h(E_h) \ , \tag{10}$$

where $\langle\mathcal{O}\rangle_0$ is its vacuum expectation value, $\langle h(\vec{k})|\mathcal{O}|h(\vec{k})\rangle$ its hadronic matrix element, $E_h^2 = m_h^2 + \vec{k}^2$, and d_h, m_h, and n_h are the hadron's spin-isospin degeneracy, mass, and thermal

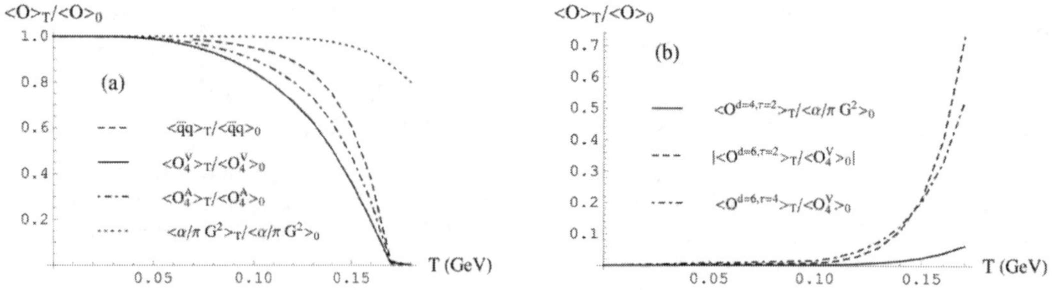

Figure 1. Temperature dependence of: quark, gluon, and axial-/vector 4-quark condensates relative to their vacuum values (left), and dimension-4 and dimension-6 non-scalar operators relative to the vacuum values of the gluon and vector 4-quark condensates (right).

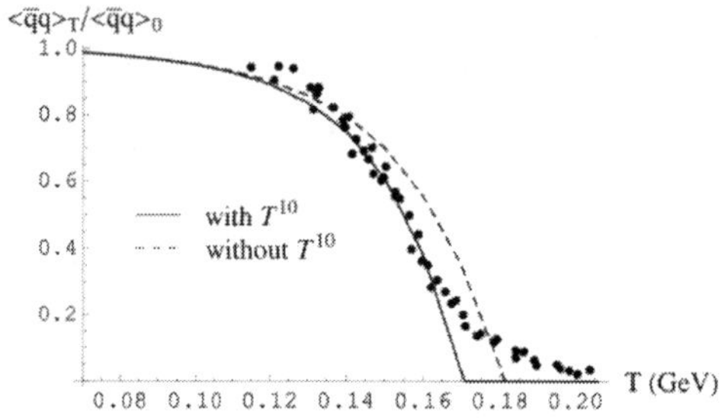

Figure 2. Temperature dependence of the quark condensate relative to its vacuum value, compared to thermal l CD data [1].

distribution function (Bose (n_b) or Fermi (n_f)), respectively. We work at zero baryon chemical potential (μ_B=0), so that anti-baryons can be absorbed into the degeneracy factor of baryons.

The HRG construction is known to reproduce the equation of state from l CD quite well for $T \leq 170\,\text{MeV}$ [32]. Importantly, it utilizes the same degrees of freedom as the in-medium ρ spectral function calculation. The resulting T-dependence of the quark, gluon and 4-quark condensates are depicted in Fig. 1 left, while the non-scalar operators are shown in Fig. 1 right.

The HRG correction to the quark condensate can be derived from the HRG partition function via $\partial \ln Z/\partial m_q$ [33, 34] and has the general structure

$$\frac{\langle \bar{q}q\rangle_T}{\langle \bar{q}q\rangle_0} = 1 - \sum_h \frac{\sigma_h}{f_\pi^2 m_\pi^2}\varrho_s^h\,, \tag{11}$$

where ϱ_s^h is the scalar density of hadron, h, and $\sigma_h = \langle h|\bar{q}q|h\rangle$ its σ-term. For the Goldstone bosons, σ_h can be calculated from current algebra while for all other hadrons, it is decomposed into a bare part (from the light valence quarks) [35] and a pion cloud contribution [36, 37]. Again, this decomposition parallels the medium effects of the ρ spectral function [38].

Our HRG calculation reproduces l CD data [1] for $T \lesssim 140\,\text{MeV}$ well, see Fig. 2. To extend this agreement, we introduced a term, αT^{10}. With $\alpha = 1.597 \cdot 10^7\,\text{GeV}^{-10}$, the quark condensate vanishes slightly above T=170 MeV. The power in T of this additional term was chosen as a balance between not affecting the low T behavior while improving the description of the high-T region of the data.

For the gluon condensate, the contributions from pions and nucleons have been evaluated in Refs. [28, 30, 39]. The HRG effect can be determined from the trace anomaly and the light quark condensate. This produces a mild dependence on temperature as shown in Fig. 1 left.

For the 4-quark condensates, current algebra again dictates the Goldstone boson contribution to the medium dependence [28]. For the non-Goldstone boson and the baryons, large N_c arguments [34, 40] suggest that their contribution can be determined by a factorization approximation which relates it to the quark condensate. We again augment the high-T dependence by a term $\beta_{V,A}T^{10}$. Since there is no guidance from lattice data, $\beta_{V,A}$ are ad usted for each channel to force the condensates to vanish at the same temperature as the quark condensate. This creates a dependence which is initially steeper than the quark condensate, cf. Fig. 1 left. The T-dependence of the chiral breaking 4-quark condensate follows from the axial-/vector ones via Eq. (9).

Hadrons in the heat bath also induce non-scalar condensates. For our CDSR analysis the relevant ones are of dimension-4 twist-2, $\langle \mathcal{O}^{d=4,\tau=2} \rangle_T$, dimension-6 twist-2, $\langle \mathcal{O}^{d=6,\tau=2} \rangle_T$, and dimension-6 twist-4, $\langle \mathcal{O}^{d=6,\tau=4} \rangle_T$. Their T dependence, adopted from Refs. [28, 29, 30], is constructed from thermal averaged powers of the momentum, with parameters related to the light-quark parton distribution and the spin-averaged (longitudinal) structure functions for each hadron [29, 41]. For hadrons with unmeasured functions, the parameters have been approximated by using the value from either the pion or nucleon with a suppression due to the valence strange-quark content. The contributions from gluonic structure functions are believed to be numerically insignificant [28, 29] and have been neglected. To compare the relative size of these contributions, the temperature dependence, shown in Fig. 1 right, is plotted as a dimensionless ratio relative to the appropriate vacuum scalar condensate (gluon for the dimension-4 operator and the vector 4-quark for the dimension-6 operator). For the dimension-6 twist-2 operator, this ratio is negative, thus the absolute value is plotted for easier comparison.

4. Finite Temperature Spectral Functions

The starting point of our study are the vacuum axial-/vector spectral functions of Ref. [12, 25]. They are constructed from three components: the ground state (ρ and a_1 peaks), a first excited state (ρ' and a_1'), and a chirally invariant (i.e., identical) continuum. The microscopic model of Ref. [42] provides the vacuum ρ spectral function, while parameterized Breit-Wigner functions are used to describe the a_1, ρ' and a_1' resonances. The availability of experimentally measured vacuum spectral functions from τ decays by ALEPH [26] allows the parameters of the a_1 and ρ' peaks to be determined by fitting these data. The a_1' contribution, on the other hand, is deduced from the WSRs, by requiring them to be satisfied. The evaluation of the vacuum CDSRs allows for the numerical determination of the vacuum gluon condensate and the 4-quark factorization parameter κ in $\langle \mathcal{O}_4^{SB} \rangle = \frac{16}{9}\kappa \langle \bar{q}q \rangle^2$. The resulting spectral functions are displayed in Fig. 3.

The changes of the condensates with temperature will induce modifications of the spectral functions. These modifications are accounted for as follows. Most importantly, to ensure that the results are consistent with dilepton data [7], the microscopic calculations for the ρ spectral function from hadronic effective theory [6] at vanishing baryon chemical potential are used. The only amendment we permit is a suppression of the vector-dominance coupling strength (as is commonly done in CDSR analyses [28, 30, 43, 44]).

For the a_1 meson, we parameterize the medium modifications of its spectral function by introducing four parameters (one for the mass, two for the width, and one for the axialvector-dominance coupling strength), which control the in-medium a_1 peak. These parameters will be scanned at each temperature to determine whether the CDSRs and WSRs can be satisfied.

A priori, the temperature dependence of the excited states is rather unconstrained. To focus the sum rules analysis on the in-medium a_1 peak, we apply the model independent low-temperature effect known as chiral mixing [45, 46] to the ρ' and a_1' states instead of introducing

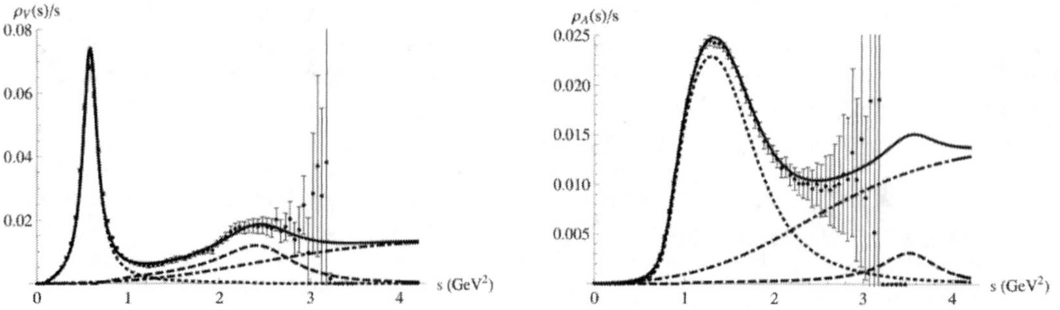

Figure 3. Vacuum spectral functions in the vector (left) and axialvector (right) channels, compared to experimental data for hadronic τ decays [26] The total spectral function in each channel (solid curve) is composed of a ground state (dotted curve), excited state (dashed curve), and a universal continuum (dot-dashed curve).

temperature dependencies to their Breit-Wigners parameters (which are hard to control). In the spirit of the HRG, both thermal pions and virtual pions from the clouds of thermal hadrons contribute to the mixing effects. Consistent with the condensates, this mixing is carried out only to linear order in the (scalar) hadron densities. In cold nuclear matter, this effect has been calculated for the pion cloud of the nucleon [47, 48]. As with the condensates, an additional T^{10} term is added to the mixing whose coefficient is determined by requiring the excited resonances become degenerate when the condensate vanishes. The addition of the virtual pions effectively approximates the extension of the mixing beyond the low-T limit, yet it does not account for finite-momentum nor finite-mass corrections from the (virtual) pions.

The chirally invariant continuum is assumed to be T independent

Lastly, the T dependence of $f_\pi(T)$ needs to be addressed. sing the Gell-Mann–Oakes–Renner relation at finite T, we relate $f_\pi(T)$ to the already-constructed quark condensate and the pion mass, whose T-dependence is approximated by the leading-order prediction from chiral perturbation theory.

To summarize this section, a microscopic model for the ρ spectral function has been supplemented with a 4-parameter ansatz for the in-medium a_1, chiral mixing for the excited states, and a weakly T-dependent pion mass from chiral perturbation theory. This setup is now deployed to the CDSRs and WSRs.

. Finite-Temperature Sum Rule Analysis

In order to critically evaluate agreement with the sum rules, quantitative criteria are needed. For the CDSR, we calculate the so-called d-value, *i.e.*, the average deviation between the LHS and RHS over a suitable Borel window [43, 49]. determined by using the same criteria as the vacuum analysis in Ref. [25]. From the analysis of Ref. [43], a d-value of 1% was found to encompass a reasonable uncertainty in the spectral function we adopt this as our figure of merit.

For the WSRs, we similarly define a deviation between each side as

$$d_{\mathrm{WSR}} = \frac{\mathrm{LHS} - \mathrm{RHS}}{\mathrm{RHS}} . \tag{12}$$

The integrands of the WSRs are oscillatory functions of s with non-trivial cancelations to yield the integrated value. This can lead to fine-tuned solutions whereby the oscillations are still large but conspire to an apparent agreement of the sum rule, with $\rho_V(s) \neq \rho_A(s)$ even close to restoration. To avoid an artificially fine-tuned solution, we define absolute-value versions of

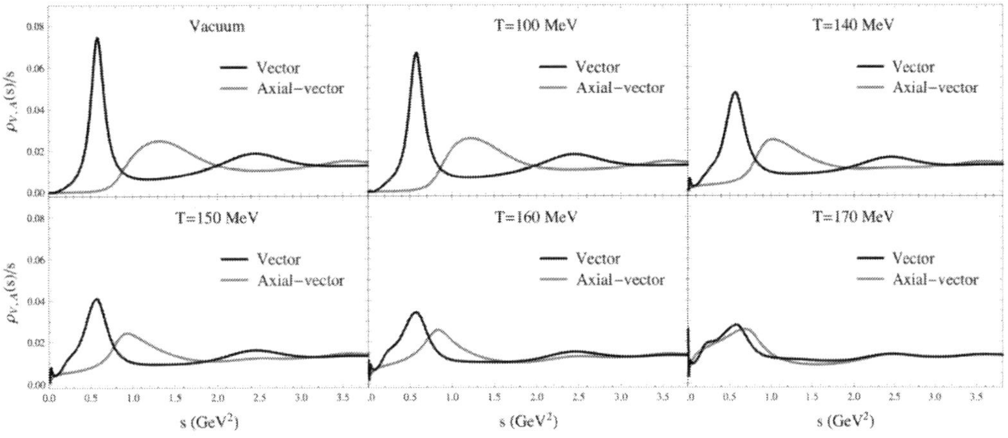

Figure 4. Finite-temperature vector (black curve) and axialvector (red curve) spectral functions following from minimizing Eq. (14) for the axialvector and d_V for the vector.

the LHS are defined by

$$\tilde{w}_n(T) \equiv \int_0^\infty ds\ s^n \ |\ \ \rho(s\ \ T)|\ . \tag{13}$$

As the spectral functions become degenerate, these moments should decrease, even without a direct relation to chiral order parameters, and thus not allow for oscillatory cancelations. We further introduce pertinent ratios, $r_n = \tilde{w}_n(T)/\tilde{w}_n(T=0)$.

Our analysis proceeds as follows. First the CDSR is evaluated for the vector channel. With a small reduction in the vector dominance coupling (up to a maximum of 7% at 170 MeV), acceptable d_V values are found for all T=0-170 MeV ranging from 0.43% to 0.67%. To satisfy the axial CDSR and the WSRs, the a_1 parameters, with a smooth temperature dependence, are found by minimizing the function

$$f = d^2_{\mathrm{WSR1}} + d^2_{\mathrm{WSR2}} + d^2_A\ . \tag{14}$$

The resulting finite-T axialvector spectral functions are shown in Fig. 4 together with the vector channel. In all cases, the deviations of WSR-1 and WSR-2 are below 0.1%, while d_A remains below 0.6%. Deviations of WSR-3 are much larger percentage-wise, but comparable to the vacuum up to $T\simeq$150 MeV. We find a monotonic decrease with T for each r_n-measure suggesting acceptable deviations even for WSR-3. We therefore conclude that our spectral functions are compatible with both CDSRs and WSRs.

To gauge the uncertainty and demonstrate the selectivity of the sum rules, we present in Fig. 5 left ranges of possible axialvector spectral functions, at a fixed representative temperature T=150 MeV, with certain constraints removed. The largest band, bordered by dashed curves, is the region spanned by requiring d_A=1% for the axialvector CDSR (the band could be larger if all spectral functions with d_A<1% were included). From this collection of spectral functions, we then select those which simultaneously satisfy WSR-1 better than 1% and arrive at the blue region bordered by dotted curves. This indicates that the axialvector CDSRs alone cannot adequately distinguish possible spectral functions. Whereby, the *combination* of the CDSR and WSRs markedly improves the selectivity of the in-medium axialvector spectral function.

An important aspect of our analysis is the systematic temperature evolution toward the restoration point at which a trivial degeneracy occurs, being compatible with our best

Journal of Physics: Conference Series **535** (2014) 012024 doi:10.1088/1742-6596/535/1/012024

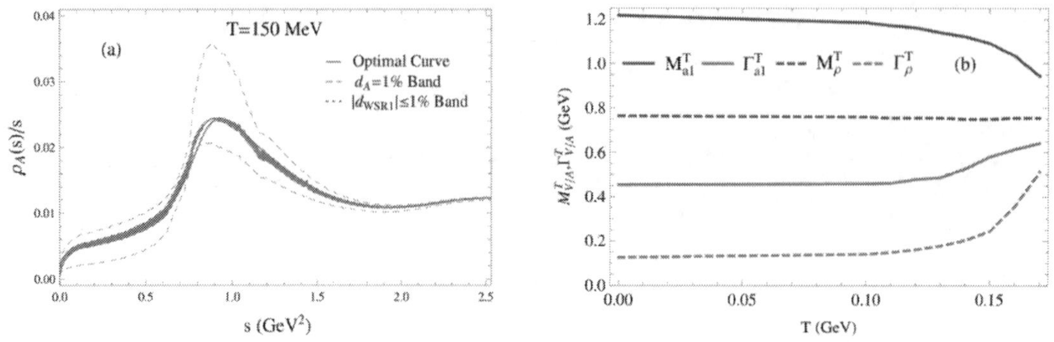

Figure . (Left panel): Allowed bands for the of axialvector spectral functions at T=150 MeV when requiring agreement with the CDSR only at d_A=1% (dashed lines), and additionally with WSR-1 at $|d_{\rm WSR1}|\leq 1\%$ (dotted lines). The solid line corresponds to a minimal f value from Eq. (14). (Right panel): Temperature dependence of the ρ and a_1 mass and width extracted from the spectral functions in Fig. 4.

estimates of the T dependent chiral order parameters and condensates. At T = 170 MeV, the condensates have the largest deviation from lattice data, and our solutions may be more illustrative of a higher temperature where the quark condensate is closer to zero. From this evolution, we can probe the temperature dependence of a_1 parameters.

We finally display in Fig. 5 right the temperature dependence of the ρ and a_1 mass and width. Two fundamental characteristics of the CD phase transition (deconfinement and chiral restoration) are exemplified here. First, the mass and width for each individual state approach each other: a sign of hadronic melting and deconfinement. Second, the two meson masses (and widths) trend towards each other (also supported by a visual inspection of Fig. 4), signaling chiral restoration. The ρ-a_1 merging is largely dictated by the WSRs, but the concrete shape close to chiral restoration is more sensitive to the CDSRs. The approach toward degeneracy at finite mass implies the chiral mass splitting to burn off , leaving an essentially persistent bare masses of $m_0 \simeq 0.8$ GeV, similar to Refs. [8, 50, 51].

. Conclusions

We have investigated implications of the ρ-melting scenario as a signal of chiral restoration in hot and dense matter. To this end, we used in-medium vector spectral functions compatible with experimental dilepton data to assess whether in-medium axialvector spectral functions could be found to simultaneously satisfy the CDSRs and WSRs. Additional input to the sum rules is the temperature dependence of pertinent condensates which was obtained from a hadron-resonance gas calculation and constrained by lattice- CD calculations. We found agreement with the vector CDSR with a small ad ustment of the vector meson dominance coupling. Additionally, by introducing four in-medium spectral parameters for the a_1, axialvector spectral functions were deduced from a combined evaluation of CDSR and WSRs over a temperature range from 0-170 MeV. A key feature of the resulting axialvector spectral functions is an a_1 mass shift towards the ρ peak through a systematic progression in temperature toward chiral degeneracy. Furthermore, the spectral melting of both ρ and a_1 resonances is suggestive for deconfinement. These insights should be sub ected to further scrutiny utilizing more comprehensive microscopic calculations of the in-medium axial-/vector spectral functions. Work in this direction is on-going.

Acknowledgments

This work is supported by the S-NSF under grant No. PH -1306359 and by the A.-v.-Humboldt Foundation (Germany).

References

[1] Borsanyi S *et al.* [Wuppertal-Budapest Collaboration] 2010 *JHEP* **1009** 073
[2] Bazavov A *et al.* 2012 *Phys. Rev.* D **85** 054503
[3] Arnaldi R *et al.* [NA60 Collaboration] 2009 *Eur. Phys. J.* C **61** 711
[4] Adamova D *et al.* [CERES NA45 Collaboration] 2008 *Phys. Lett.* B **666** 425
[5] Geurts F *et al.* [STAR Collaboration] 2013 *Nucl. Phys.* **A904-905** 217c
[6] Rapp R and Wambach J 1999 *Eur. Phys. J.* A **6** 415
[7] Rapp R 2013 *PoS CPOD* **2013** 008
[8] Urban M, Buballa M and Wambach J 2002 *Phys. Rev. Lett.* **88** 042002
[9] Urban M, Buballa M and Wambach J 2002 *Nucl. Phys.* A **697** 338
[10] Parganli a D, Giacosa F and Rischke D H, 2010 *Phys. Rev.* D **82** 054024
[11] Hohler P M and Rapp R 2014 *Phys. Rev.* D **89** 125013
[12] Hohler P M and Rapp R 2014 *Phys. Lett.* B **731** 103
[13] Weinberg S 1967 *Phys. Rev. Lett.* **18** 507
[14] Das T, Mathur V S and kubo S 1967 *Phys. Rev. Lett.* **19** 859
[15] Kapusta J I and Shuryak E V 1994 *Phys. Rev.* D **49** 4694
[16] Shifman M A, Vainshtein A I and akharov V I 1979 *Nucl. Phys.* B **147** 385
[17] Shifman M A, Vainshtein A I and akharov V I 1979 *Nucl. Phys.* B **147** 448
[18] Marco E, Hofmann R and Weise W 2002 *Phys. Lett.* B **530** 88
[19] Holt N P M, Hohler P M and Rapp R 2013 *Phys. Rev.* D **87** 076010
[20] Ayala A, Dominguez C A, Loewe M and hang Y *Preprint* arXiv:1405.2228
[21] Dominguez C A, Loewe M and hang Y 2012 *Phys. Rev.* D **86** 034030
[22] Ayala A, Dominguez C A, Loewe M and hang Y 2012 *Phys. Rev.* D **86** 114036
[23] Hilger T, Kampfer B and Leupold S 2011 *Phys. Rev.* C **84** 045202
[24] Hilger T, Thomas R, Kampfer B and Leupold S 2012 *Phys. Lett.* B **709** 200
[25] Hohler P M and Rapp R 2012 *Nucl. Phys.* A **892** 58
[26] Barate R *et al.* [ALEPH Collaboration] 1998 *Eur. Phys. J.* C **4** 409
[27] Ackerstaff K *et al.* [PAL Collaboration] 1999 *Eur. Phys. J.* C **7** 571
[28] Hatsuda T, Koike Y and Lee S H 1993 *Nucl. Phys.* B **394** 221
[29] Leupold S and Mosel U 1998 *Phys. Rev.* C **58** 2939
[30] schocke S, Pavlenko P and Kämpfer B 2002 *Eur. Phys. J.* A **15** 529
[31] Nakamura K *et al.* (Particle Data Group) 2010 *J. Phys.* G **37** 075021
[32] Karsch F, Redlich K and Taw k A 2003 *Eur. Phys. J.* C **29** 549
[33] Gerber P and Leutwyler H 1989 *Nucl. Phys.* B **321** 387
[34] Leupold S 2006 *J. Phys.* G **32** 2199
[35] Gasser J, Leutwyler H and Sainio M E 1991 *Phys. Lett.* B **253** 252
[36] Jameson I, Thomas A W and Chanfray G 1992 *J. Phys.* G **18** L159
[37] Birse M C and McGovern J A 1992 *Phys. Lett.* B **292** 242
[38] Rapp R 2013 *J. Phys. Conf. Ser.* **420** 012017
[39] Cohen T D, Furnstahl R J and Griegel D K 1992 *Phys. Rev.* C **45** 1881
[40] Leupold S 2005 *Phys. Lett.* B **616** 203
[41] Choi S, Hatsuda T, Koike Y and Lee S H 1993 *Phys. Lett.* B **312** 351
[42] Urban M, Buballa M, Rapp R and Wambach J 1998 *Nucl. Phys.* A **641** 433
[43] Leupold S, Peters W and Mosel U 1998 *Nucl. Phys.* A **628** 311
[44] Leupold S 2001 *Phys. Rev.* C **64** 015202
[45] Dey M, Eletsky V L and Ioffe B L 1990 *Phys. Lett.* B **252** 620
[46] Steele J V, Yamagishi H and ahed I 1996 *Phys. Lett.* B **384** 255
[47] Chanfray G, Delorme J and Ericson M 1998 *Nucl. Phys.* A **637** 421
[48] Krippa B 1998 *Phys. Lett.* B **427** 13
[49] Leinweber D B 1997 *Ann. Phys.* **254** 328
[50] Pisarski R D *Preprint* hep-ph 9503330
[51] Pisarski R D 1995 *Phys. Rev.* D **52** 3773

Journal of Physics: Conference Series **535** (2014) 012025 doi:10.1088/1742-6596/535/1/012025

A Comparison of HBT Measurements for $d + Au$ and $Au + Au$ collision systems at $\sqrt{s_{NN}} = 200$ GeV

Alex Mwai for the PHENIX Collaboration

Department of Chemistry, Stony Brook University, Stony Brook, NY 11794-3800

E-mail: alexmwai@bnl.gov

Abstract. Two-pion Bose-Einstein correlations are a valuable tool for studying the space-time extent of emission sources in $p + A$, $d + A$, and $A + A$ collisions. In recent work, PHENIX has extracted the 3-D HBT radii as a function of centrality and transverse-pair mass (m_T) for the $d + Au$ and $Au + Au$ collision systems. A comparison of the radii for both systems indicate strong similarities in the detailed dependencies on centrality and m_T, suggestive of important final-state re-scattering effects in the reaction dynamics for the $d + Au$ and $Au + Au$ systems. The measurements also point to a smaller freeze-out size and system lifetime in $d + Au$ as compared to $Au + Au$.

1. Introduction

A double-ridge structure has recently been reported for high multiplicity 2-D hadron correlations in $p + p$ and $p + A$ systems at the LHC [4, 1, 27, 22]. The observation of this structure suggests that the $p + p$ and $p + A$ systems might be large enough and be sufficiently long-lived for significant medium effects to form, such as those previously seen in $A + A$ collisions [10, 5].

Two theories have been presented as to the origin of the double-ridge structure: The Color Glass Condensate (CGC) framework argues that the structure is as a result of quantum interference effects between correlated gluons due to gluon saturation at small impact parameter [23, 24]. The 3+1-D viscous hydrodynamic model, on the other hand, suggests that the collective effects observed could be attributed to fluctuations in the initial state that are then carried through the final-state hydrodynamic evolution [20, 32].

Hanbury Brown-Twiss (HBT) interferometry is a useful technique for studying the space-time extent of pion-emitting sources [25, 18]. The HBT measurements dependence on transverse mass and collision centrality in $A + A$ systems have been extensively studied, and are known to show characteristics that point to the influence of final-state re-scattering effects [12, 8, 2]. Similar HBT measurements for $p + p$ and $p + A$ systems can therefore provide a useful and independent check of proposed models. Specifically, observation of common trends in $A + A$ and $p + A$ systems for the HBT measurements would be an indication of the strong role that final-state re-scattering effects has in $p + A$ systems [21]. In these proceedings, we present two-pion HBT measurement results for the $d + Au$ collision system at $\sqrt{s_{NN}} = 200$ GeV using the PHENIX detector. The measurements are studied for their dependence on the transverse mass as well as the collision geometry. These results are also compared with similar measurements done for the $Au + Au$ collision system at $\sqrt{s_{NN}} = 200$ GeV, as well as the $\sqrt{s_{NN}} = 2.76$ TeV $Pb + Pb$ collision system results from LHC.

2. Analysis

This analysis is done using data collected by the PHENIX detector. The $Au + Au$ results are from the 2007 running period while the $d + Au$ results are from the 2008 running period. In total, 3.5 billion events are utilized for $Au + Au$ and 1.8 billion events for $d + Au$. The collision vertex was determined using the Zero Degree Calorimeter (ZDC) and restricted to $|z| < 30$ cm along the beam direction. The collision centrality classes were selected from the distribution of the charge deposited in the Beam Beam Counter (BBC). Charged particle tracking was done using both the Drift Chamber (DC) and the Pad Chambers (PC), with the 3-D trajectories of the particles reconstructed from a model based on hits in the two subsystems [11].

PID in this anaysis was done using the Time-of-Flight detector (ToF) and the Electromagnetic Calorimeter (EMC). Both detectors provide for very good timing resolution ($\sigma_t \sim 110$ psec and $\sigma_t \sim 400$ psec respectively) and sufficient azimuthal angle coverage ($\Delta\phi < \pi/2$) [14, 15]. Charged pions were identified to within 2 standard deviations of their mass squared distribution peak to ensure purity.

The two-particle correlaion function is defined as $C_2(q) = A(q)/B(q)$. Here, $A(q)$ is the same charge pair distribution in relative momentum for pions from the same event while $B(q)$ is a background distribution obtained from pairing same charge pions from different events, but with similar binning in vertex and centrality as the $A(q)$ distribution. Pairs in both distributions were built in vertex bins of 3 cm and centrality bins of 5%. The relative momentum, q, is defined as $\vec{p}_1 - \vec{p}_2$ where $\vec{p}_1$ and $\vec{p}_2$ are the momenta for the particles in the pair. To remove spurious correlations due to effects of track splitting as well as detector resolution limitations, cuts were applied in DC, PC, and EMC to both the $A(q)$ and $B(q)$ distributions. These cuts were rigorously studied for each centrality selection, and applied for each collision system independently. The 3-D correlation functions were projected in the Bertsch-Pratt parameterization [16, 31] such that q_{long} is projected along the beam direction, q_{out} is parallel to the average transverse momentum of the pair (k_T), and q_{side} is orthogonal to q_{out} and q_{long}. k_T is defined as $(\vec{p}_{T,1} + \vec{p}_{T,2})/2$ where $\vec{p}_{T,1}$ and $\vec{p}_{T,2}$ are the transverse momenta of the particles. Figure 1 shows a selection of the correlation functions for $Au + Au$ and $d + Au$ at comparable number of participants (N_{part}) selections (15.7 $\pm$ 1.6 and 16.7 $\pm$ 1.1 respectively) and an average k_T of 0.39 GeV/c.

The obtained correlation functions were studied for their dependence on the expansion dynamics of the system in the form of the transverse mass given as $m_T = \sqrt{k_T + m_\pi}$ where m_π is the pion mass. The correlation functions were also studied for their dependence in the collision centrality in the form of N_{part} and the initial transverse size ($\bar{R}$). These two quantities were determined for each centrality selection using Monte-Carlo Glauber calculations [29, 30] . $\bar{R}$ is obtained from the root-mean-square widths of density distributions in 2-D, σ_x, σ_y, and defined as $1/\bar{R} = \sqrt{(1/\sigma_x^2) + (1/\sigma_y^2)}$ [17]. Since $\bar{R}$ is proportional to τ, the expansion time of the system, it is a more natural scaling variable for the system size as measured in HBT as compared to N_{part} [36, 28]. To allow for a more quantitative comparison of the two systems, the correlation functions were fit with the Sinyukov fit function [19, 37]. This fit function allows for the extraction of the HBT radii while compensating for

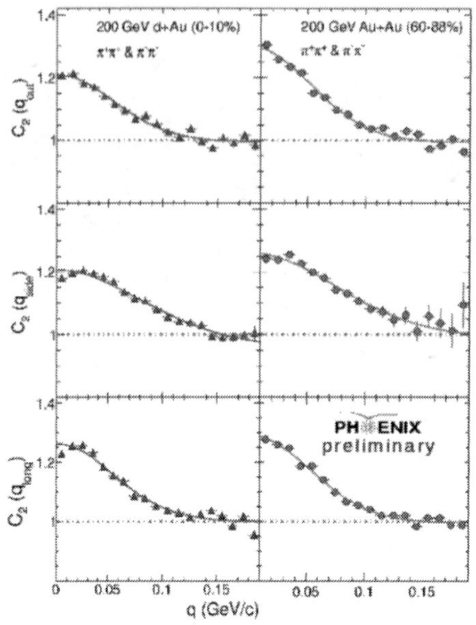

Figure 1. Correlation functions in q for $Au + Au$ and $d + Au$ at comparable N_{part} values and the same average k_T of 0.39 GeV/c.

Final State Interactions (FSI), which in this study are dominated by Coulomb pair repulsion at small q.

$$C_2(q) = N[(\lambda(1 + G(q)))F_c + (1 - \lambda)]$$
$$G(q) \cong exp(-R^2_{side}q^2_{side} - R^2_{out}q^2_{out} - R^2_{long}q^2_{long}) \quad (1)$$

N is a normalization factor and λ is a measure of the chaoticity. F_c is a correction for the FSI. R_{side} carries information about the geometrical transverse size of the system. R_{out} carries information about both the transverse size and the particle emission duration of the system, $\Delta\tau$, while from R_{long}, we can extract information about the lifetime of the system, τ.

3. Results

3.0.1. m_T dependence Figure 2 shows the m_T dependence of R_{out}, R_{side}, and R_{long} for $Au + Au$ and $d + Au$ collision systems at similar N_{part} values. A decreasing trend with an increase in m_T is evident for both collision systems. This trend has previously been shown to be characteristic of an expanding source [26]. To further extract information about the geometrical size (R_{geom}) and lifetime of the system (τ), blastwave model fits [33] were made to the curves for R_{side} and R_{long} with the fit functions:

$$R_{side} = R_{geom}/\sqrt{(1 + \beta^2(m_T/T))}, \quad (2)$$

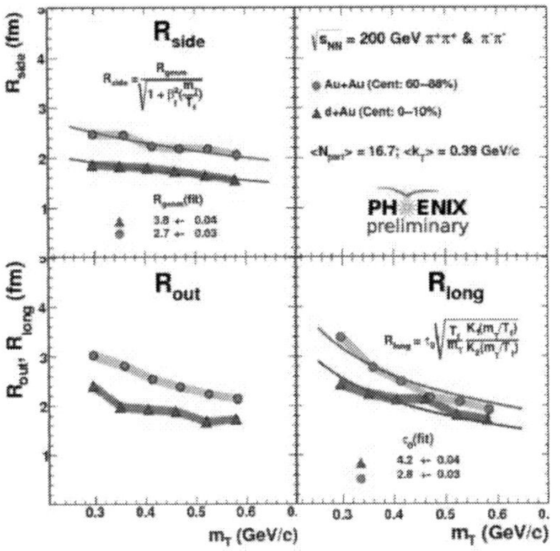

Figure 2. R_{out}, R_{side}, and R_{long} dependence on m_T for central $d+Au$ and peripheral $Au+Au$. Dashed lines are blastwave fits to the curves.

	$d+Au$	$Au+Au$
R_{geom} (fm)	2.7±0.03±0.2(sys)	3.8±0.04±0.2(sys)
τ (fm/c)	2.8±0.03±0.4(sys)	4.2±0.04±0.3(sys)

Table 1. R_{geom} and τ values for $d+Au$ and $Au+Au$ obtained from blastwave model fits to curves for R_{side} and R_{long} in figure 2.

$$R_{long} = \tau\sqrt{(T/m_T)[(K_2(m_T/T))/(K_1(m_T/T))]} \tag{3}$$

T (0.118 ± 0.02 GeV and 0.123 ± 0.02 GeV for $d+Au$ and $Au+Au$ respectively), the freeze-out temperature and β (0.42 ± 0.03 and 0.38 ± 0.08 c for $d+Au$ and $Au+Au$ respectively), the surface velocity are the input parameters. T and β are obtained from blastwave fits to the p_T spectra for identified charged hadrons [6] while K_2 and K_1 are modified Bessel functions. The values given in table 1 show that $d+Au$ has both a smaller transverse size and system lifetime as compared to $Au+Au$.

The smaller size of the $d+Au$ system as compared to the $Au+Au$ can also be seen in the freeze-out volume which is proportional to the product $R_{out} \times R_{side} \times R_{long}$. Figure 3 shows $R_{out} \times R_{side} \times R_{long}$ dependence on m_T for $d+Au$ and $Au+Au$ systems. While $d+Au$ indicates a smaller freeze-out volume as compared to $Au+Au$, the fall-off with m_T for both systems is comparable.

R_{out}/R_{side} has traditionally been used as a proxy for $\Delta\tau$ for a pion-emitting source [13, 35, 34, 7]. Figure 4 shows the R_{out}/R_{side} dependence on m_T for the $d+Au$ and $Au+Au$ systems. The ratio is found to be flat and close to unity, suggesting that the two systems both exhibit a very short emission duration.

Journal of Physics: Conference Series **535** (2014) 012025 doi:10.1088/1742-6596/535/1/012025

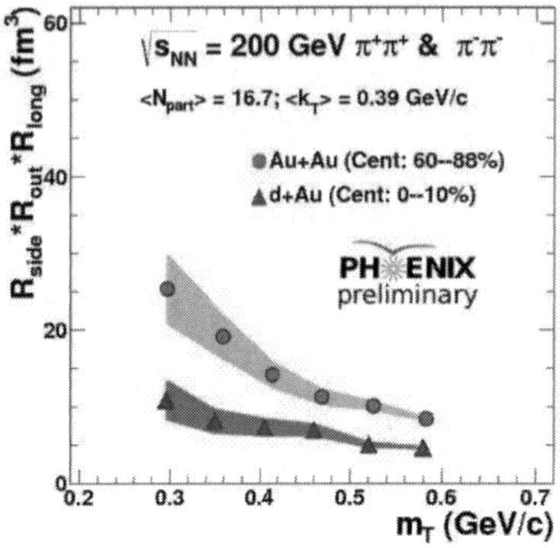

Figure 3. $R_{out} \times R_{side} \times R_{long}$ dependence on m_T for $d + Au$ and $Au + Au$ systems.

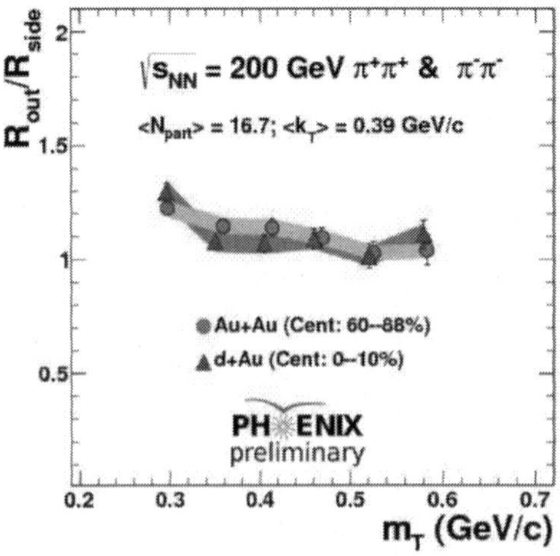

Figure 4. The R_{out}/R_{side} ratio dependence on m_T for central $d + Au$ and peripheral $Au + Au$ collisions.

Journal of Physics: Conference Series **535** (2014) 012025 doi:10.1088/1742-6596/535/1/012025

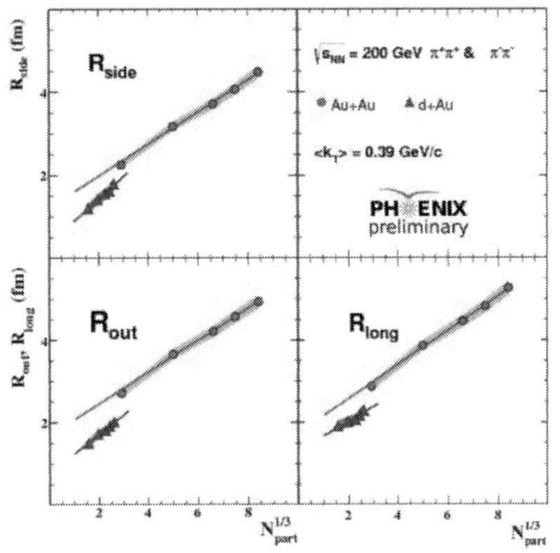

Figure 5. N_{part} dependence of R_{out}, R_{side}, and R_{long} for the $d + Au$ and $Au + Au$ collision systems at $k_T \approx 0.39$ GeV/c. Dashed line are linear fits to the data.

3.0.2. Collision geometry dependence To investigate the collision geometry dependence, R_{out}, R_{side}, and R_{long} were studied for their dependence on N_{part} and $\overline{R}$.

N_{part} dependence The N_{part} dependence for both $d + Au$ and $Au + Au$ collision systems was done at $k_T \approx 0.39$ GeV/c. Figure 5 shows this dependence. A similar linear increase in both R_{out} and R_{side} is observed for both collision systems. R_{long} shows a slight difference in slope which could be due to differences in the longitudinal dynamics for $d + Au$ and $Au + Au$. These results suggest that there is a common strong correlation between the transverse size and the initial geometry in both collision systems. The results also indicate that the dependence of the transverse expansion rate with collision geometry is similar in $d + Au$ and $Au + Au$ systems.

$\overline{R}$ dependence Figure 6 shows the $\overline{R}$ dependence of R_{side} for the $d+Au$ and $Au+Au$ collision systems, as well as ALICE $Pb + Pb$ at 2.76 TeV and STAR $Au + Au$ at 200 GeV [3, 9]. For all systems, R_{side} shows a linear relationship with $\overline{R}$. $d + Au$ and $Au+Au$ collision systems have similar slopes while the $Pb+Pb$ results demonstrate a larger slope which could be attributed to the stronger expansion rate expected at 2.76 TeV. A good match between the 200 GeV $Au+Au$ PHENIX and STAR measurements is also seen in the comparison. These results strongly emphasize the role of final-state re-scattering effects in $d + Au$ which is typical of a hydrodynamic evolution scenario.

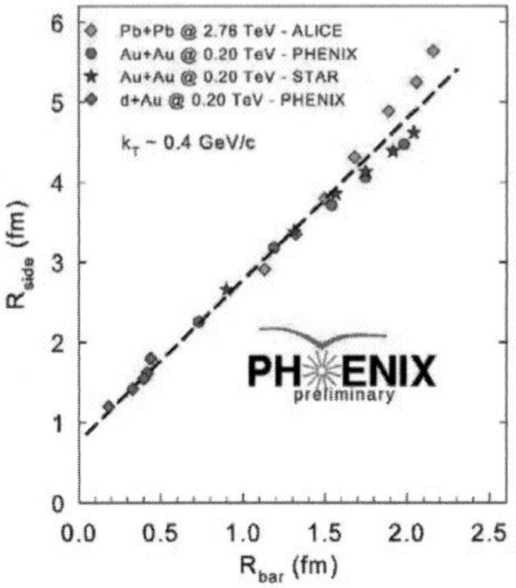

Figure 6. $\overline{R}$ dependence of R_{side} for $d + Au$, $Au + Au$, and $Pb + Pb$ collision systems at $k_T \approx 0.39$ GeV/c [3, 9].

4. Summary and Conclusion

We have, in this proceedings, presented new PHENIX results on HBT measurements for the $d + Au$ and $Au + Au$ collision systems at $\sqrt{s_{NN}} = 200$ GeV. The HBT radii for the two systems show comparable trends both for their m_T and collision centrality dependence. The $d + Au$ system was also found to have both a smaller size and shorter lifetime as compared to the $Au + Au$ system. R_{side} in the $d + Au$ and $Au + Au$ systems, as well as $Pb + Pb$ at $\sqrt{s_{NN}} = 2.76$ TeV, shows a linear relationship with the initial transverse size but with a larger slope for the $Pb + Pb$ collision system. The difference in slope could be due to the larger expansion rate at the higher LHC energy. In conclusion, these results indicate that final-state re-scattering effects play an important role in the $d + Au$ collision system.

References

[1] Georges Aad et al. Observation of Associated Near-side and Away-side Long-range Correlations in $\sqrt{s_{NN}}$=5.02 TeV Proton-lead Collisions with the ATLAS Detector. *Phys.Rev.Lett.*, 110:182302, 2013.

[2] K. Aamodt et al. Two-pion Bose-Einstein correlations in central Pb-Pb collisions at $sqrt(s_N N) = 2.76$ TeV. *Phys.Lett.*, B696:328–337, 2011.

[3] K. Aamodt et al. Two-pion Bose-Einstein correlations in central Pb-Pb collisions at $sqrt(s_N N) = 2.76$ TeV. *Phys.Lett.*, B696:328–337, 2011.

[4] Betty Abelev et al. Long-range angular correlations on the near and away side in p-Pb collisions at $\sqrt{s_{NN}} = 5.02$ TeV. *Phys.Lett.*, B719:29–41, 2013.

[5] B.I. Abelev et al. Long range rapidity correlations and jet production in high energy nuclear collisions. *Phys.Rev.*, C80:064912, 2009.

[6] B.I. Abelev et al. Systematic Measurements of Identified Particle Spectra in pp, d^+ Au and Au+Au Collisions from STAR. *Phys.Rev.*, C79:034909, 2009.

[7] J. Adams et al. Azimuthally sensitive HBT in Au + Au collisions at s(NN)**(1/2) = 200-GeV. *Phys.Rev.Lett.*, 93:012301, 2004.

[8] J. Adams et al. Pion interferometry in Au+Au collisions at S(NN)**(1/2) = 200-GeV. *Phys.Rev.*, C71:044906, 2005.

[9] J. Adams et al. Pion interferometry in Au+Au collisions at S(NN)**(1/2) = 200-GeV. *Phys.Rev.*, C71:044906, 2005.

[10] A. Adare et al. Dihadron azimuthal correlations in Au+Au collisions at s(NN)**(1/2) = 200-GeV. *Phys.Rev.*, C78:014901, 2008.

[11] K. Adcox et al. PHENIX central arm tracking detectors. *Nucl.Instrum.Meth.*, A499:489–507, 2003.

[12] S.S. Adler et al. Bose-Einstein correlations of charged pion pairs in Au + Au collisions at s(NN)**(1/2) = 200-GeV. *Phys.Rev.Lett.*, 93:152302, 2004.

[13] S.S. Adler et al. Bose-Einstein correlations of charged pion pairs in Au + Au collisions at s(NN)**(1/2) = 200-GeV. *Phys.Rev.Lett.*, 93:152302, 2004.

[14] M. Aizawa et al. PHENIX central arm particle ID detectors. *Nucl.Instrum.Meth.*, A499:508–520, 2003.

[15] L. Aphecetche et al. PHENIX calorimeter. *Nucl.Instrum.Meth.*, A499:521–536, 2003.

[16] G.F. Bertsch. Pion Interferometry as a Probe of the Plasma. *Nucl.Phys.*, A498:173C–180C, 1989.

[17] R.S. Bhalerao, Jean-Paul Blaizot, Nicolas Borghini, and Jean-Yves Ollitrault. Elliptic flow and incomplete equilibration at RHIC. *Phys.Lett.*, B627:49–54, 2005.

[18] David H. Boal, Claus-Konrad Gelbke, and Byron K. Jennings. Intensity interferometry in subatomic physics. *Rev. Mod. Phys.*, 62:553–602, Jul 1990.

[19] M.G. Bowler. Coulomb corrections to Bose-Einstein correlations have been greatly exaggerated. *Phys.Lett.*, B270:69–74, 1991.

[20] Piotr Bozek. Collective flow in p-Pb and d-Pd collisions at TeV energies. *Phys.Rev.*, C85:014911, 2012.

[21] Piotr Bozek and Wojciech Broniowski. Size of the emission source and collectivity in ultra-relativistic p-Pb collisions. *Phys.Lett.*, B720:250–253, 2013.

[22] Serguei Chatrchyan et al. Observation of long-range near-side angular correlations in proton-lead collisions at the LHC. *Phys.Lett.*, B718:795–814, 2013.

[23] Kevin Dusling and Raju Venugopalan. Azimuthal collimation of long range rapidity correlations by strong color fields in high multiplicity hadron-hadron collisions. *Phys.Rev.Lett.*, 108:262001, 2012.

[24] Kevin Dusling and Raju Venugopalan. Comparison of the color glass condensate to dihadron correlations in proton-proton and proton-nucleus collisions. *Phys.Rev.*, D87(9):094034, 2013.

[25] R. Hanbury Brown and R. Q. Twiss. A Test of a New Type of Stellar Interferometer on Sirius. *Nature*, 178(4541):1046–1048, November 1956.

[26] M. Herrmann and G.F. Bertsch. Source dimensions in ultrarelativistic heavy ion collisions. *Phys.Rev.*, C51:328–338, 1995.

[27] Vardan Khachatryan et al. Observation of Long-Range Near-Side Angular Correlations in Proton-Proton Collisions at the LHC. *JHEP*, 1009:091, 2010.

[28] Roy A. Lacey, Yi Gu, X. Gong, D. Reynolds, N.N. Ajitanand, et al. Is anisotropic flow really acoustic? 2013.

[29] Roy A. Lacey, Rui Wei, N.N. Ajitanand, and A. Taranenko. Initial eccentricity fluctuations and their relation to higher-order flow harmonics. *Phys.Rev.*, C83:044902, 2011.

[30] Michael L. Miller, Klaus Reygers, Stephen J. Sanders, and Peter Steinberg. Glauber modeling in high energy nuclear collisions. *Ann.Rev.Nucl.Part.Sci.*, 57:205–243, 2007.

[31] S. Pratt. Pion Interferometry of Quark-Gluon Plasma. *Phys.Rev.*, D33:1314–1327, 1986.

[32] Guang-You Qin and Berndt MÃŒller. Elliptic and triangular flow anisotropy in deuteron-gold collisions at RHIC and proton-lead collisions at the LHC. *Phys.Rev.*, C89:044902, 2014.

[33] Fabrice Retiere and Michael Annan Lisa. Observable implications of geometrical and dynamical aspects of freeze out in heavy ion collisions. *Phys.Rev.*, C70:044907, 2004.

[34] Dirk H. Rischke. Hydrodynamics and collective behavior in relativistic nuclear collisions. *Nucl.Phys.*, A610:88C–101C, 1996.

[35] Dirk H. Rischke and Miklos Gyulassy. The Time delay signature of quark - gluon plasma formation in relativistic nuclear collisions. *Nucl.Phys.*, A608:479–512, 1996.

[36] Edward Shuryak and Ismail Zahed. High-multiplicity pp and pA collisions: Hydrodynamics at its edge. *Phys.Rev.*, C88(4):044915, 2013.

[37] Yu. Sinyukov, R. Lednicky, S.V. Akkelin, J. Pluta, and B. Erazmus. Coulomb corrections for interferometry analysis of expanding hadron systems. *Phys.Lett.*, B432:248–257, 1998.

Journal of Physics: Conference Series **535** (2014) 012026 doi:10.1088/1742-6596/535/1/012026

Particle production and final state effects in nuclear collisions

Charles Gale[1], Sangyong Jeon[1], Björn Schenke[2], Prithwish Tribedy[3] and Raju Venugopalan[2]

[1] Department of Physics, McGill University, 3600 University Street, Montreal, Quebec, H3A 2T8, Canada

[2] Physics Department, Brookhaven National Laboratory, Upton, NY 11973, USA

[3] Variable Energy Cyclotron Centre, 1/AF Bidhan Nagar, Kolkata 700064, India

E-mail: bschenke@bnl.gov

Abstract. We discuss various features of multi-particle production in high-energy nuclear collisions within the IP-Glasma model. We demonstrate that for some observables in heavy-ion collisions the effects of final state interactions governed by fluid-dynamics are essential. In smaller systems, like proton-lead collisions, the same model fails to describe the azimuthal anisotropy of produced particles. This failure can be due to neglected initial state correlations or the lack of a detailed description of the fluctuating spatial structure of the proton at high energies, or both.

1. Introduction

For the physical interpretation of observables in heavy-ion, proton/deuteron-heavy-ion and (high-multiplicity) proton+proton collisions, it is essential to develop a sophisticated description of the multi-particle production mechanism and event-by-event fluctuations. In the high energy limit the color-glass-condensate (CGC) framework [1] is the proper effective theory of quantum-chromodynamics (QCD) that provides such description.

One particular implementation of the color-glass condensate picture is the IP-Glasma model [2, 3]. It combines the IP-Sat dipole model [4, 5, 6] with the classical dynamics of gluon fields created in the collision [7, 8, 9]. One of its advantages over more fundamental approaches is its ability to model the spatial distribution, or impact parameter dependence, of hadrons and nuclei. With parameters of the initial lumpy gluon distributions constrained by inclusive and diffractive DIS data from e+p scattering at HERA, the IP-Glasma model correctly describes the bulk features of various systems over a wide range of energies [10].

After briefly introducing the IP-Glasma model and the employed relativistic fluid dynamic simulation MUSIC in Section 2, we highlight various important features of the model that manifest in observables in different collision systems and compare to the well known Monte-Carlo (MC) Glauber model. These range from the correct description of the event-by-event distributions of flow coefficients v_n in heavy-ion collisions (Section 3), to the correlation of multiplicity and geometry in ultra-central collisions of deformed nuclei (Section 4) to the prediction of relative system sizes in p+p, p+A, and heavy-ion collisions (Section 5). We close by discussing shortcomings of the model that become prominent in p+A collisions where the predicted v_n coefficients are significantly smaller than the experimental data (Section 6).

2. IP-Glasma + MUSIC

The IP-Glasma model relates the deeply inelastic scattering (DIS) constrained nuclear dipole cross-sections to the initial classical dynamics of highly occupied gluon Glasma fields after the nuclear collision. Given an initial distribution of color charges in the high energy nuclear wave-functions, the IP-Glasma framework computes the strong early time multiple scattering of gluon fields by event-by-event solutions of Yang-Mills equations. It is more general and accurate relative to "k_T-factorization" models often used in the small x literature, especially for the soft dynamics that characterizes hydrodynamic flow, because it does not assume at least one nucleus to be dilute and non-linear effects in the interactions of strong gluon fields are taken into account.

The IP-Glasma model naturally includes the effect of several sources of quantum fluctuations: fluctuating distributions of nucleons in the nuclear wave-functions and intrinsic fluctuations of the color charge distribution. This, combined with Lorentz contraction, results in lumpy transverse projections of the gluon field configurations that vary event to event. The scale of this lumpiness is given on average by the nuclear saturation scale Q_s which corresponds to distance scales smaller than the nucleon size [6].

A detailed description of the IP-Glasma model can be found in [2, 3, 10]. For the results shown in this work, the same model parameters as in [10] were used.

The IP-Glasma model provides the gluon fields' energy-momentum tensor $T^{\mu\nu}$ at every transverse position and at a given time after the collision, from which the energy density ε and flow velocities u^μ are extracted via the relation $u_\mu T^{\mu\nu} = \varepsilon u^\nu$. This provides the initial conditions for fluid dynamic simulations. In the results presented below, the viscous part of the energy momentum tensor is set to zero at the initial time of the fluid dynamic simulation. This is done because the gluon field $T^{\mu\nu}$ is very anisotropic (the longitudinal pressure is approximately zero) and without a mechanism for isotropization, which possibly could be provided by fluctuation triggered instabilities in 3+1 dimensional classical statistical simulations [11, 12], we assume instant isotropization (and thermalization) at the time of switching to fluid dynamics.

We employ the viscous relativistic fluid dynamic simulation MUSIC [13, 14, 15], which is a 3+1 dimensional simulation. However, because the initial conditions from the IP-Glasma model are boost-invariant it is used in its 2+1 dimensional mode.

3. Event-by-event flow distributions in heavy-ion collisions

Flow harmonics v_n are defined via the Fourier decomposition of the azimuthal particle distribution

$$dN/d\phi \propto 1 + 2\sum_{n=1}^{\infty} v_n \cos(n - \phi_n) \qquad (1)$$

where ϕ_n represents the phase (event plane) of the nth-order harmonic.

Their event-by-event distributions in Pb+Pb collisions at $\sqrt{s} = 2.76$ TeV have been measured by the ATLAS collaboration [16]. Comparisons of the experimental data to event-by-event fluid dynamic calculations have demonstrated that only IP-Glasma initial conditions can describe the distributions over the entire centrality range [17, 18, 19]. In Fig. 1 we present the maximally peripheral bins reported by the ATLAS collaboration for distributions of v_2, v_3, and v_4. In addition to the scaled v_n distributions we show scaled eccentricity ε_n distributions determined from the initial energy density distribution in the transverse plane. They show fairly good agreement with the v_n distributions, in particular for $n = 3$. However, differences appear in the large v_n-tails for $n = 2$ and $n = 4$. This observation underlines the importance of final state fluid dynamic evolution and its non-linearities that allow a coupling of different n components.

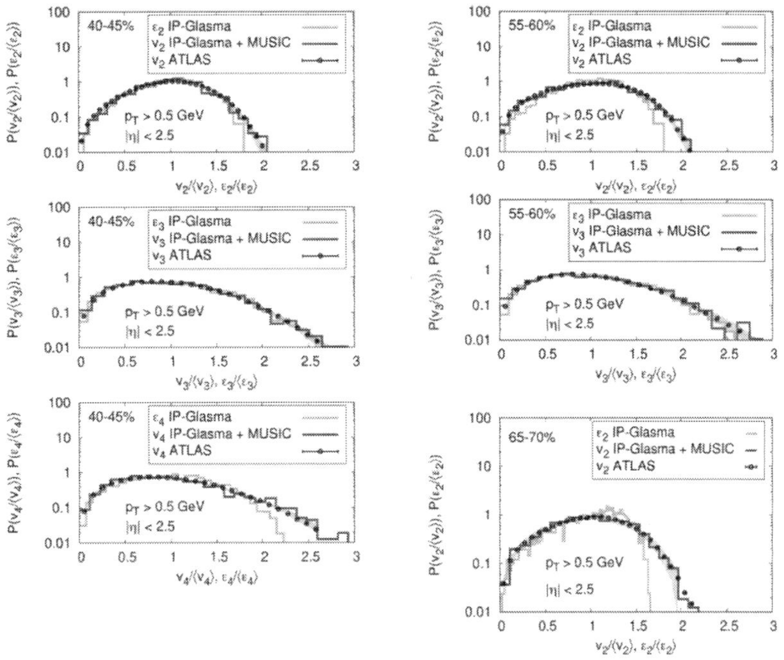

Figure 1. Event-by-event distributions of v_n in peripheral collisions. Shown are comparisons of the IP-Glasma+MUSIC calculations (of v_n and ε_n) to experimental data from the ATLAS collaboration [16].

The results indicate that the fluctuations relevant for event-by-event fluctuations of flow harmonics are very well reproduced in the IP-Glasma model. These are both multiplicity fluctuations that enter via centrality selection, and geometry fluctuations. The former are more strongly affected by color charge fluctuations than the latter. The length scales of the color charge fluctuations are short (of order $1/Q_s$) and are thus not very important for geometric fluctuations probed by v_2, v_3 or v_4. Higher harmonics would be needed to resolve such length scales, but viscous effects dampen them strongly.

4. Multiplicity-geometry correlations in collisions of deformed nuclei

Ultra-central collisions of deformed nuclei such as uranium allow to study the mechanism of particle production via correlations of the multiplicity and the geometry [20]. In particular, after selecting very central events based on the number of spectator neutrons in the forward (and backward) direction, one can correlate the measured v_2 with the multiplicity.

A simple two-component Glauber model [21] predicts for collisions of two prolate ^{238}U nuclei that the eccentricity of the interaction region decreases with increasing multiplicity for very central events. This is due to the contribution of the number of collisions $N_{\rm coll}$ to the multiplicity in this model. If the uranium nuclei collide with their elongated direction aligned with the beam line (so called tip-tip collisions) the number of collisions is larger than if they collide with the shorter direction aligned with the beam line (side-side or body-body collisions). At the same time, tip-tip collisions produce an almost perfectly round interaction region, while body-body collisions generate larger eccentricities on average because of the deformation. For collisions of

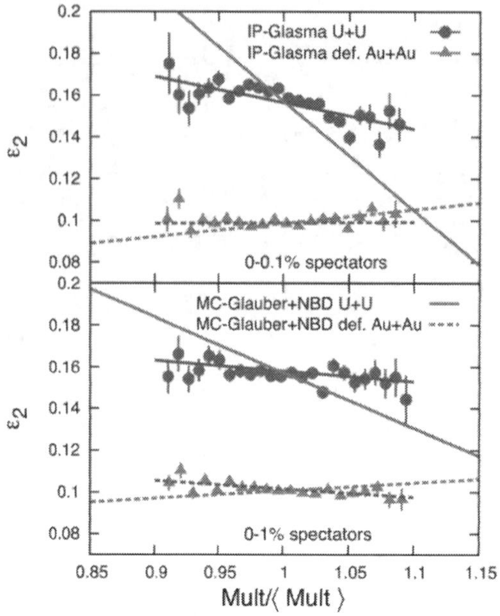

Figure 2. Correlation between the multiplicity (scaled by the mean multiplicity in the given bin) and the eccentricity of the interaction region in the transverse plane. In the IP-Glasma model a weak anti-correlation is found for both U+U and Au+Au collisions, while in the two component MC-Glauber model a stronger anti-correlation is found for U+U collisions and a positive correlation for Au+Au collisions. Figure from [22].

oblate gold nuclei the argument above turns around and a positive correlation between v_2 and multiplicity is predicted [20].

In the IP-Glasma model there is no explicit dependence of the multiplicity on $N_{\rm part}$ and $N_{\rm coll}$. There is, however a weak logarithmic dependence on the orientation of deformed nuclei that is related to the relevant scale Q_s and the running of the strong coupling constant. In [22] it was shown that this leads to a weaker anti-correlation of ε_2 and multiplicity in U+U collisions, and to a very weak anti-correlation (as opposed to a positive correlation) in Au+Au collisions. We demonstrate this in Fig. 2.

5. System sizes in various collision systems

Another interesting aspect, in particular when interpreting various experimental observations as emerging from final state collective dynamics, is the study of the system size in different collision systems. We define the initial size of the system $r_{\rm max}$ as the (angle-averaged) radius where the energy density reaches a minimal threshold $\varepsilon_{\rm min}$. The choice of $\varepsilon_{\rm min}$ only affects the overall normalization of $r_{\rm max}$; it does not affect the dependence of $r_{\rm max}$ on the number of charged particles $N_{\rm ch}$ [23].

In this way we can study within the IP-Glasma model how the system size depends on the collision system for a given multiplicity. We compare $r_{\rm max}$ (for $\varepsilon_{\rm min} = \Lambda_{\rm QCD}^4 = (0.2\,{\rm GeV})^4$) as a function of the third root of the charged particle multiplicity in p+p, p+Pb, and Pb+Pb collisions in Fig. 3. The lower end of the band is the initial size, the upper end the maximal size during fluid dynamic evolution. We find that initial sizes at a given multiplicity in p+p and p+Pb collisions are similar, while Pb+Pb collisions are initially larger. When applying fluid

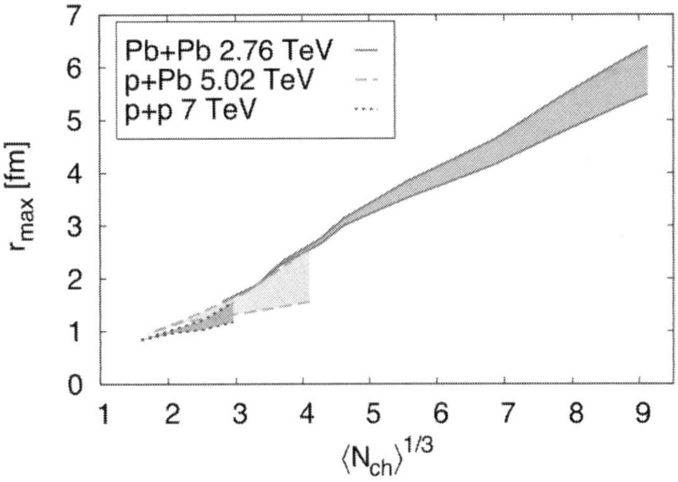

Figure 3. $r_{\max}$ determined using the IP-Glasma model and fluid dynamic expansion. The lower end of the band indicates the size of the initial state, the upper end the maximal value of $r_{\max}$ during the hydrodynamic evolution. Figure from [19].

dynamics, this leads to a more violent expansion of the smaller (higher energy-density) system, as can be seen when comparing the width of the band of p+Pb and Pb+Pb collisions at e.g. $\langle N_{\rm ch}\rangle^{1/3} = 4$.

6. Hints towards complex geometries in high-multiplicity p+A collisions

In [19] it was established that the IP-Glasma+MUSIC model does a good job in describing experimental data from heavy-ion collisions out to fairly peripheral centrality bins. One can then ask how the model does in describing data from p+A collisions that produce similar multiplicities. Fig. 4 shows the multiplicity dependence of v_2 and v_3 in peripheral Pb+Pb collisions and (high-multiplicity) p+Pb collisions with comparable multiplicity. While the agreement with experimental data from the CMS collaboration [24] is fairly good in the Pb+Pb case, for p+Pb collisions, however, v_2 is significantly under-predicted. v_3 agrees for the lower multiplicity studied, but has a rather flat multiplicity dependence and underestimates the experimental data largely at the higher multiplicities.

The reason for this disagreement can be two-fold [19]. On the one hand, we are neglecting all initial state correlations that lead to a v_2 [25, 26, 27, 28, 29] and when including fluctuations should also produce odd harmonics. On the other hand, the description of the proton in the IP-Glasma model is somewhat over-simplified. The proton shape is approximated by a sphere, and any deviations from that are due to the small scale color charge fluctuations. The interaction region in p+Pb collisions is dominated by the shape of the smaller projectile, i.e., the proton. This leads to initial geometries in p+Pb collisions that have very small eccentricities and thus lead to small v_n coefficients via fluid dynamic expansion. If there is indeed a significant contribution to v_n coefficients in p+Pb collisions from collective effects, this indicates that the shape of the proton fluctuates significantly more than assumed in the IP-Glasma model. One could envision a description where the small x gluon distributions are still concentrated around large x valence quark positions, which would lead to much larger eccentricities and fluctuations.

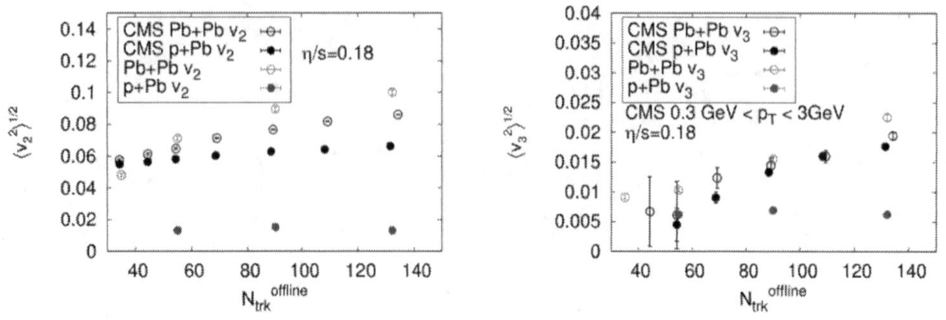

Figure 4. Multiplicity dependence of the rms v_2 (left) and v_3 (right) in Pb+Pb (open symbols) and p+Pb collisions (filled symbols) from the IP-Glasma+MUSIC model (red) compared to experimental data (black) by the CMS collaboration [24]. Figure adapted from [19].

In this case p+Pb collisions could be used to determine the shape and fluctuations of gluon distributions in the proton at high energies.

7. Conclusions

We have presented various important features of the IP-Glasma+MUSIC model, including the correct description of event-by-event flow harmonics in heavy-ion collisions, the prediction of geometry-multiplicity correlations in ultra-central collisions of deformed nuclei, and the multiplicity dependence of the system size in various collision systems. We have further demonstrated that within the IP-Glasma model, the measured azimuthal anisotropy of produced particles in p+Pb collisions is not well reproduced. We argue that this may be due to missing initial state correlations or the lack of a detailed enough description of the fluctuating gluon distribution in the proton. These details are most likely irrelevant for (not too peripheral) heavy-ion collisions, because initial state correlations can easily be destroyed by the strong final state interactions, and the fluctuations of nucleon positions dominate the geometry. The detailed sub-structure of each nucleon is much less important than in a p+A or p+p collision.

An improved description of the proton's spatial structure is highly desirable to determine the origin of the large azimuthal anisotropies measured in p+A (and d+A) collisions.

Acknowledgments

This work was supported under DOE Contract No. DE-AC02-98CH10886.

[1] Gelis F, Iancu E, Jalilian-Marian J and Venugopalan R 2010 *Ann.Rev.Nucl.Part.Sci.* **60** 463–489
[2] Schenke B, Tribedy P and Venugopalan R 2012 *Phys. Rev. Lett.* **108** 252301
[3] Schenke B, Tribedy P and Venugopalan R 2012 *Phys. Rev.* **C86** 034908
[4] Bartels J, Golec-Biernat K J and Kowalski H 2002 *Phys. Rev.* **D66** 014001
[5] Kowalski H and Teaney D 2003 *Phys. Rev.* **D68** 114005
[6] Kowalski H, Lappi T and Venugopalan R 2008 *Phys. Rev. Lett.* **100** 022303
[7] Krasnitz A and Venugopalan R 2000 *Phys. Rev. Lett.* **84** 4309–4312
[8] Krasnitz A and Venugopalan R 2001 *Phys. Rev. Lett.* **86** 1717–1720
[9] Lappi T 2003 *Phys. Rev.* **C67** 054903
[10] Schenke B, Tribedy P and Venugopalan R 2014 *Phys.Rev.* **C89** 024901 (*Preprint* 1311.3636)
[11] Berges J, Boguslavski K, Schlichting S and Venugopalan R 2014 *Phys.Rev.* **D89** 074011 (*Preprint* 1303.5650)
[12] Epelbaum T and Gelis F 2013 *Phys.Rev.Lett.* **111** 232301 (*Preprint* 1307.2214)

[13] Schenke B, Jeon S and Gale C 2010 *Phys. Rev.* **C82** 014903
[14] Schenke B, Jeon S and Gale C 2011 *Phys. Rev. Lett.* **106** 042301
[15] Schenke B, Jeon S and Gale C 2011 *Phys. Rev.* **C85** 024901
[16] Aad G *et al.* (ATLAS Collaboration) 2013 *JHEP* **1311** 183 (*Preprint* 1305.2942)
[17] Renk T and Niemi H 2014 *Phys.Rev.* **C89** 064907 (*Preprint* 1401.2069)
[18] Gale C, Jeon S, Schenke B, Tribedy P and Venugopalan R 2013 *Phys.Rev.Lett.* **110** 012302 (*Preprint* 1209.6330)
[19] Schenke B and Venugopalan R 2014 (*Preprint* 1405.3605)
[20] Wang H and Sorensen P (STAR Collaboration) 2014 (*Preprint* 1406.7522)
[21] Kharzeev D and Nardi M 2001 *Phys. Lett.* **B507** 121–128
[22] Schenke B, Tribedy P and Venugopalan R 2014 *Phys. Rev. C* **89** 064908
[23] Bzdak A, Schenke B, Tribedy P and Venugopalan R 2013 (*Preprint* 1304.3403)
[24] Chatrchyan S *et al.* (CMS Collaboration) 2013 *Phys.Lett.* **B724** 213–240 (*Preprint* 1305.0609)
[25] Dumitru A, Dusling K, Gelis F, Jalilian-Marian J, Lappi T and Venugopalan R 2011 *Phys. Lett.* **B697** 21–25 (*Preprint* 1009.5295)
[26] Dusling K and Venugopalan R 2012 *Phys.Rev.Lett.* **108** 262001 (*Preprint* 1201.2658)
[27] Dusling K and Venugopalan R 2013 *Phys.Rev.* **D87** 051502 (*Preprint* 1210.3890)
[28] Dusling K and Venugopalan R 2013 *Phys.Rev.* **D87** 054014 (*Preprint* 1211.3701)
[29] Dusling K and Venugopalan R 2013 *Phys.Rev.* **D87** 094034 (*Preprint* 1302.7018)

Measurement of D-meson production in pp, p–Pb, and Pb–Pb collisions at the LHC with the ALICE detector

Chitrasen Jena, for the ALICE Collaboration
Università degli Studi di Padova and INFN Sezione di Padova
E-mail: `jena@pd.infn.it`

Abstract. Heavy quarks are a powerful probe for investigating the properties of the Quark-Gluon Plasma created in heavy-ion collisions, since they are produced in initial hard scattering processes and experience all the stages of the medium evolution. ALICE has measured the production of D^0, D^+, D^{*+}, and D_s^+ mesons at central rapidity in their hadronic decay channels in various collision systems and energies. We present recent results for D-meson production measured by the ALICE Collaboration in pp collisions at $\sqrt{s} = 7$ and 2.76 TeV, Pb–Pb collisions at $\sqrt{s_{NN}} = 2.76$ TeV and p–Pb collisions at $\sqrt{s_{NN}} = 5.02$ TeV.

1. Introduction

The measurement of heavy-flavour production in various collision systems and energies offers the opportunity to investigate the properties of the hot and dense QCD matter, the Quark-Gluon Plasma (QGP), created in high energy heavy-ion collisions. Due to their large masses, charm and beauty quarks are produced predominantly in hard parton scattering processes in the early stage of the collision. They interact with the medium created in the collision and lose energy via both inelastic (medium-induced gluon radiation, or radiative energy loss) [1, 2] and elastic (collisional energy loss) [3, 4] processes. Theoretical calculations for energy loss predict a dependence on the colour charge and mass of the parton propagating through the medium. In QCD, quarks have a smaller colour coupling factor with respect to gluons, so that the energy loss for quarks is expected to be smaller than for gluons. The dead-cone effect should reduce small-angle gluon radiation for heavy quarks with moderate energy-over-mass values and hence the radiative energy loss for heavy quarks is expected to be smaller than for light ones [5]. In addition, collisional energy loss is expected to be reduced for heavier quarks, because the spatial diffusion coefficient that regulates the momentum transfer with the constituents of the medium is expected to scale as $1/m_q$ [6].

A sensitive observable to characterize the effect of the dense medium on the measured heavy-flavour hadron yields is the nuclear modification factor, defined as

$$R_{AA}(p_T) = \frac{1}{\langle T_{AA}\rangle} \frac{dN_{AA}/dp_T}{d\sigma_{pp}/dp_T}, \tag{1}$$

where $\langle T_{AA}\rangle$ is the average nuclear overlap function estimated using the Glauber model [7], dN_{AA}/dp_T is the transverse momentum differential yield measured in nucleus-nucleus collisions and $d\sigma_{pp}/dp_T$ is the p_T-differential production cross section in pp collisions. In-medium energy loss results in a suppression, $R_{AA} < 1$, of hadron yields at moderate to high transverse momentum. The dependence of the energy loss on the parton species and mass can be investigated by comparing the nuclear modification factors of hadrons with charm (R^D_{AA}) and beauty (R^B_{AA}) with that of pions (R^π_{AA}), mostly originating from gluon fragmentation at LHC energies. The expected dependence of the energy loss on the parton species should determine a difference in their R_{AA}. However, it is important to note that the comparison of heavy-flavour hadron and pion R_{AA} cannot be interpreted directly as a comparison of charm, beauty, and gluon energy losses, due to the different parton fragmentation functions and slopes of the p_T-differential cross section. For example, the theoretical calculations in Ref. [8] shows similar R_{AA} for charmed hadrons and pions despite the different energy loss of c quarks, light quarks and gluons. Moreover, at low p_T, a significant fraction of pions are produced from soft processes, not hard-scatterings.

The measurement of anisotropy in the azimuthal distribution of particle momenta provides further insight into the properties of the QGP. In heavy-ion collisions, anisotropic patterns of particle production originate from the initial anisotropy in the spatial distribution of the nucleons participating in the collision. The azimuthal anisotropy of produced particles is characterized by the Fourier coefficients $v_n = \langle \cos(n(\varphi - \Psi_{RP})) \rangle$, where n is the order of the harmonic, φ the azimuthal angle of the particle and Ψ_{RP} is the azimuthal angle of the reaction plane. The reaction plane is spanned by the vector of the impact parameter and the beam direction. For non-central collisions, the overlap region of the colliding nuclei has a lenticular shape and the azimuthal anisotropy is dominated by the second Fourier coefficient v_2, commonly denoted elliptic flow. At low and intermediate p_T, the v_2 of heavy-flavour hadrons is sensitive to the degree of thermalisation of charm quarks in the medium [9], while at higher p_T it allows the study of the path-length dependence of in-medium parton energy loss [10].

The measurement of charm and beauty production cross sections in pp collisions at the LHC provides an important test of perturbative QCD (pQCD) calculations and serves as an essential reference for understanding the results from heavy-ion collisions [11]. In addition, heavy-flavour production as a function of charged-particle multiplicity in pp collisions could provide insight into the relevance of multiple hard scattering processes. Initial-state effects, such as the modification of parton distribution functions in the nucleus and momentum broadening due to parton scattering in the nucleus, affect heavy-quark production and cannot be accounted for with a reference from pp data. It is therefore necessary to study p–Pb collisions, where an extended hot and dense strongly-interacting medium is not expected to form, to quantify these initial-state effects.

In these proceedings, we will discuss the ALICE measurements of D-meson production in pp collisions at $\sqrt{s} = 7$ TeV, Pb–Pb collisions at $\sqrt{s_{NN}} = 2.76$ TeV, and p–Pb collisions at $\sqrt{s_{NN}} = 5.02$ TeV.

2. D-meson decay reconstruction with ALICE

The D-meson decays are reconstructed in the central rapidity region utilizing the tracking and particle identification capabilities of the ALICE central barrel detectors. The central barrel detectors are surrounded by a large solenoidal magnet, which provides a magnetic field of 0.5 T along the beam direction. A detailed description of the ALICE apparatus is given in Ref. [12].

The measurement of charm production is performed by reconstructing D^0, D^+, D^{*+} and D^+_s charmed hadrons via their hadronic decays $D^0 \to K^-\pi^+$(with branching ratio, BR, of $3.88 \pm 0.05\%$), $D^+ \to K^-\pi^+\pi^+$(BR of $9.13 \pm 0.19\%$), $D^{*+} \to D^0\pi^+$(BR of $67.7 \pm 0.5\%$) with $D^0 \to K^-\pi^+$ and $D^+_s \to \phi\pi^+ \to K^+K^-\pi^+$(BR of $2.28 \pm 0.12\%$) and their charge conjugates [13]. The D^0, D^+, and D^+_s mesons have mean proper decay lengths $c\tau \approx 123$, 312, and 150 μm, respectively [13]. Therefore, their secondary decay vertices are typically displaced by a few hundred μm from the primary interaction vertex. The analysis strategy for the D^0, D^+, and D_s^+ is based on the reconstruction and selection of secondary vertex topologies with significant separation from the primary vertex. In the case of the D^{*+}, which decays strongly, the secondary vertex topology of the daughter D^0 is reconstructed. The topological reconstruction of the decays allows for an efficient rejection of the combinatorial background from uncorrelated tracks. The Time Projection Chamber (TPC) and the Inner Tracking System (ITS) detectors, in particular the two innermost ITS layers equipped with Silicon Pixel Detectors (SPD), provide a resolution of the track impact parameter of about 60 μm for tracks with $p_T = 1$ GeV/c in Pb–Pb collisions. In order to further suppress the combinatorial background, measurements of the particle time-of-flight from the collision point to the Time Of Flight (TOF) detector and the specific energy loss in the TPC gas are used to identify kaons and pions. Finally, the signal yield is extracted by fitting the invariant mass distribution with a Gaussian function for the signal and an exponential function for the background shape for D^0, D^+, and D_s^+. The D^{*+} signal is calculated by fitting the invariant-mass difference $\Delta m = m_{D^{*+}} - m_{D^0}$ with a function that consists of a Gaussian term describing the signal and the term $a\sqrt{\Delta m - m_\pi} \cdot e^{b(\Delta m - m_\pi)}$ for the background, where m_π is the charged-pion mass, and a and b are free parameters. The raw yields extracted with the fit are corrected for acceptance and efficiency using Monte Carlo simulations based on Pythia [14] and HIJING event generators [15].

In order to extract the cross section of prompt D mesons, the contribution of D mesons from B decays is subtracted. This B feed-down contribution is evaluated using the B-meson cross section from the FONLL pQCD calculations [16], the $B \to D + X$ decay kinematics from the EvtGen package [17], the efficiencies for prompt and feed-down D mesons from simulations, and a hypothesis on the R_{AA} (R_{pPb}) of D mesons arising from B-meson decays in Pb–Pb (p–Pb) collisions. Details of the analysis procedure in pp, Pb–Pb, and p–Pb collisions are reported in Refs. [18], [11], and [19], respectively.

3. Results in pp collisions

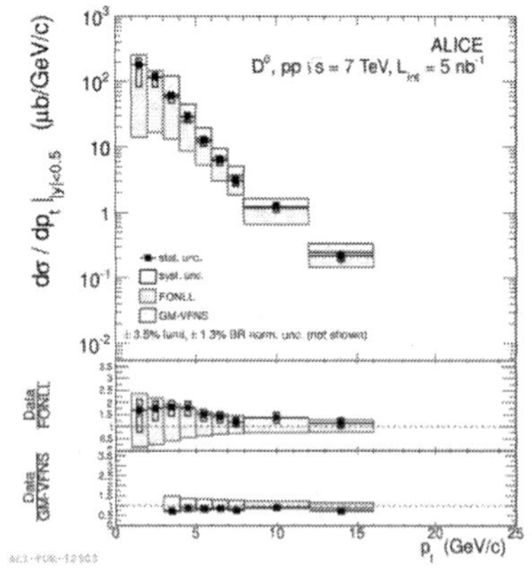

Figure 1. p_T-differential cross section of prompt D^0 in pp collisions at $\sqrt{s} = 7$ TeV, compared with results from FONLL [16] and GM-VFNS [21] theoretical calculations [18].

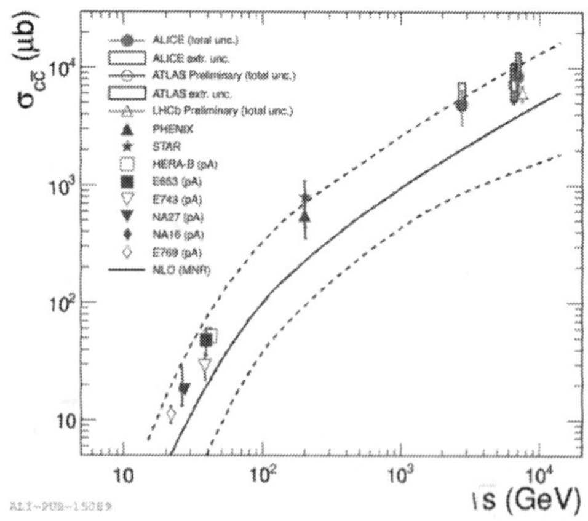

Figure 2. Total charm cross section measured in pp collisions with ALICE as a function of $\sqrt{s}$, in comparison with other experimental results and MNR [25] calculations [23].

The measurement of prompt D-meson production in pp collisions at $\sqrt{s} = 7$ TeV was performed using a data sample collected in 2010 with a minimum-bias trigger. The analysed data correspond to an integrated luminosity of 5 nb^{-1}. The measured p_T-differential cross section for prompt D^0, D^+, D^{*+}, and D_s^+ mesons are published in Ref. [18, 20]. Figure 1 shows the differential cross section of prompt D^0 production in the p_T range $1 < p_T < 16$ GeV/c. The measured cross sections are described within uncertainties by pQCD calculations such as FONLL [16], GM-VFNS [21] and k_T-factorization at LO [22]. The data are at the upper edge of the uncertainty band for FONLL and k_T-factorization model, but at the lower edge for GM-VFNS.

The D-meson production cross section has also been measured in pp collisions at $\sqrt{s} =$ 2.76 TeV [23]. This data sample was collected in 2011 with an integrated luminosity of 1.1 nb^{-1}, much smaller than that recorded at $\sqrt{s} = 7$ TeV. Therefore, these data were not used as the pp reference for the Pb–Pb results, but only to verify the validity of an energy scaling of the cross section from centre-of-mass energy 7 TeV to 2.76 TeV using FONLL [24].

The measured cross sections for pp collisions at $\sqrt{s} = 7$ TeV and 2.76 TeV were extrapolated to full phase space using FONLL calculations, in order to extract the total $c\bar{c}$ cross section in pp collisions at these energies. The extrapolated results are shown in Fig. 2, along with the results from other experiments and a next-to-leading-order pQCD calculation (MNR [25]). The ALICE result at $\sqrt{s} = 7$ TeV is in agreement with those from the ATLAS [26] and LHCb [27] Collaborations, and both of the ALICE points follow the trends exhibited by the NLO predictions and other experimental results spanning three orders of magnitude in energy.

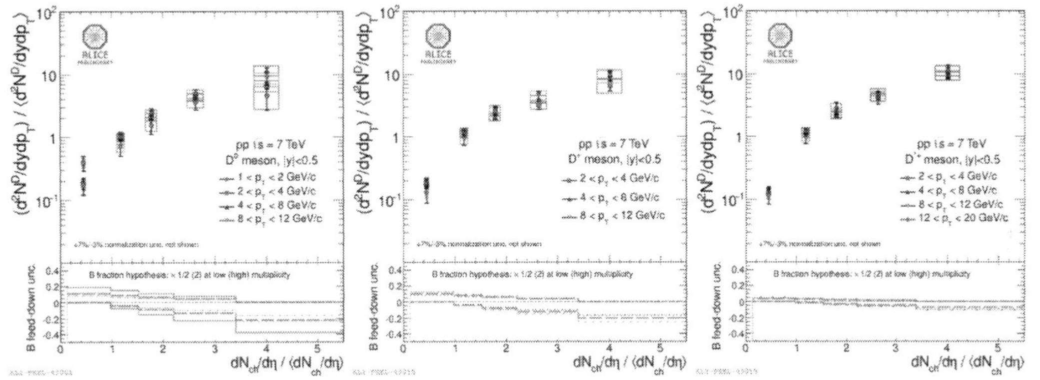

Figure 3. Preliminary results on self-normalized yields of D^0, D^+ and D^{*+} as a function of the relative multiplicity of charged particles produced in pp collision at $\sqrt{s} = 7$ TeV for several p_T intervals.

The production of D mesons was also studied in several p_T intervals as a function of the multiplicity of charged particles generated in the collision. The multiplicity estimator used in this analysis is the number of SPD tracklets N_{trk} (combination of two hits in the two SPD layers) in $|\eta| < 1.0$. Since the pseudorapidity coverage of the SPD changes with the z position of the interaction vertex, a correction to the measured N_{trk} is applied event-by-event. The D-meson yields are extracted in five intervals of multiplicity. Figure 3 shows the yield of D^0, D^+ and D^{*+} in a given multiplicity interval divided by the value integrated over multiplicity as a function of relative charged-particle multiplicity (in units of the average) for different p_T intervals in pp collisions at $\sqrt{s} = 7$ TeV. The results for D^0, D^+ and D^{*+} are compatible within uncertainties and show an increase of the D-meson yields with the charged-particle multiplicity of the event. A similar trend has also been observed for inclusive J/ψ measured with ALICE [28]. These results may suggest that either D-meson production in pp collisions is connected with a strong hadronic activity, or with multiple hard parton interactions.

4. Results in Pb–Pb collisions

The Pb–Pb data at $\sqrt{s_{NN}} = 2.76$ TeV were collected in 2010 and 2011, using both minimum-bias and centrality triggers. In 2010, the data were collected with a minimum-bias trigger ($L_{int} = 2.12\ \mu b^{-1}$), while in 2011 an online selection based on the VZERO signal amplitude was used to enhance the sample of central (0–10%, $L_{int} = 28\ \mu b^{-1}$) and mid-central (10–50%, $L_{int} = 6\ \mu b^{-1}$) collisions. In order to obtain the pp reference spectrum at $\sqrt{s} = 2.76$ TeV, the high-statistics 7 TeV data were scaled to 2.76 TeV using FONLL. The nuclear modification factors of D^0, D^+ and D^{*+} mesons measured in the range $2 < p_T < 16$ GeV/c using the 2010 data sample were reported in Ref. [11] and the v_2 results from the 2011 data sample were reported in Ref. [29, 30].

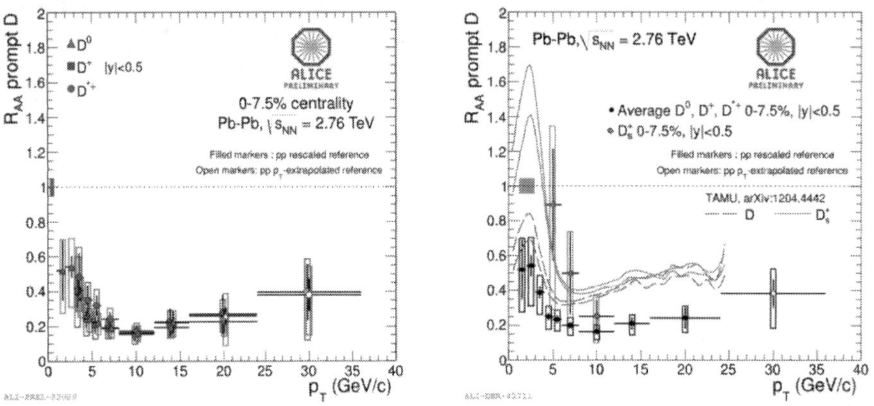

Figure 4. Left panel: R_{AA} of D^0, D^+, and D^{*+} mesons in central (0–7.5%) Pb–Pb collisions at $\sqrt{s_{NN}} = 2.76$ TeV. Right panel: Average R_{AA} of D^0, D^+, and D^{*+} mesons compared with that of D_s^+ mesons. Theoretical model predictions for R_{AA} of non-strange D mesons and the D_s^+ meson are shown as lines [31].

The R_{AA} of D^0, D^+, D^{*+} and D_s^+ mesons was measured in most-central (0–7.5%) Pb–Pb collisions in a wider transverse momentum range $1 < p_T < 36$ GeV/c, using 2011 data. The left panel of Fig. 4 shows the R_{AA} measurements for D^0, D^+, and D^{*+}. They are consistent within statistical uncertainties and show a strong suppression (factor of 4–5) for $p_T > 5$ GeV/c. The right panel of Fig. 4 shows R_{AA} of the D_s^+ meson in the range $4 < p_T < 12$ GeV/c, compared with the average R_{AA} of non-strange D mesons. A suppression of the D_s^+ meson is observed for $8 < p_T < 12$ GeV/c, in agreement within the uncertainties with that of non-strange D mesons in the same p_T range. Theoretical calculations predict that D_s^+ meson yield should be less suppressed at low and intermediate p_T in comparison to non-strange D mesons, if the dominant mechanism for D-meson formation is via recombination of charm quarks with light quarks in the medium [31, 32]. One of these theoretical predictions for R_{AA} of non-strange D mesons and D_s^+ mesons is shown in the right panel of Fig. 4. Present data hint at the central value of R_{AA} of D_s^+ mesons being higher than that of non-strange D mesons at low p_T, however a more precise measurement is needed to draw a firm conclusion as they are compatible within current uncertainties.

The comparison of the centrality dependence of R_{AA} of non-strange D mesons and of J/ψ from B-meson decay measured with CMS [33] is shown in the left panel of Fig. 5. The p_T ranges for D mesons ($8 < p_T < 16$ GeV/c) and J/ψ from B-meson decays ($6.5 < p_T < 30$ GeV/c)

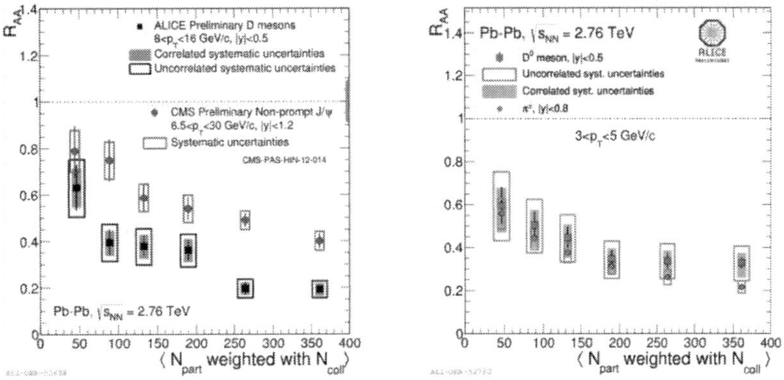

Figure 5. Average $R_{\rm AA}$ of $\rm D^0$, $\rm D^+$ and $\rm D^{*+}$ mesons as a function of centrality in Pb–Pb collisions at $\sqrt{s_{\rm NN}} = 2.76$ TeV, compared with $R_{\rm AA}$ of non-prompt $\rm J/\psi$ mesons measured with CMS (left panel) [33] and charged pions measured with ALICE (right panel)

were chosen such that the kinematics of D and B mesons were similar. These results indicate a stronger suppression of charm hadrons than beauty hadrons in central Pb-Pb collisions, as expected from the predicted mass dependence discussed in section 1. The right panel of Fig. 5 shows the comparison of the centrality dependence of $R_{\rm AA}$ of D mesons and charged pions for $3 < p_{\rm T} < 5$ GeV/c. The magnitude of the suppression is similar for D mesons and pions, which is consistent with some of the calculations discussed in section 1. A hint of a difference in central collisions is suggested but more precise measurements are needed to draw a conclusion on the difference between D-meson and pion suppression due to the mass and colour-charge dependence of the energy loss.

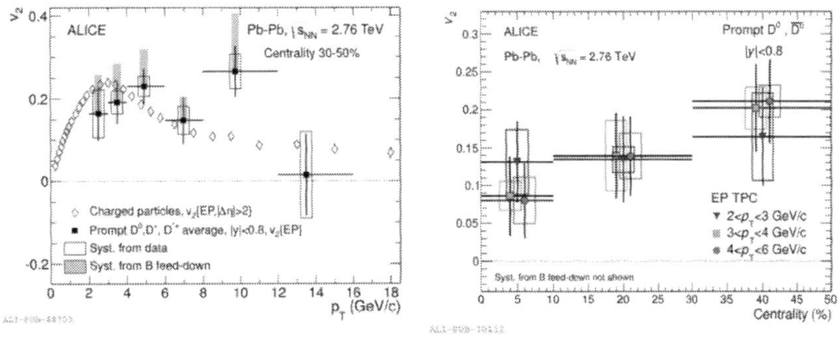

Figure 6. Left panel: Average v_2 of $\rm D^0$, $\rm D^+$ and $\rm D^{*+}$ mesons as a function of $p_{\rm T}$ for Pb–Pb collisions at $\sqrt{s_{\rm NN}} = 2.76$ TeV in the 30–50% centrality range, compared with the v_2 of charged particles [29]. Right panel: v_2 of $\rm D^0$ in three $p_{\rm T}$ intervals as a function of centrality. The data points are displaced horizontally for two of the $p_{\rm T}$ intervals for better visibility [30].

The v_2 of prompt D mesons at mid-rapidity was measured in the three centrality classes 0–10%, 10–30% and 30–50% using the event-plane method [29,30]. The average v_2 as a function of $p_{\rm T}$ for $\rm D^0$, $\rm D^+$ and $\rm D^{*+}$ mesons in the 30–50% centrality class is shown in the left panel of Fig. 6, compared with the measured v_2 of charged particles. The D-meson v_2 is larger than zero with 5σ significance in the range $2 < p_{\rm T} < 6$ GeV/c. A positive v_2 is also observed for $p_{\rm T} > 6$ GeV/c, which most likely originates from the path-length dependence of the in-medium partonic energy loss, although the large uncertainties do not allow a firm conclusion. The measured D-meson v_2 is comparable in magnitude with that of charged particles, which are mostly light-flavour hadrons. This result suggests that low-momentum charm quarks take part in the collective motion of the system. The centrality dependence of v_2 of D^0 mesons is shown in the right panel of Fig. 6 for three p_T intervals in the range $2 < p_T < 6$ GeV/c. A decreasing trend of v_2 towards more central collisions is observed, as expected because of the decreasing initial geometrical anisotropy.

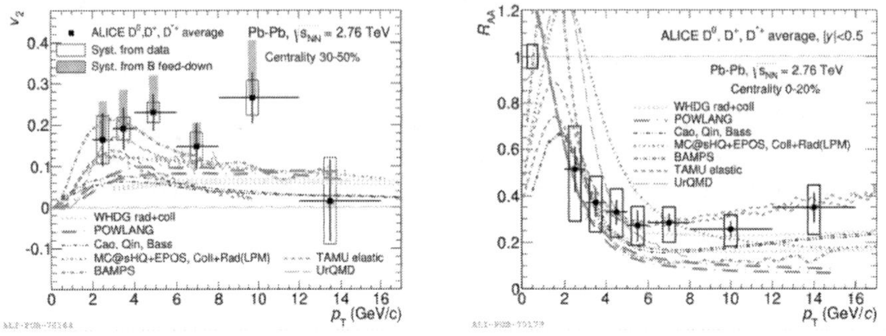

Figure 7. Left panel: Model comparisons for average v_2 of D^0, D^+ and D^{*+} in the 30–50% centrality class (left panel), average R_{AA} of D^0, D^+ and D^{*+} in the 0–20% centrality class (right panel) [34–42].

A number of theoretical model calculations are currently available for the elliptic flow coefficient v_2 and the nuclear modification factor R_{AA} of heavy-flavour hadrons [34–42]. Figure 7 shows the comparison of these models to measurements of the average v_2 of D^0, D^+ and D^{*+} in the 30–50% centrality class and average R_{AA} of D^0, D^+ and D^{*+} in the 0–20% centrality class. The anisotropy is qualitatively described by the models that include both charm quark energy loss in a geometrically anisotropic medium and mechanisms that transfer the elliptic flow of the medium to charm quarks during the system expansion. The measured nuclear modification factor is qualitatively described by the models including in-medium parton energy loss. The model comparison for R_{AA} and v_2 shows that some models describe either the R_{AA} or the v_2, but a simultaneous description of both observables remains a challenge.

5. Results in p–Pb collisions

The p–Pb data at $\sqrt{s_{NN}} = 5.02$ TeV were collected in 2013 with a minimum-bias trigger. The analysed data sample corresponds to an integrated luminosity of 48.6 μb^{-1}. The D-meson production cross section and the nuclear modification factor R_{pPb} were measured in p–Pb collisions at $\sqrt{s_{NN}} = 5.02$ TeV [19]. The left panel of Fig. 8 shows the p_T-differential production cross sections of prompt D^0, D^+, D^{*+}, and D_s^+ mesons and the right panel of the same figure shows the production cross section as a function of rapidity for D^0 in $2 < p_T < 5$ GeV/c, $5 < p_T < 8$ GeV/c and $8 < p_T < 16$ GeV/c. In the centre of mass system of the p–Pb collisions, the measurement covers the rapidity range $-0.96 < y < 0.04$. Within the current statistical and systematic uncertainties, no evidence of a dependence of the production cross section on rapidity is observed within this window. The measured p_T, y-differential cross section is compatible with predictions obtained by scaling the y-differential cross section calculated for pp collisions with FONLL [16] by the Pb mass number and by a R_{pPb} estimated, as a function of p_T, on the basis of the MNR [25] calculation and EPS09 nuclear PDF parametrizations [43].

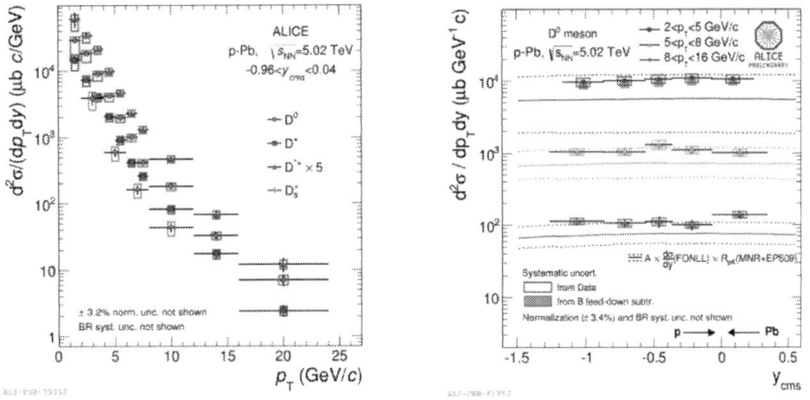

Figure 8. Left panel: p_T-differential inclusive production cross section of prompt D^0, D^+, D^{*+}, and D_s^+ mesons in p–Pb collisions at $\sqrt{s_{NN}} = 5.02$ TeV [19]. Right panel: p_T, y-differential production cross section of prompt D^0 mesons in p–Pb collisions as a function of the rapidity in the centre-of-mass system for three different p_T ranges. The lines represent expectations based on pQCD calculations including the EPS09 parametrization of nuclear PDFs.

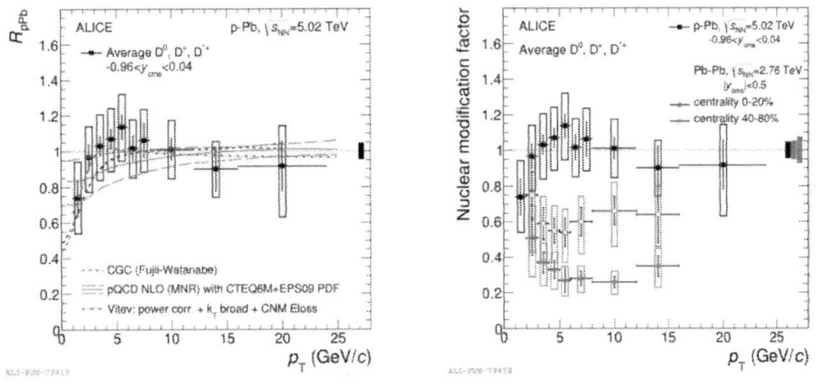

Figure 9. Average R_{pPb} of D^0, D^+ and D^{*+} mesons as a function of p_T at $\sqrt{s_{NN}} = 5.02$ TeV compared to model calculations [25, 43] (left panel), and D-meson R_{AA} in central (0–20%) and in semi-peripheral (40–80%) Pb-Pb collisions at $\sqrt{s_{NN}} = 2.76$ TeV (right panel) [19].

The p_T-differential cross sections were used to compute the nuclear modification factor R_{pPb}, defined as

$$R_{pPb}(p_T) = \frac{d\sigma_{pPb}/dp_T}{A \times (d\sigma_{pp}/dp_T)}, \tag{2}$$

where σ_{pPb} is the cross section in p–Pb collisions, σ_{pp} is the cross section in pp collisions at the same centre-of-mass energy, and A is the mass number of the Pb nucleus. The reference pp cross sections at $\sqrt{s} = 5.02$ TeV were obtained by a FONLL based energy scaling of the p_T-differential cross sections measured at $\sqrt{s} = 7$ TeV [24].

The R_{pPb} measurements for all D-meson species, including D_s^+, are consistent, and compatible with unity within the uncertainties in the measured p_T range [19]. The average R_{pPb} of D^0, D^+ and D^{*+} was calculated using the relative statistical uncertainties as weights. The left panel of

Fig. 9 shows the average $R_{\rm pPb}$ of D mesons as a function of $p_{\rm T}$, compared to model calculations. It can be seen from the present measurement that the cold nuclear matter effects are smaller than the uncertainties for $p_{\rm T} \gtrsim 3$ GeV/c. Predictions based on NLO pQCD calculations (MNR [25]) of D-meson production, including the EPS09 [43] nuclear modification of the CTEQ6M PDF [44] and calculations based on the Color Glass Condensate [45] can describe the measurement considering only initial state effects. Data are also well described by a calculation including energy loss in cold nuclear matter, nuclear shadowing and $k_{\rm T}$-broadening [46, 47]. Within uncertainties, the experimental results do not favour one prediction over the other.

In the right panel of Fig. 9, the average $R_{\rm AA}$ of prompt D mesons in central (0–20%) and in semi-peripheral (40–80%) Pb–Pb collisions at $\sqrt{s_{\rm NN}} = 2.76$ TeV [11] is shown along with the average $R_{\rm pPb}$ of prompt D mesons in p–Pb collisions at $\sqrt{s_{\rm NN}} = 5.02$ TeV. Although the centre-of-mass energies of the two systems differ, there is a clear suppression in Pb–Pb collisions compared with p–Pb for $p_{\rm T} \gtrsim 4$ GeV/c. This indicates that final-state effects, such as partonic energy loss in the medium, play a far more dominant role in the suppression of D-meson production at higher $p_{\rm T}$ than initial-state effects, such as nuclear shadowing.

6. Conclusions and Outlook

Measurements of the production of D mesons at mid-rapidity in pp, Pb–Pb and p–Pb collisions performed with ALICE have been presented. The $p_{\rm T}$-differential cross sections of D mesons in pp collisions are described within uncertainties by theoretical predictions based on perturbative QCD, such as FONLL, GM-VFNS and $k_{\rm T}$-factorization at LO. D-meson production was studied as a function of the charged particle multiplicity in pp collisions at $\sqrt{s} = 7$ TeV. An increase of the D-meson yield as a function of multiplicity is observed, which suggests a connection of D-meson production to a strong hadronic activity or to multiple hard parton interactions.

A strong suppression of prompt D mesons is observed at high $p_{\rm T}$ in central Pb-Pb collisions at $\sqrt{s_{\rm NN}} = 2.76$ TeV. The suppression is of similar magnitude for all non-strange D mesons. Data are not conclusive on a possible enhancement of the $\rm D_s^+$ yield relative to non-strange D mesons at low $p_{\rm T}$. The observed difference in the suppression of D mesons and non-prompt $\rm J/\psi$ from B-meson decays measured with CMS at high $p_{\rm T}$ in central collisions is consistent with predictions of a mass hierarchy in the energy loss of heavy quarks.

A v_2 significantly larger than zero was measured for D mesons at low $p_{\rm T}$ in semi-peripheral Pb-Pb collisions, suggesting that charm quarks take part in the collective expansion of the medium. The measured $R_{\rm AA}$ and v_2 of D mesons were compared with several theoretical model calculations. Although some models can describe measurements of $R_{\rm AA}$ and v_2 separately, it is challenging for models to describe simultaneously the large suppression of D mesons in central collisions and their azimuthal anisotropy in non-central collisions. $R_{\rm pPb}$ for D mesons as a function of $p_{\rm T}$ in p–Pb collisions is consistent with unity and also compatible with models including initial-state effects. This suggests that the suppression visible at high $p_{\rm T}$ in Pb–Pb collisions primarily is due to the final-state effects rather than initial-state effects.

After a shutdown of close to 2 years, the LHC will resume operations in 2015 with an energy almost a factor of two higher than in the previous running period and it is expected to deliver a factor 5–10 larger integrated luminosity for Pb–Pb collisions prior to the second long shutdown. The increased statistics resulting from this improvement will improve the precision of D-meson measurements with ALICE and shed light on the intriguing questions opened by the present data.

References

[1] Gyulassy M and Plumer M 1990 *Phys. Lett. B* **243** 432
[2] Baier R, Dokshitzer Y L, Mueller A H, Peigne S and Schiff D 1997 *Nucl. Phys. B* **484** 265
[3] Thoma M H and Gyulassy M 1991 *Nucl. Phys. B* **351** 491

Journal of Physics: Conference Series **535** (2014) 012027 doi:10.1088/1742-6596/535/1/012027

[4] Braaten E and Thoma M H 1991 *Phys. Rev. D* **44** 1298; 1991 *Phys. Rev. D* **44** 2625
[5] Dokshitzer Y L and Kharzeev D E 2001 *Phys. Lett. B* **519** 199
[6] Van Hees H, Greco V and Rapp R 2006 *Phys. Rev. C* **73** 034913
[7] Glauber R J 1959 *Lectures in Theoretical Physics* Vol. **1** 315
[8] Djordjevic M 2014 *Phys. Rev. Lett.* **112** 042302
[9] Ollitrault J Y 1992 *Phys. Rev. D* **46** 229
[10] Gyulassy M, Vitev I and Wang X N 2001 *Phys. Rev. Lett.* **86** 2537
[11] Abelev B *et al.* (ALICE Collaboration) 2012 *JHEP* **1209** 112
[12] Aamodt K *et al.* (ALICE Collaboration) 2008 *JINST* **3** S08002
[13] Beringer J *et al.* (Particle Data Group) 2012 *Phys. Rev. D* **86** 010001
[14] Sjöstrand T, Mrenna S, Skands P 2006 *JHEP* **0605** 026
[15] Wang X N and Gyulassy M 1991 *Phys. Rev. D* **44** 3501
[16] Cacciari M, Greco M and Nason P 1998 *JHEP* **9805** 007
[17] Lange D 2001 *Nucl. Instrum. Meth. A* **462** 152
[18] Abelev B *et al.* (ALICE Collaboration) 2012 *JHEP* **1201** 128
[19] Abelev B *et al.* (ALICE Collaboration) 2014 *Preprint* arXiv:1405.3452 [nucl-ex]
[20] Abelev B *et al.* (ALICE Collaboration) 2012 *Phys. Lett. B* **718** 279
[21] Kniehl B A *et al.* 2012 *Eur. Phys. J. C* **72** 2082
[22] Maciula R and Szczurek A, 2013 *Phys. Rev. D* **87** 094022
[23] Abelev B *et al.* (ALICE Collaboration) 2012 *JHEP* **1207** 191
[24] Averbeck R *et al.* 2011 *Preprint* arXiv:1107.3243 [hep-ph]
[25] Mangano M, Nason P and Ridolfi G 1992 *Nucl. Phys. B* **373** 295
[26] ATLAS Collaboration 2011 PHYS-PUB-2011-012; 2011 ATLAS-CONF-2011-017
[27] LHCb Collaboration 2010 LHCb-CONF-2010-013
[28] Abelev B *et al.* (ALICE Collaboration) 2012 *Phys. Lett. B* **712** 165175
[29] Abelev B *et al.* (ALICE Collaboration) 2013 *Phys. Rev. Lett.* **111** 102301
[30] Abelev B *et al.* (ALICE Collaboration) 2014 *Preprint* arXiv:1405.2001 [nucl-ex]
[31] He M, Fries R J and Rapp R 2013 *Phys. Rev. Lett.* **110** 112301
[32] Kuznetsova I and Rafelski J 2007 *Eur. Phys. J. C* **51** 113
[33] CMS Collaboration 2012 CMS-PAS-HIN-12-014
[34] Horowitz W A and Gyulassy M 2011 *J. Phys. G* **38** 124114
[35] Alberico W M *et al.* 2011 *Eur. Phys. J. C* **71** 1666
[36] Cao S, Qin G Y and Bass S A 2013 *Phys. Rev. C* **88** 044907
[37] Gossiaux P B, Bierkandt R and Aichelin J 2009 *Phys. Rev. C* **79** 044906
[38] Gossiaux P B *et al.* 2010 *J. Phys. G* **37** 094019
[39] Fochler O *et al.* 2011 *J. Phys. G* **38** 124152
[40] He M, Fries R J and Rapp R 2012 *Phys. Rev. C* **86** 014903
[41] Lang T *et al.* 2012 *Preprint* arXiv:1211.6912 [hep-ph]
[42] Lang T *et al.* 2012 *Preprint* arXiv:1212.0696 [hep-ph]
[43] Eskola K J, Paukkunen H and Salgado C A 2009 *JHEP* **0904** 065
[44] Stump D *et al.* 2003 *JHEP* **0310** 046
[45] Fujii H and Watanabe K 2013 *Preprint* arXiv:1308.1258 [hep-ph]
[46] Sharma R, Vitev I and Zhang B W 2009 *Phys. Rev. C* **80** 054902
[47] Sharma R and Vitev I 2013 *Phys. Rev. C* **87** 044905

Journal of Physics: Conference Series **535** (2014) 012028 doi:10.1088/1742-6596/535/1/012028

Hydrodynamic flow in proton-nucleus collisions at the LHC

K. Werner[(a)], **B. Guiot**[(a)], **Iu. Karpenko**[(b,c)], **T. Pierog**[(d)]

[(a)] SUBATECH, University of Nantes – IN2P3/CNRS– EMN, Nantes, France
[(b)] Bogolyubov Institute for Theoretical Physics, Kiev 143, 03680, Ukraine
[(c)] FIAS, Johann Wolfgang Goethe Universitaet, Frankfurt am Main, Germany
[(d)]Karlsruhe Inst. of Technology, KIT, Campus North, Inst. f. Kernphysik, Germany

Abstract. There is little doubt that hydrodynamic flow has been observed in heavy ion collisions at the LHC and RHIC, mainly based on results on azimuthal anisotropies, but also on particle spectra of identified particles, perfectly compatible with hydrodynamic expansions. Surprisingly, in p-Pb collisions one observes a very similar behavior. So do we see flow even in p-Pb? We will try to answer this question.

Collective hydrodynamic flow seems to be well established in heavy ion (HI) collisions at energies between 200 and 2760 AGeV, whereas p-p and p-nucleus (p-A) collisions are often considered to be simple reference systems, showing "normal" behavior, such that deviations of HI results with respect to p-p or p-A reveal "new physics". Surprisingly, the first results from p-Pb at 5 TeV on the transverse momentum dependence of azimuthal anisotropies and particle yields are very similar to the observations in HI scattering [1, 2].

Do we see radial flow in p-Pb collisions? In order to answer this question, we will employ the EPOS3 approach [3], well suited for this problem, since it provides within a unique theoretical scheme the initial conditions for a hydrodynamical evolution in p-p, p-A, and HI collisions. The initial conditions are generated in the Gribov-Regge multiple scattering framework. An individual scattering is referred to as Pomeron, identified with a parton ladder, eventually showing up as flux tubes (also called strings). Each parton ladder is composed of a pQCD hard process, plus initial and final state linear parton emission. Our formalism is referred to as "Parton based Gribov Regge Theory" and described in very detail in [4]. Based on these initial conditions, we performed already ideal hydrodynamical calculations (EPOS2) [5, 6, 7, 8] to analyse HI and p-p scattering at RHIC and LHC. In EPOS3 we add two major improvements: a more sophisticated treatment of nonlinear effects in the parton evolution by considering individual (per Pomeron) saturation scales [9], and a 3D+1 viscous hydrodynamical evolution. There are also changes in our core-corona procedure, which amounts to separate the initial energy of the flux tubes into a part which constitutes the initial conditions for hydro (core) and the particles which leave the "matter". This is crucial as well in proton-nucleus collisions (as in all other collision types).

To understand the results discussed later in this paper, we show in fig. 1 the effect of flow on identified particle spectra, by comparing p_t distributions from pure string decay to spectra from a pure hydrodynamic evolution. In case of string fragmentation, heavier particles are strongly suppressed compared to lighter ones, but the shapes are not so different. This picture changes completely in the fluid case: The heavier the particle, the more it gets shifted to higher p_t. This

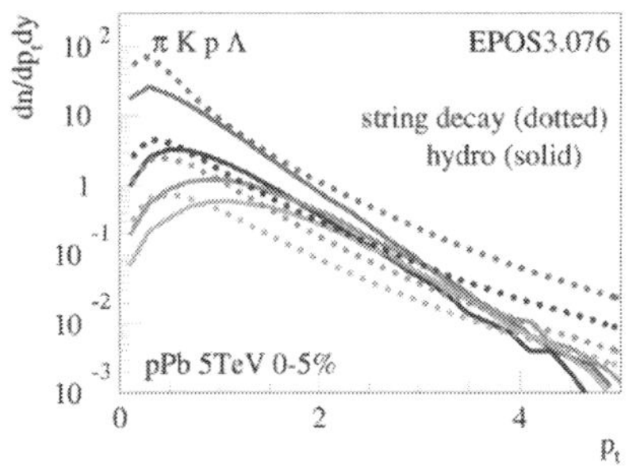

Figure 1. (Color online) Identified particle spectra as a function of p_t, for central (0-5%) p-Pb collisions at 5.02 TeV. We show results for particle production from string decay, i.e. EPOS without hydro (dotted curves), and particle production from pure hydro, without corona (solid lines). In both cases, we show (from top to bottom) pions, kaons, protons, and lambdas.

is a direct consequence of the fact that the particles are produced from fluid cells characterized by radial flow velocities, which gives more transverse momentum to heavier particles.

There are few other studies of hydrodynamic expansion in proton-nucleus systems. In [10], fluctuating initial conditions based on the so-called Monte Carlo Glauber model (which is actually a wounded nucleon model) are employed, followed by a viscous hydrodynamical evolution. Also [11] uses fluctuating initial conditions, here based on both Glauber Monte Carlo and Glasma initial conditions. Finally in [12], ideal hydrodynamical calculations are performed, starting from smooth Glauber model initial conditions.

In the following, we will compare experimental data on identified particle production with our simulation results (referred to as EPOS3), and in addition to some other models, as there are QGSJETII [13], AMPT [14], and EPOS LHC [15]. The QGSJETII model is also based on Gribov-Regge multiple scattering, but there is no fluid component. The main ingredients of the AMPT model are a partonic cascade and then a hadronic cascade, providing in this way some "collectivity". EPOS LHC is a tune (using LHC data) of EPOS1.99. As all EPOS1 models, it contains flow, put in by hand, parametrizing the collective flow at freeze-out. Finally, the approach discussed in this paper (EPOS3) contains a full viscous hydrodynamical simulation. So it is interesting to compare these four models, since they differ considerably concerning the implementation of flow, from full hydrodynamical flow in EPOS3 to no flow in QGSJETII.

The CMS collaboration published a detailed study [1] of the multiplicity dependence of (normalized) transverse momentum spectra in p-Pb scattering at 5.02 TeV. The multiplicity (referred to as N_{track}) counts the number of charged particles in the range $|\eta| < 2.4$. In fig. 2, we compare experimental data [1] for pions (black symbols) with the simulations from QGSJETII (upper left figure), AMPT (upper right), EPOS LHC (lower left), and EPOS3 (lower right). The different curves in each figure refer to different centralities, with mean values (from bottom to top) of 8, 84, 160, and 235 charged tracks. They are shifted relative to each other by a constant amount. Concerning the models, QGSJETII is the easiest to discuss, since here there are no flow features at all, and the curves for the different multiplicities are identical. The data, however, show a slight centrality dependence: the spectra get somewhat harder with increasing multiplicity. The other models, AMPT, EPOS LHC, and EPOS3 are close to the data.

In figs. 3, 4, we compare experimental data [1] for kaons and protons (black symbols) with the simulations. The experimental shapes of the p_t spectra change considerably, getting much harder with increasing multiplicity. In QGSJETII, having no flow, the curves for the different multiplicities are identical. The AMPT model shows some (but too little) change

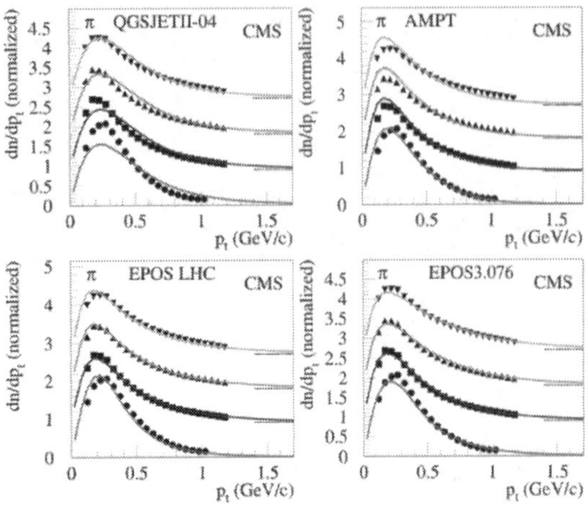

Figure 2. (Color online) Transverse momentum spectra of pions in p-Pb scattering at 5.02 TeV, for four different multiplicity classes with mean values (from bottom to top) of 8, 84, 160, and 235 charged tracks.

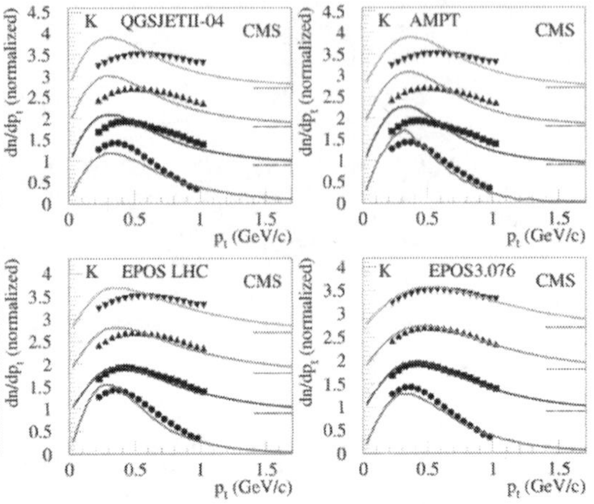

Figure 3. (Color online) Same as fig. 2, but for kaons. We show data from CMS [1] (symbols) and simulations from QGSJETII, AMPT, EPOS LHC, and EPOS3, as indicated in the figures.

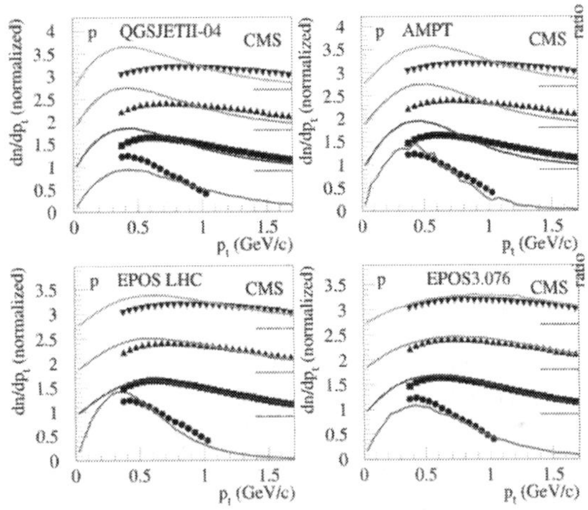

Figure 4. (Color online) Same as fig. 2, but for protons.

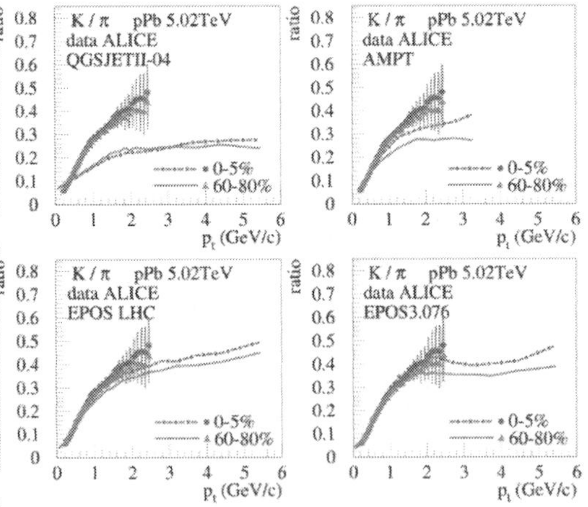

Figure 5. (Color online) Kaon over pion ratio as a function of transverse momentum in p-Pb scattering at 5.02 TeV, for the 0-5% highest multiplicity (red dashed-dotted lines, circles) and 60-80% (green solid lines, triangles).

with multiplicity. EPOS LHC goes into the right direction, whereas EPOS3 gives a reasonable description of the data. **It seems that hydrodynamical flow helps considerably to reproduce these data.**

Also ALICE [2] has measured identified particle production for different multiplicities in p-Pb

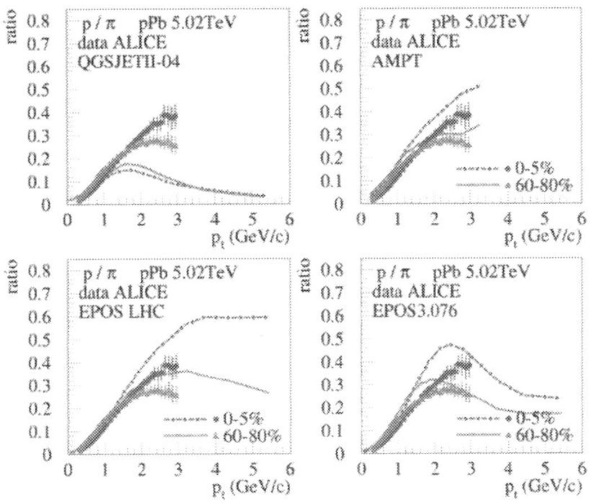

Figure 6. (Color online) Same as fig. 5, but proton over pion ratio.

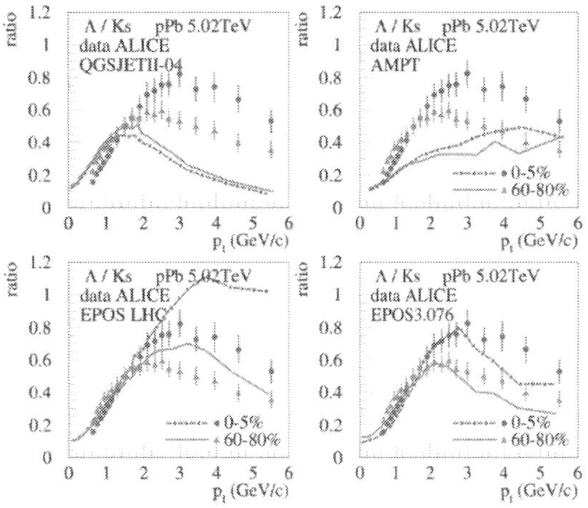

Figure 7. (Color online) Same as fig. 5, but Λ over K_s ratio.

scattering at 5.02 TeV. Here, multiplicity counts the number of charged particles in the range $2.8 < \eta_{\rm lab} < 5.1$. It is useful to study the multiplicity dependence, best done by looking at ratios. In fig. 5, we show the pion over kaon (K/π) ratio as a function of transverse momentum in p-Pb scattering at 5.02 TeV, for high multiplicity (red dashed-dotted lines, circles) and low multiplicity events (green solid lines, triangles), comparing data from ALICE [2] (symbols) and simulations from QGSJETII, AMPT, EPOS LHC, and EPOS3 (lines). In all models, as in the data, there is little multiplicity dependence. However, the QGSJETII model is considerably below the data, for both high and low multiplicity events. AMPT is slightly below, whereas EPOS LHC and EPOS3 do a reasonable job. Concerning the proton over pion (p/π) ratio, fig. 6, again QGSJETII is way below the data, for both high and low multiplicity events, whereas the three other models show the trend correctly, but being slightly above the data. Most interesting are the lambdas over kaon (Λ/K_s) ratios, as shown in fig. 7, because here a wider transverse momentum range is considered, showing a clear peak structure with a maximum around 2-3 GeV/c and a slightly more pronounced peak for the higher multiplicities. QGSJETII and AMPT cannot (even qualitatively) reproduce this structure. EPOS LHC shows the right trend, but the peak is much too high for the high multiplicities. EPOS3 is close to the data.

To summarize these ratio plots (keeping in mind that the QGSJETII model has no flow, AMPT "some" flow, EPOS LHC a parametrized flow, and EPOS3 hydrodynamic flow): Flow seems to help considerably. However, from the Λ/K_s ratios, we conclude that EPOS LHC uses a too strong radial flow for high multiplicity events. The hydrodynamic flow employed in EPOS3 seems to get the experimental features reasonably well. Crucial is the core-corona procedure discussed earlier: there is more core (compared to corona) in more central collisions, but the centrality (or multiplicity) dependence is not so strong, and there is already an important core (=flow) contribution in peripheral events.

Finally, we sketch very briefly results on elliptical flow v_2 obtained from dihadron correlations, showing ALICE results [16, 17] and EPOS3 simulation, see ref. [18] for details. In fig. 8. we plot v_2 as a function of p_t. Clearly visible in data and in the simulations: a separation of the results for the tywo hadron species: in the p_t range of 1-1.5 GeV/c, the proton result is clearly below the pion one. Within our fluid dynamical approach, the above results are nothing but a "mass splitting". The effect is based on an asymmetric (mainly elliptical) flow, which translates into the corresponding azimuthal asymmetry for particle spectra. Since a given velocity translates

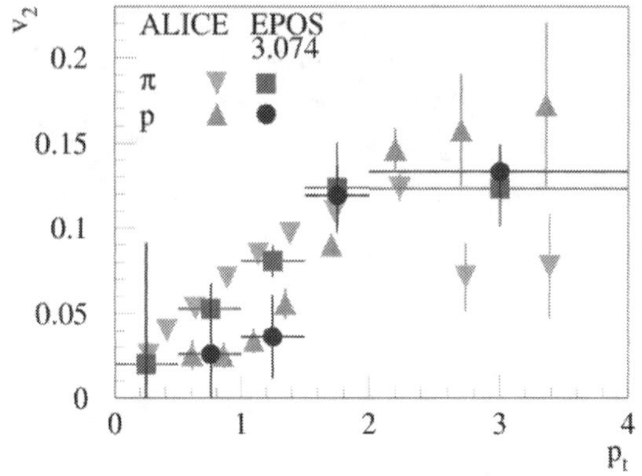

Figure 8. (Color online) Elliptical flow coefficients v_2 for pions and protons. We show ALICE results (triangles) and EPOS3 simulations (squares and circles). Pions appear red and yellow, protons blue and green.

into momentum as $m_A\gamma v$, with m_A being the mass of hadron type A, flow effects show up at higher values of p_t for higher mass particles.

To summarize : Comparing experimental data on identified particle production to various Monte Carlo generators, we conclude that hydrodynamical flow seems to play an important role in p-Pb scattering.

[1] CMS collaboration, arXiv:1307.3442
[2] ALICE collaboration, arXiv:1307.6796
[3] K. Werner et al., to be published
[4] H. J. Drescher, M. Hladik, S. Ostapchenko, T. Pierog and K. Werner, Phys. Rept. 350, 93, 2001
[5] K. Werner, Iu. Karpenko, T. Pierog, M. Bleicher, K. Mikhailov, Phys. Rev. C 82, 044904 (2010)
[6] K. Werner, Iu.Karpenko, M. Bleicher, T.Pierog, S. Porteboeuf-Houssais, arXiv:1203.5704, Phys. Rev. C 85 (2012) 064907
[7] K. Werner, Iu. Karpenko, T. Pierog, M. Bleicher, K. Mikhailov, arXiv:1010.0400, Phys. Rev. C 83, 044915 (2011)
[8] K. Werner, Iu. Karpenko, T. Pierog, arXiv:1011.0375, Phys. Rev. Lett. 106, 122004,2011
[9] L. McLerran, R. Venugopalan, Phys. Rev. D 49 (1994) 2233; L. McLerran, R. Venugopalan, Phys. Rev. D 49 (1994) 3352; L. McLerran, R. Venugopalan, Phys. Rev. D 50 (1994) 2225.
[10] P. Bozek, W. Broniowski, arXiv:1304.3044
[11] A. Bzdak, B. Schenke, P. Tribedy, R. Venugopalan, arXiv:1304.3403
[12] G.Y. Qin, B. Mueller, arXiv:1306.3439
[13] S. Ostapchenko, Phys. Rev. D74 (2006) 014026; S.Ostapchenko, Phys.Rev. D83 (2011) 014018
[14] Z.-W. Lin, C. M. Ko, B.-A. Li, B. Zhang and S. Pal, Phys. Rev. C 72, 064901 (2005).
[15] T. Pierog, Iu. Karpenko, J.M. Katzy, E. Yatsenko, K. Werner, arXiv:1306.5413.
[16] ALICE collaboration, Phys.Lett. B719 (2013) 29-41, arXiv:1212.2001
[17] ALICE collaboration, preprint CERN-PH-EP-2013-115arXiv:1307.4379
[18] K. Werner, M. Bleicher, B. Guiot, Iu. Karpenko, T. Pierog, arXiv:1307.4379 [nucl-th]

Measurements of heavy-flavour decay leptons in pp, p-Pb, and Pb-Pb collisions with ALICE

Sarah LaPointe on behalf of the ALICE Collaboration
Istituto Nazionale di Fisica Nucleare, Via P. Giuria 1, 10125 Torino
E-mail: s.lapointe@cern.ch

Abstract.
We present measurements of electrons and muons originating from the semi-leptonic decays of heavy-flavour hadrons at central and forward rapidities with ALICE in pp ($\sqrt{s}$ = 2.76 and 7 TeV), p-Pb ($\sqrt{s_{\rm NN}}$ = 5.02 TeV), and Pb-Pb ($\sqrt{s_{\rm NN}}$ = 2.76 TeV) collisions. The measured $p_{\rm T}$-differential production cross sections in pp collisions, the nuclear modification factors in p-Pb and Pb-Pb, and the v_2 measurements in Pb-Pb collision are disgusted. In addition, the angular correlation distribution between heavy-flavour decay electrons and charged particles, measured in p-Pb collisions, is shown. The results are compared to their corresponding theoretical predictions.

1. Introduction

The main objective of ALICE is to investigate the strongly interacting, partonic medium that forms from high energy collisions of heavy ions [1]. At the CERN LHC heavy quarks, i.e. charm and beauty, are abundantly produced [2, 3] and, because of their large masses, they predominately originate from hard parton scattering processes in the initial phase of the collision. Thus their production occurs early relative to the expected formation time of the Quark-Gluon Plasma (QGP). Additionally, in-QGP annihilation and production of heavy quark pairs is negligible, thus it is expected that they sample each stage of the QGP phase.

While propagating through the medium, heavy quarks are expected to lose energy via elastic scatterings and radiative processes. An experimental observable sensitive to the energy loss is the nuclear modification factor ($R_{\rm AA}$), which quantifies the modification of heavy-flavour transverse momentum ($p_{\rm T}$) distributions in Pb-Pb collisions, relative to pp collisions. $R_{\rm AA}$ is expected to be equal to unity in the absence of initial and final state medium effects. Another experimental observable of interest is v_2, which is the second Fourier coefficient of the azimuthal distribution of particles in the transverse plane with respect to the orientation of the reaction plane. Studying the v_2 of heavy-flavour decay particles can provide knowledge of path-length dependent quark energy loss at high transverse momentum, while at low transverse momentum it offers information on the degree of thermalization of heavy quarks within the medium. An additional, relevant measurement is the azimuthal correlation of heavy-flavour decay electrons and charged hadrons, which can be utilized to estimate the relative contribution of charm and beauty hadrons to the measured heavy-flavour decay electron yield in pp collisions, while also providing a tool to examine the production and fragmentation of heavy quarks in other systems.

The measurement of heavy-flavour production in pp collisions provides the necessary reference for the measurements carried out in heavy-ion collisions, while it also provides essential tests of perturbative Quantum Chromodynamics (pQCD). However, any modification observed in Pb-Pb arises from either initial or final state effects and the pp reference alone cannot allow to decipher the origin. Therefore p-Pb collisions are necessary to disentangle initial state effects, such as the modification of parton distributions in nuclei or energy loss in cold nuclear matter, from final state effects in Pb-Pb collisions.

In the following, the ALICE measurements of electrons and muons originating from the semi-leptonic decays of heavy-flavour hadrons in pp and Pb-Pb, and electrons in p-Pb collisions are presented. The measured p_T-differential production cross sections, nuclear modification factors, v_2, and electron-charged particle angular correlation distributions are shown and compared to the respective theoretical predictions.

2. Heavy-flavour measurements with electrons and muons in ALICE

The heavy-flavour particle detection performance of ALICE is detailed in [4]. Concerning the measurements presented here, the detectors utilized are located in the central rapidity ($|\eta| < 0.9$) and forward rapidity ($-4 < \eta < -2.5$) regions. Open heavy-flavour hadrons are measured via their semi-electronic decay channels in the central rapidity region and in the forward region through their semi-muonic decay channels. Track reconstruction in the central region is based on the Inner Tracking System and the Time Projection Chamber (TPC). Electrons are identified with TPC and Time-of-Flight (TOF) detectors and at higher momentum using the ElectroMagnetic Calorimeter (EMCal) and the Transition Radiation Detector. The electron sample includes electrons from the decay of heavy-flavour hadrons and from background sources, dominated by photon conversions in detector material and Dalitz decays. The background is estimated using either an e^+e^- invariant mass method or a Monte Carlo (MC) calculation of the contributions based on measured π^0 and η meson cross sections [5]. In addition, electrons from decays of beauty hadrons can be identified statistically using a track displacement based selection which exploits the relatively long lifetime of B mesons ($c\tau \sim 500\mu$m). In the forward region, muon tracking and identification is provided by the muon spectrometer. The dominant background sources for the muon measurement are from the decays of pions and kaons. They are estimated in pp collisions using a MC simulation, while in Pb-Pb collisions they are extrapolated from the measured pion and kaon yields at central rapidity.

3. Measurements in pp collisions

The p_T-differential production cross section of electrons from heavy-flavour hadron decays in pp collisions at $\sqrt{s} = 7$ TeV at central rapidity ($|y| < 0.5$) and in the p_T range 0.5–8 GeV/c [5] is shown in Fig. 1 (left). The measurement is compared to the complementary result of the ATLAS experiment [6], measured in the rapidity interval $|y| < 2$ (excluding $1.37 < |y| < 1.52$), which spans the p_T range 7–26 GeV/c. Shown in Fig. 1 (right) is the p_T-differential production cross section of muons from heavy-flavour hadron decays in pp collisions at $\sqrt{s} = 7$ TeV at forward rapidty ($2.5 < y < 4$) and in the p_T range 2–12 GeV/c [7]. Also shown in both panels of Fig. 1 is a comparison of the corresponding pQCD FONLL predictions [8] in the respective rapidity intervals, along with the ratio between data and FONLL. Within the experimental and theoretical uncertainties, both the electron and muon data are well described by the predictions.

In pp collisions at $\sqrt{s} = 7$ TeV electrons from beauty-hadron decays have been measured with ALICE [9]. The relatively long lifetime of beauty hadrons was exploited by selecting on the track impact parameter (d_0), which is the projection, in the transverse plane, of the charged track distance of closest approach to the interaction vertex. This selection enhances the relative contribution of electrons from beauty-hadron decays in the total heavy-flavour decay electron sample.

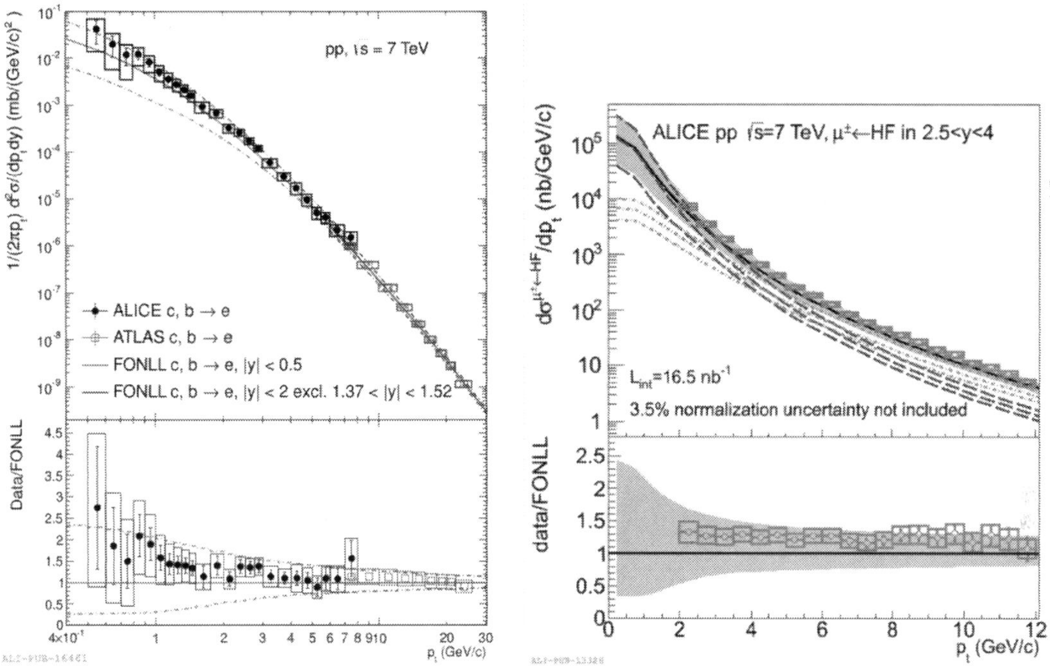

Figure 1. Left: p_T-differential production cross section of electrons from heavy-flavour hadron decays measured in pp collisions at $\sqrt{s} = 7$ TeV with ALICE in $|y| < 0.5$ and ATLAS in $|y| < 2$, excluding $1.37 < |y| < 1.52$ [6]. The solid (dashed) curve shows the corresponding FONLL prediction (uncertainty). Right: p_T-differential production cross section of muons from heavy-flavour hadron decays measured in pp collisions at $\sqrt{s} = 7$ TeV in $2.5 < y < 4$. The solid curve (band) is the corresponding FONLL prediction (uncertainty). The dashed (dot–dashed) curve is the FONLL prediction of muons from charm (beauty) hadron decays. Both figures show the corresponding ratio between the data and the FONLL central prediction in the lower panels.

Heavy-flavour decay leptons have been measured in pp collisions at $\sqrt{s} = 2.76$ TeV, the reference energy for Pb-Pb collisions. Relative to the $\sqrt{s} = 7$ TeV sample, the running time for this data sample was short, and the uncertainties of the electron p_T-differential cross section from the 7 TeV sample were significantly lower compared to those at 2.76 TeV. Hence the cross section at $\sqrt{s} = 7$ TeV was scaled down to 2.76 TeV using the FONLL prediction and used as the pp reference for Pb-Pb measurements. The resulting scaled p_T-differential cross section was compared to the one measured at $\sqrt{s} = 2.76$ TeV and it was found to be in agreement, validating the use of the scaling method [10]. The heavy-flavour decay muon p_T-differential production cross section in pp collisions at $\sqrt{s} = 2.76$ TeV in the p_T range 2–10 GeV/c and $2.5 < y < 4$ has been measured [11]. With a statistical precision better than that of the scaled 7 TeV p_T-differential cross section it is thus used as the pp reference directly.

In pp collisions at $\sqrt{s} = 2.76$ TeV electrons from beauty-hadron decays have been measured [12]. The primary analysis technique used the track impact parameter as a selection criterion. A second method discriminates beauty from charm production using the distribution of the azimuthal angle between heavy-flavour decay electrons and charged hadrons, $\Delta\varphi$. Due to the different decay kinematics of charm and beauty hadrons, the width of the peak in the $\Delta\varphi$ distribution around zero is larger for beauty than for charm hadron decays and from this the relative beauty contribution to the measured heavy-flavour decay electron p_T-differential production cross section is extracted. Fig. 2 shows the p_T-differential production cross section of electrons from beauty-hadron decays measured in pp collisions at $\sqrt{s} = 2.76$ TeV. The result in the p_T range 1–8 GeV/c and $|y| < 0.8$ was supplied by the impact-parameter based analysis, while the p_T interval 8–10 GeV/c was taken from the electron-hadron azimuthal correlation based analysis, measured in $|y| < 0.7$. The p_T-differential cross section is compared to three pQCD calculations [8, 13, 14] and the data and theory are consistent within the experimental and theoretical uncertainties.

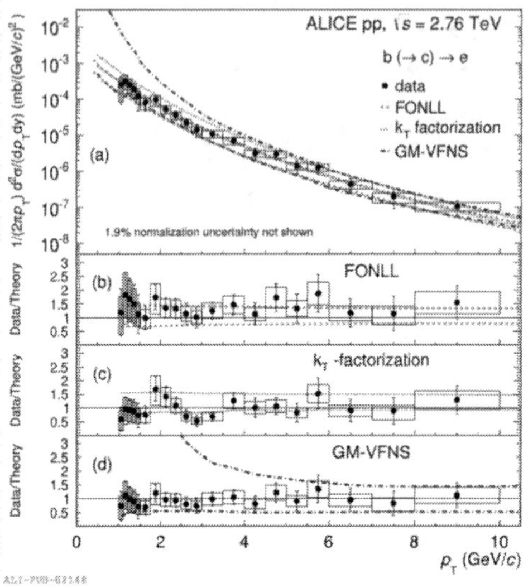

Figure 2. (a) p_T-differential production cross section of electrons from beauty-hadron decays measured in pp collisions at $\sqrt{s} = 2.76$ TeV with ALICE. The dashed, dotted, and dot-dashed curves represent the FONLL [8], k_T-factorization [14], and GM-VFNS [13] predictions, respectively. (b)-(d) Ratios of the data and the central prediction of the pQCD calculations. For all panels, the error bars (boxes) represent the statistical (systematic) uncertainties.

4. Measurements in Pb-Pb and p-Pb collisions

In the following, the nuclear modification factors (R_{AA} and R_{pPb}), the ν_2 coefficient of electrons and muons from heavy-flavour hadron decays measured in Pb-Pb collisions, and the electron-hadron angular correlations measured in p-Pb collisions are presented and compared to the most current theoretical predictions.

The nuclear modification factor is defined as:

$$R_{AA}(p_T) = \frac{1}{\langle T_{AA}\rangle}\frac{dN_{AA}/dp_T}{d\sigma_{pp}/dp_T}; \qquad R_{pPb}(p_T) = \frac{1}{A}\frac{d\sigma_{pPb}/dp_T}{d\sigma_{pp}/dp_T} \tag{1}$$

where dN_{AA}/dp_T is the yield in Pb-Pb collisions, $d\sigma_{pp}/dp_T$ is the differential cross section in pp collisions, and $\langle T_{AA}\rangle$ is the average nuclear overlap function, calculated using the Glauber model [15]. For the calculation in p-Pb collisions the p_T-differential cross section is taken as the numerator and $\langle T_{AA}\rangle$ is replaced by A, the mass number of the Pb nucleus.

In Pb-Pb collisions the p_T-differential yield of heavy-flavour decay electrons in the range 3 $< p_T <$ 18 GeV/c is measured at central rapidity. The measurement extends above the upper

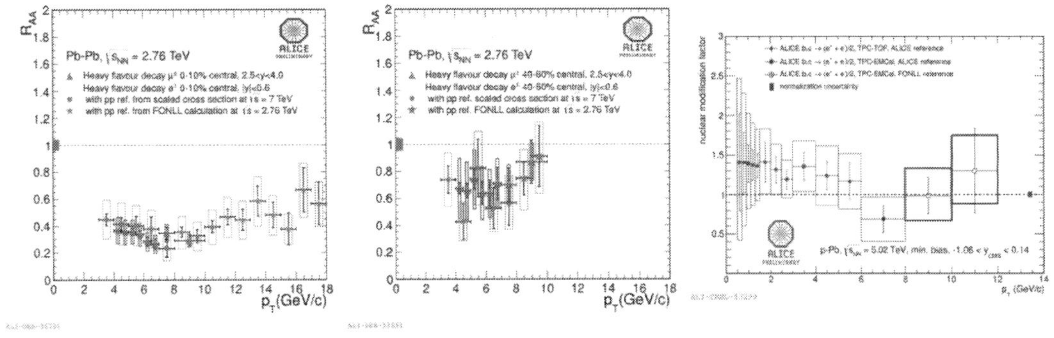

Figure 3. Left: Nuclear modification factor (R_{AA}) of electrons (circle and star symbols, depending on the pp reference) and muons [7] (triangles) from heavy-flavour hadron decays as a function of p_{T} in Pb-Pb collisions at $\sqrt{s_{\mathrm{NN}}} = 2.76$ TeV for the centrality class 0–10 %. Center: R_{AA} of electrons measured in the 40–80 % and muons in the 40–50 % centrality classes. Right: R_{pPb} of electrons from the decay of heavy-flavour hadrons as a function of p_{T}. The circle symbols represent the electrons identified using TPC-TOF, while those identified using TPC-EMCal are shown as square symbols. For all panels, the muons are measured in the forward rapidity region ($2.5 < y < 4$), while the electrons are observed at central rapidity ($|y| < 0.6$).

limit (8 GeV/c) of the measurement in pp at $\sqrt{s} = 7$ TeV, hence the FONLL calculation was used as the reference for $p_{\mathrm{T}} > 8$ GeV/c. The heavy-flavour decay muons are measured in the forward region in the range $4 < p_{\mathrm{T}} < 10$ GeV/c. Fig. 3 (left) shows the measured R_{AA} of the heavy-flavour decay electrons and muons for the centrality class 0–10 %. Both the muons and electrons reach a suppression factor 3–4 for $p_{\mathrm{T}} > 5$ GeV/c, which is where pQCD calculations predict that leptons from beauty-hadron decays start to dominate the p_{T}-differential cross section in pp collisions. This suggests that not only charm, but also beauty quarks are experiencing significant in-medium energy loss. Also shown in Fig. 3 (center) is the resulting R_{AA} measured in less central collisions, muons in the 40–80 % and electrons in the 40–50 % centrality class. The observation of less modification of the p_{T} spectrum is expected, as the medium formed in peripheral collisions should be less dense that that formed in a central collision.

Heavy-flavour decay electrons were measured in minimum bias p-Pb collisions at $\sqrt{s_{\mathrm{NN}}} = 5.02$ TeV in the rapidity interval $|\eta| < 0.6$. Using the TPC and the TOF for electron identification the measurement spans the p_{T} range 0.5–6 GeV/c, while electrons identified utilizing the TPC and the EMCal were measured in the p_{T} range 2–12 GeV/c. In the overlap region these results are in agreement, with the TPC–TOF based measurement featuring smaller uncertainties. The nuclear modification factor R_{pPb} is shown in Fig. 3 (right). For $p_{\mathrm{T}} < 8$ GeV/c the p-Pb p_{T}-differential cross section is compared to the pp reference measured at $\sqrt{s} = 7$ TeV and scaled to 5.02 TeV, while above 8 GeV/c the FONLL calculation was used as the reference. Although the experimental uncertainties are large, this result suggests small cold nuclear matter effects and indicates that the observed suppression observed in Pb-Pb collisions is a dense medium effect.

For non-central heavy-ion collisions, an azimuthal anisotropy in momentum space can be observed in the measured final state particle distributions. The particle distribution can be expanded in a Fourier series [16]

$$E\frac{\mathrm{d}^3N}{\mathrm{d}p^3} = \frac{1}{2\pi}\frac{\mathrm{d}^2N}{p_{\mathrm{T}}\mathrm{d}p_{\mathrm{T}}\mathrm{d}y}\left(1 + 2\sum_{n=1}^{\infty} \nu_n \cos[n(\varphi - \Psi_{RP})]\right) \quad (2)$$

Here ν_2 is the second Fourier coefficient, often referred to as "elliptic flow", φ is the particle

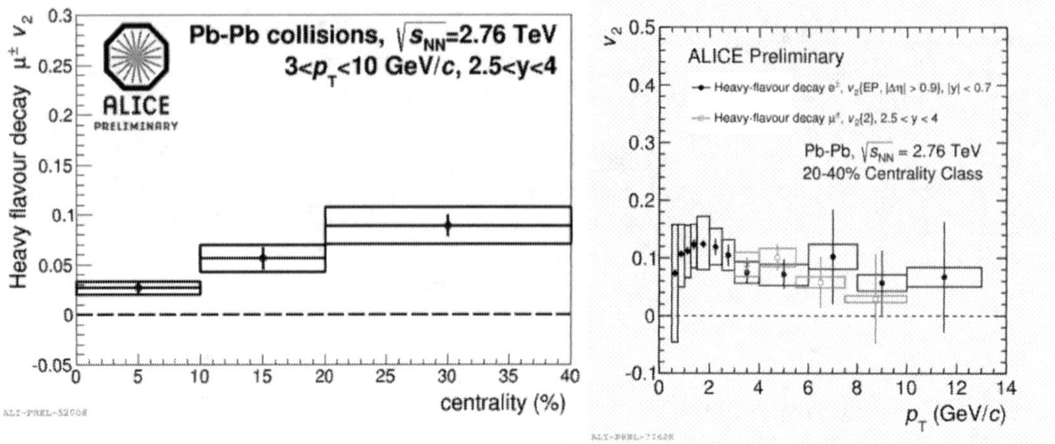

Figure 4. Left: p_T-integrated ν_2 of heavy-flavour decay muons as a function of collision centrality measured in Pb-Pb collisions at $\sqrt{s_{NN}}$ = 2.76 TeV in the p_T range 3–10 GeV/c in $2.5 < y < 4$. Right: Comparison of the measured ν_2 as a function of p_T in Pb-Pb collisions in the centrality class 20–40 % of the heavy-flavour decay muons (open squares) measured in the forward region ($2.5 < y < 4$) and electrons (full circles) measured at central rapidity ($|y| < 0.7$).

azimuthal angle, Ψ_{RP} is the reaction plane azimuthal angle. The reaction plane is defined by the beam direction and the collision impact parameter.

Fig. 4 (left) shows the p_T-integrated ($3 < p_T < 10$ GeV/c) heavy-flavour decay muon ν_2 as a function of centrality. From central (0–10 %) to semi-central (20–40 %) collisions ν_2 increases. The near-zero ν_2 for the most central collisions is consistent with the expectation that the initial geometrical anisotropy vanishes. The comparison of the heavy-flavour decay electron ν_2 as a function of p_T measured in $|y| < 0.7$ for the centrality class 20–40%, to the heavy-flavour decay muons measured in the same centrality class but in the range $2.5 < y < 4$ is shown in Fig. 4 (right). The values of heavy-flavour decay lepton ν_2 measured in the two rapidity regions are similar.

In Fig. 5 the measured electron R_{AA} (left) and ν_2 (right) are compared to five parton transport models: BAMPS, which includes collisional in-medium energy loss and mimics the radiative processes by increasing the elastic cross section [17], BAMPS + rad., where this version includes radiative processes [18], TAMU, which includes collisional energy loss, with resonance formation and dissociation in an evolving hydrodynamic medium [19], POWLANG, which is based on the Langevin equation with collisional energy loss in an expanding, deconfined medium [20], and MC@sHG+EPOS, which includes radiative and collisional energy loss in an expanding medium based on the EPOS model. The comparison to the models suggests that re-scatterings transfer to heavy quarks information on the azimuthal anisotropy of the medium.

In p-Pb collisions at $\sqrt{s_{NN}}$ = 5.02 TeV the angular correlation between heavy-flavour decay electrons and charged particles has been measured in three multiplicity classes. The multiplicity was measured with the V0A detector ($2.8 < \eta < 5.1$) in the Pb hemisphere. A typical two-particle angular correlation distribution is the difference in azimuth ($\Delta\varphi$) and in pseudorapidity ($\Delta\eta$) between a trigger particle and associated particle, normalized to the number of trigger particles. Here the trigger particle is the electron and the associated are the charge particles produced in the event. The electron sample contains a background contribution which is subtracted via the e^+e^- invariant mass method, so as to obtain the correlation between heavy-flavour decay electrons and charged particles. In Fig 6 (left) the measured angular correlation,

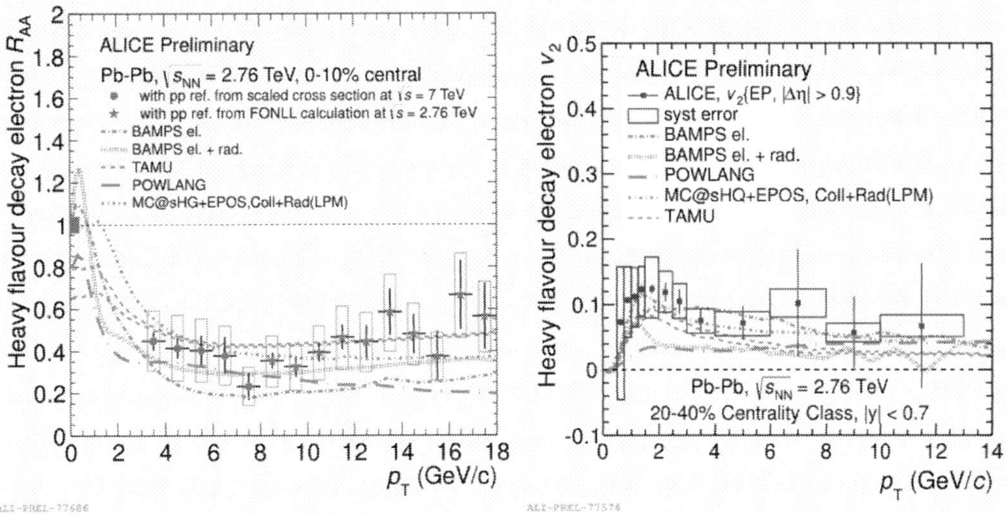

Figure 5. Left: Heavy-flavour decay electron $R_{\rm AA}$ as a function of $p_{\rm T}$ measured in the 0–10 % centrality class in $|y| < 0.6$. The circles (stars) indicate where the pp reference is taken as the cross section scaled from $\sqrt{s} = 7$ TeV (FONLL calculation at $\sqrt{s} = 2.76$ TeV). The results are compared to four parton transport models: BAMPS [17], BAMPS+rad. [18], TAMU [19], POWLANG [20], and MC@sHQ+EPOS, Coll+Rad(LPM) [21]. Right: Heavy-flavour decay electron v_2 as a function of $p_{\rm T}$ measured in the 20-40 % centrality class in $|y| < 0.6$. The results are compared to the aforementioned models.

after the projection on the $\Delta\varphi$ axis, for three multiplicity classes in p-Pb collisions at $\sqrt{s_{\rm NN}}$ =5.02 TeV and for pp collisions at $\sqrt{s} = 7$ TeV is shown, where the heavy-flavour decay electron is in the $p_{\rm T}$ range of 1-2 GeV/c. The distribution observed in the low multiplicity events has a similar shape on the near ($\frac{-\pi}{2} < \Delta\varphi < \frac{\pi}{2}$) and away ($\frac{\pi}{2} < \Delta\varphi < \frac{3\pi}{2}$) side to that obtained from pp collisions. However, a clear modification of the correlation function on both the near and away side is visible in high multiplicity p-Pb collisions compared to pp and low multiplicity p-Pb collisions. $I_{\rm CP}$, the ratio between the per-trigger yield in high and low multiplicity collisions, can be used to quantify the observed modification. In Fig 6 (center) the near side $I_{\rm CP}$ is shown as a function of the electron $p_{\rm T}$ and demonstrates that for low $p_{\rm T}$ electrons there is an unexpected enhancement of associated particles.

To further characterize the modification, the correlation distribution in $\Delta\varphi\Delta\eta$ measured in the low multiplicity event class was subtracted from that in the high multiplicity event class to reduce the contribution from jets. The result is shown in Fig 6 (right) and suggests a long range correlation in $\Delta\eta$, also observed for various hadron-hadron angular correlations measured in p-Pb collisions at $\sqrt{s_{\rm NN}} = 5.02$ TeV [22, 23]. The 'double ridge' structure can be described by either initial state effects, e.g. parton saturation in the nucleus [24], or by final state effects, e.g. hydrodynamic expansion in small systems [25, 26]. Regardless of which process(es) is(are) responsible for the 'double ridge', it seems that heavy-flavour hadrons are also affected by it.

5. Summary

Electrons and muons, which originate from the semi-leptonic decays of heavy-flavour hadrons, have been measured at mid-rapidity and forward rapidity in pp, Pb-Pb, and p-Pb collisions. The $p_{\rm T}$-differential production cross sections in pp collisions are well described by pQCD calculations. In Pb-Pb collisions, the nuclear modification factor indicates a strong suppression of the heavy-

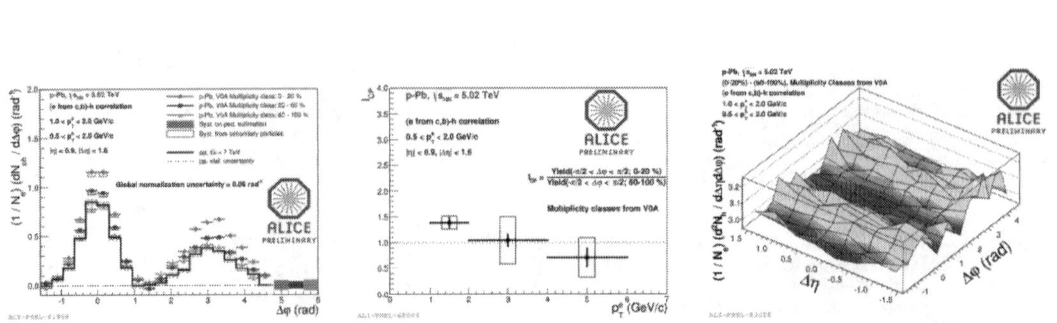

Figure 6. Left: Distribution of the azimuthal difference between heavy-flavour decay electrons and hadrons measured in the low (triangles), intermediate (squares), and high (circles) multiplicity classes in p-Pb collisions at $\sqrt{s_{\rm NN}} = 5.02$ TeV. The resulting distribution measured in pp collisions at $\sqrt{s} = 7$ TeV is shown as the black line. Center: The ratio between the near-side yields in high (0–20%) and low (60–100%) multiplicity p-Pb collisions ($I_{\rm CP}$) for three electron $p_{\rm T}$ ranges. Right: Difference between the two-particle correlation distributions measured in high and low multiplicity p-Pb collisions, in $\Delta\varphi$ $\Delta\eta$.

flavour yield at high $p_{\rm T}$ in the central collisions, which points to significant energy loss of heavy quarks in the produced medium. v_2 measurements of heavy-flavour decay leptons in Pb-Pb collisions indicate a positive v_2 in semi-central collisions, suggesting re-scatterings transfer information on the azimuthal anisotropy of the medium to heavy quarks. In p-Pb collisions, $R_{\rm pPb}$ is consistent with small cold nuclear matter effects, demonstrating that the observed suppression in central Pb-Pb collisions is mainly a final state effect, due to in-medium energy loss. In high multiplicity p-Pb collisions the measurement of angular correlation between heavy-flavour decay electrons and charged particles reveals a double ridge structure, hinting that the mechanism responsible for the long range correlation observed for light-flavour hadrons may also affect heavy-quark dynamics.

References

[1] Aamodt K *et al.* (ALICE Collaboration) 2008 *JINST* **3** S08002
[2] Abelev B *et al.* (ALICE Collaboration) 2012 *JHEP* **1207** 191
[3] Abelev B *et al.* (ALICE Collaboration) 2012 *JHEP* **1211** 065
[4] Alessandro B *et al.* (ALICE Collaboration) 2006 *J. Phys. G* **32** 1295
[5] Abelev B *et al.* (ALICE Collaboration) 2012 *Phys.Rev.* **D86** 112007
[6] Aad G *et al.* (ATLAS Collaboration) 2012 *Phys.Lett.* **B707** 438–458
[7] Abelev B *et al.* (ALICE Collaboration) 2012 *Phys.Lett.* **B708** 265–275
[8] Cacciari M, Frixione S, Houdeau N, Mangano M L, Nason P *et al.* 2012 *JHEP* **1210** 137
[9] Abelev B *et al.* (ALICE Collaboration) 2013 *Phys.Lett.* **B721** 13–23
[10] Averbeck R, Bastid N, del Valle Z C, Crochet P, Dainese A *et al.* 2011 (*Preprint* 1107.3243)
[11] Abelev B *et al.* (ALICE Collaboration) 2012 *Phys. Rev. Lett.* **109**(11) 112301
[12] Abelev B *et al.* (ALICE Collaboration) 2014 (*Preprint* 1405.4144)
[13] Bolzoni P and Kramer G 2013 *Nucl.Phys.* **B872** 253–264 (*Preprint* 1212.4356)
[14] Maciula R and Szczurek A 2013 (*Preprint* 1301.3033)
[15] Aamodt K *et al.* (ALICE Collaboration) 2008 *Phys. Rev. Lett.* **106** S08002
[16] Poskanzer A M and Voloshin S A 1998 *Phys. Rev. C* **58** 1671
[17] Uphoff J, Fochler O, Xu Z and Greiner C 2012 *Phys.Lett.* **B717** 430–435
[18] Uphoff J, Fochler O, Xu Z and Greiner C 2014 *J.Phys.Conf.Ser.* **509** 012077
[19] He M, Fries R J and Rapp R 2014 (*Preprint* 1401.3817)
[20] Alberico W, Beraudo A, De Pace A, Molinari A, Monteno M *et al.* 2013 *Eur.Phys.J.* **C73** 2481
[21] Nahrgang M, Aichelin J, Gossiaux P B and Werner K 2013 (*Preprint* 1305.6544)
[22] Abelev B *et al.* (ALICE Collaboration) 2013 *Phys.Lett.* **B719** 29–41
[23] Abelev B *et al.* (ALICE Collaboration) 2013 *Phys.Lett.* **B726** 164–177
[24] Dusling K and Venugopalan R 2013 *Phys.Rev.* **D87** 094034
[25] Werner K, Karpenko I and Pierog T 2011 *Phys.Rev.Lett.* **106** 122004
[26] Bozek P and Broniowski W 2014 (*Preprint* 1403.6042)

Journal of Physics: Conference Series **535** (2014) 012030 doi:10.1088/1742-6596/535/1/012030

Fluctuations of conserved charges on the lattice and in heavy ion collisions

S. Borsanyi[1], Z. Fodor[1,2,3], S. D. Katz[2,4], S. Krieg[1,3], C. Ratti[5,6], K. K. Szabo[1,3]

[1] Department of Physics, Wuppertal University, Gaussstr. 20, D-42119 Wuppertal, Germany
[2] Inst. for Theoretical Physics, Eötvös University, Pázmány P. sétány 1/A, H-1117 Budapest, Hungary
[3] Jülich Supercomputing Centre, Forschungszentrum Jülich, D-52425 Jülich, Germany
[4] MTA-ELTE "Lendület" Lattice Gauge Theory Research Group, Pázmány P. sétány 1/A, H-1117 Budapest, Hungary
[5] Dip. di Fisica, Università di Torino and INFN, Sezione di Torino via Giuria 1, I-10125 Torino, Italy
[6] Department of Physics, University of Houston, Houston, TX 77204, USA

E-mail: ratti@to.infn.it

Abstract. We present our latest results for fluctuations of electric charge and baryon number, simulated on the lattice in a system of 2+1 dynamical quark flavors at the physical quark masses and continuum extrapolated. In order to extract the chemical freeze-out temperature and chemical potential, we compare our results to the moments of multiplicity distribution of the corresponding conserved charges, measured in heavy ion collision experiments by the STAR collaboration. Consistency between the freeze-out parameters obtained through different conserved charges is discussed.

1. Introduction

Fluctuations of conserved charges have been the object of intense theoretical and experimental activity in the last few years. Several studies have identified fluctuations as ideal observables to spot the change from crossover to first-order phase transition in the QCD phase diagram [1, 2, 3]. Consequently the beam energy scan program at RHIC, in search for the QCD critical point, has recently published efficiency corrected results for these observables, in the collision-energy range $\sqrt{s} = 7.7 - 200$ GeV [4, 5]. Besides, it has been proposed to use fluctuations of conserved charges to extract the chemical freeze-out conditions of a heavy-ion collision from first principles, by comparing their experimental value to results of lattice QCD simulations at finite temperature and (small) chemical potential [6, 7, 8]. Such comparison assumes that, in the evolution of a heavy-ion collision, it is possible to identify two lines in the (T, μ_B) plane. The first one, called chemical freeze-out, is the point at which all inelastic interactions between particles cease: conserved charges are frozen at this time. The second one is the so-called kinetic freeze-out, at lower temperatures, at which all elastic interactions cease: the momentum of the particles is frozen at this time.

The chemical freeze-out has been extensively studied in the past, by means of a Statistical Hadronization Model (SHM) fit to the particle yields and ratios measured in experiments

Journal of Physics: Conference Series **535** (2014) 012030 doi:10.1088/1742-6596/535/1/012030

[9, 10, 11, 12]. By decreasing the collision energy, the freeze-out chemical potential increases; repeating the analysis for a series of beam energies provides a freeze-out curve in the (T, μ) plane. Our lattice analysis relies on the same assumption as the SHM, namely that the experimentally measured fluctuations can be described in terms of a thermally equilibrated system. However, the advantage is a first principle determination of the freeze-out parameters, without having to rely on a model which can lead to different results when a different hadronic spectrum is included [13, 14].

In comparing lattice results to experimental ones, one has to make sure that possible sources of non-thermal fluctuations are kept under control. The data published by the STAR collaboration are corrected for several effects: the centrality-bin-width correction method minimizes effects due to volume variation because of finite centrality bin width; the finite reconstruction efficiency of the detector is corrected for, based on binomial probability distribution [15]; spallation protons, coming from the interaction with the beam-pipe, are removed by appropriate cuts in p_T [5]. Effects due to kinematic cuts have been investigated in the Hadron Resonance Gas model and found to be small [16, 17]. Interactions in the hadronic phase and non-equilibrium effects might in principle affect fluctuations [18, 19]; the analysis that we are performing can be considered as a test for the validity of the equilibrium scenario: we are checking whether the freeze-out parameters yielded by different conserved charges (electric charge and baryon number) are consistent with each other. In particular, while the freeze-out temperature might be flavor-dependent [20], the chemical potentials as a function of the collision energy should be the same for all species. Therefore, in our approach we are able to check whether the three conserved charges B, Q, S have a common freeze-out surface, or whether the chemical freeze-out is a more involved scenario which allows different surfaces for different conserved charges. The present level of precision reached by lattice QCD results, obtained at physical quark masses and continuum-extrapolated, allows to perform this check for the first time [21].

One more caveat in our analysis is worth mentioning: while lattice QCD can only address conserved charges, experimental fluctuations are limited to a sub-set of the full hadron spectrum; in particular, the net-proton multiplicity distribution is measured, as opposed to the lattice net-baryon number fluctuations. Recently it was shown that, once the effects of resonance feed-down and isospin randomization are taken into account [22, 23], the net-proton and net-baryon number fluctuations are numerically very similar [24].

In this manuscript we give an overview of the analysis that we performed in [21]: we systematically compare our continuum-extrapolated results on fluctuations of electric charge and baryon number to the most recent, efficiency corrected measurements by the STAR collaboration [4, 5]. This allows us to extract an upper limit for the chemical freeze-out temperature, as well as freeze-out chemical potentials for the four highest energies at RHIC. The independent analyses of electric charge and baryon number yield consistent results, thus validating our approach and the assumptions on which it is based.

2. Results

The definition of the fluctuations of baryon number, electric charge and strangeness is

$$\chi_{lmn}^{BSQ} = \frac{\partial^{l+m+n}(p/T^4)}{\partial(\mu_B/T)^l \partial(\mu_S/T)^m \partial(\mu_Q/T)^n}; \tag{1}$$

they are related to the moments of the multiplicity distributions of the corresponding conserved charges. The following, volume-independent ratios, are usually introduced, in order to get rid of the extra volume unknown parameter:

$$\begin{aligned} \chi_3/\chi_2 = S\sigma \quad &; \quad \chi_4/\chi_2 = \kappa\sigma^2 \\ \chi_1/\chi_2 = M/\sigma^2 \quad &; \quad \chi_3/\chi_1 = S\sigma^3/M\,. \end{aligned} \tag{2}$$

Journal of Physics: Conference Series **535** (2014) 012030 doi:10.1088/1742-6596/535/1/012030

The chemical potentials μ_B, μ_Q and μ_S are not independent: a relationship between them is obtained, in order to match the experimental situation. The finite baryon density in the system is due to light quarks only, since it is generated by the nucleon stopping in the collision region. The strangeness density $\langle n_S \rangle$ is then equal to zero for all collision energies, as a consequence of strangeness conservation. Besides, the electric charge and baryon-number densities are related, in order to match the isospin asymmetry of the colliding nuclei: $\langle n_Q \rangle = Z/A \langle n_B \rangle$. $Z/A = 0.4$ represents a good approximation for Pb-Pb and Au-Au collisions. As a consequence, μ_Q and μ_S depend on μ_B so that these conditions are satisfied. This is achieved by Taylor-expanding the densities in these three chemical potentials up to μ_B^3 [7]:

$$\begin{aligned} \mu_Q(T, \mu_B) &= q_1(T)\mu_B + q_3(T)\mu_B^3 + ... \\ \mu_S(T, \mu_B) &= s_1(T)\mu_B + s_3(T)\mu_B^3 + ... \end{aligned} \tag{3}$$

Our continuum extrapolated results for the functions $q_1(T)$, $q_3(T)$, $s_1(T)$, $s_3(T)$ were shown in [8]. The quantities that we consider to extract the freeze-out T and μ_B, are the ratios $R_{31}^B = \chi_3^B/\chi_1^B$ and $R_{12}^B = \chi_1^B/\chi_2^B$ respectively, at values of (μ_B, μ_Q, μ_S), which satisfy the pyhsical conditions discussed in the previous paragraph.

The first terms of their Taylor expansion around $\mu_B = 0$ read:

$$\begin{aligned} R_{31}^B(T,\mu_B) &= \frac{\chi_3^B(T,\mu_B)}{\chi_1^B(T,\mu_B)} = \frac{\chi_4^B(T,0) + \chi_{31}^{BQ}(T,0)q_1(T) + \chi_{31}^{BS}(T,0)s_1(T)}{\chi_2^B(T,0) + \chi_{11}^{BQ}(T,0)q_1(T) + \chi_{11}^{BS}(T,0)s_1(T)} + \mathcal{O}(\mu_B^2) \\ R_{12}^B(T,\mu_B) &= \frac{\chi_1^B(T,\mu_B)}{\chi_2^B(T,\mu_B)} = \frac{\chi_2^B(T,0) + \chi_{11}^{BQ}(T,0)q_1(T) + \chi_{11}^{BS}(T,0)s_1(T)}{\chi_2^B(T,0)}\frac{\mu_B}{T} + \mathcal{O}(\mu_B^3). \end{aligned} \tag{4}$$

From the above equations, it is straightforward to identify R_{31}^B as an ideal thermometer, since its leading order is independent of μ_B. Once the freeze-out temperature has been extracted from R_{31}^B, it can then be plugged into R_{12}^B: the latter can then be used to extract the freeze-out chemical potential (notice that our results for R_{12}^B are obtained up to NLO in μ_B).

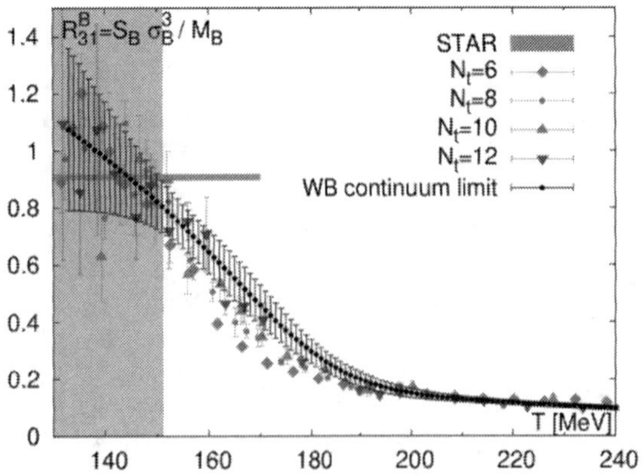

Figure 1. R_{31}^B: the colored symbols show finite-N_t lattice QCD results. The continuum extrapolation is represented by black points (from Ref. [8]). The dark-orange band shows the recent experimental measurement by the STAR collaboration [4]: it was obtained by averaging the 0-5% and 5-10% data at the four highest energies ($\sqrt{s}$ = 27, 39, 62.4, 200 GeV). The green-shaded area shows the valid temperature range.

A similar expansion in μ_B holds also for the fluctuations of electric charge [8]: the leading order in χ_3^Q/χ_1^Q is independent of μ_B, while the LO in χ_2^Q/χ_1^Q is linear in μ_B. The present precision reached by the experimental results on χ_3^Q/χ_1^Q does not allow an independent determination of T_f. However, we are able to extract independent estimates of the freeze-out chemical potential from R^B_{12} and R^Q_{12}.

In Fig. 1 we show the comparison between the lattice results for $\chi^B_3(T,\mu_B)/\chi^B_1(T,\mu_B)$ and the experimental measurement of $S_p\sigma^3_p/M_p$ by the STAR collaboration [4]. The latter is obtained by averaging the 0-5% and 5-10% centrality data, at the four highest energies ($\sqrt{s} = 27$, 39, 62.4, 200 GeV). Since the curvature of the phase diagram is small around $\mu_B = 0$ [25], it is safe to assume that this average allows to determine the freeze-out temperature for small μ_B. The green-shaded area shows the valid temperature range: due to the uncertainty on the lattice results in the low-temperature regime, at the moment it is only possible to extract an upper value for the freeze-out temperature: the freeze-out takes place at a temperature $T_f \leq 151$ MeV, which is somewhat lower than expected from previous analyses based on preliminary, not efficiency corrected data [26] (allowing for a two-sigma deviation for the lattice simulations and the experimental measurements, the highest possible T_f is 155 MeV). In Refs. [27, 28] we have published the lattice determination of the transition temperature from various chiral observables in the range 147-157 MeV. For the minimum of the speed of sound we found 145(5) MeV in [29]. The discussed freeze-out temperature is thus in the cross-over region around or slightly below the central value.

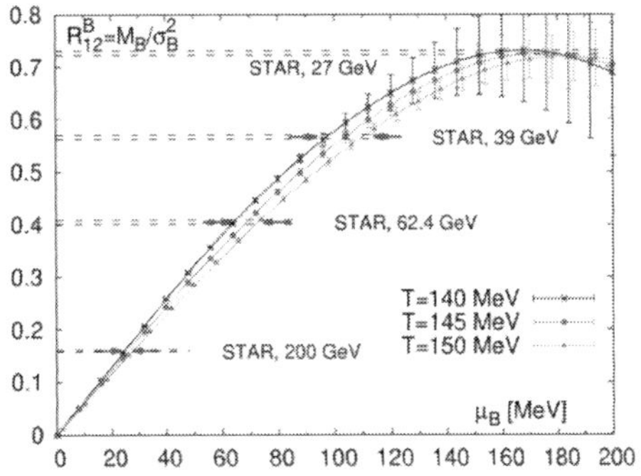

Figure 2. R^B_{12} as a function of μ_B. The three points correspond to the STAR data for M_p/σ^2_p at collision energies $\sqrt{s} = 39$, 62.4, 200 GeV and centrality 0-10%, from Ref. [4] (the $\sqrt{s} = 27$ GeV point is also shown, but the non-monotonicity of the lattice results at $\mu_B \geq 130$ MeV does not allow a determination of μ_B from it). The arrows show the extracted values for μ_B at freeze-out.

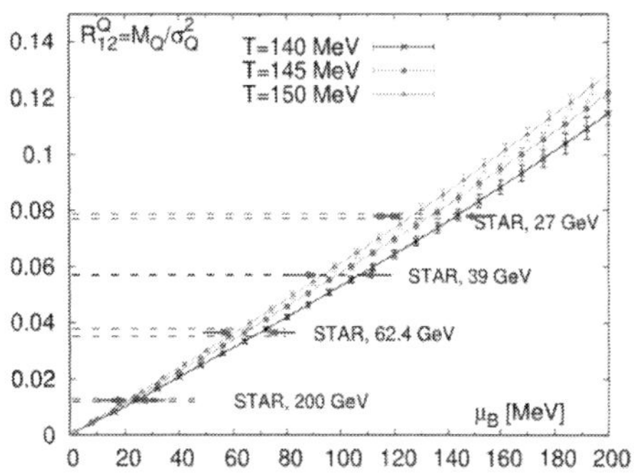

Figure 3. R^Q_{12} as a function of μ_B. The four points correspond to the STAR data for M_Q/σ^2_Q at $\sqrt{s} = 27$, 39, 62.4, 200 GeV and centrality 0-10%, from Ref. [5]. In both figures, the colored symbols correspond to the lattice QCD results in the continuum limit, for the range ($140 \leq T_f \leq 150$) MeV. The arrows show the extracted values for μ_B at freeze-out.

We can now determine the freeze-out chemical potential μ_B, by comparing the lattice results for R^B_{12} as function of the chemical potential, and in the temperature range ($140 \leq T_f \leq 150$) MeV to the experimental results for M_p/σ^2_p published by the STAR collaboration in Ref. [4, 30]. This procedure is shown in Fig. 2. In Fig. 3 we repeat the analysis for R^Q_{12} vs M_Q/σ^2_Q published by the STAR collaboration in Ref. [5]. The two quantities allow for an independent determination of μ_B from electric charge and baryon number: the corresponding values are listed in Table 1, and shown in Fig. 4. Consistency between the two values of baryon-chemical potential is found for all collision energies (notice that the non-monotonicity of the lattice results for R^B_{12} at

$\mu_B \geq 130$ MeV does not allow a determination of μ_B from this observable at $\sqrt{s} = 27$ GeV). We also compare the chemical potentials in Table 1 to those found earlier in statistical fits [11, 9]: we find a remarkable agreement (see Fig. 4).

Table 1. Freeze-out μ_B vs. $\sqrt{s}$, for the four highest-energy STAR measurements. The μ_B values and error-bars in this table have been obtained under the assumption that 140 MeV $\leq T_f \leq$ 150 MeV. This uncertainty dominates the overall errors. Other (minor) sources of uncertainty are the lattice statistics and the experimental error.

$\sqrt{s}[GeV]$	μ_B^f [MeV] (from B)	μ_B^f [MeV] (from Q)
200	25.8±2.7	22.8±2.6
62.4	69.7±6.4	66.6±7.9
39	105±11	101±10
27	-	136±13.8

Consistency between the two values of baryon-chemical potential is found for all collision energies (the non-monotonicity of the lattice results for R_{12}^B at $\mu_B \geq 130$ MeV does not allow a determination of μ_B from this observable at $\sqrt{s} = 27$ GeV). We also compare the chemical potentials in Table 1 to those found earlier in statistical fits [11, 9]: we find a remarkable agreement (see Fig. 4).

Note that, in spite of the remarkable agreement between the chemical potentials obtained from the lattice and SHM analyses, for the freeze-out temperature statistical models typically yield a somewhat higher value: e.g. 164 MeV in Refs. [32, 9]. The latest fit to particle ratios from the ALICE collaboration [35] yields a temperature $T_{ch} \simeq 156$ MeV. However, it was pointed out that the protons are overestimated by this temperature: they would be optimally reproduced with a temperature consistent with the upper limit that we find in our analysis. Towards the lower end in temperature range we find also Ref. [33] with $T_f = 155 \pm 8$ MeV and $\mu_B = 25 \pm 1$ MeV, at $\sqrt{s} = 200$ GeV.

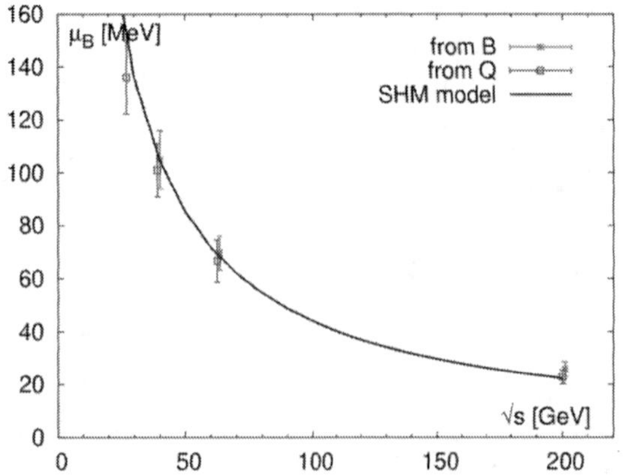

Figure 4. Freeze-out chemical potential μ_B as a function of the collision energy. The red stars show the μ_B obtained by fitting R_{12}^B, the blue squares have been obtained by fitting R_{12}^Q. The black curve comes from the Statistical Hadronization Model analysis of Refs. [9].

Our analysis finds a consistency between the freeze-out chemical potentials obtained from electric charge and baryon number, if we assume an agreement in the corresponding freeze-out temperatures. If the freeze-out can be described by the same temperature and chemical potentials for charge and protons, then one can define a combined observable: $R_{12}^{Q}/R_{12}^{B} = [M_Q/\sigma_Q^2]/[M_B/\sigma_B^2]$. Here, the volume factor of the charge and baryon (proton) measurements cancel separately. This ratio of ratios is an ideal thermometer: it is far easier to obtain both for lattice and experiment since it does not involve skewness or kurtosis which are obviously affected by larger error bars; besides, the lattice curve shows a high sensitivity to the temperature, thus allowing a more precise determination of T_f. A comparison between lattice results and experimental measurement for this observable is shown in Fig. 5. Contrary to the skewness thermometer, here we see a clear monotonic temperature dependence without the hardly controllable lattice errors at low temperatures. This allows for the identification of a narrow temperature band, instead of an upper limit.

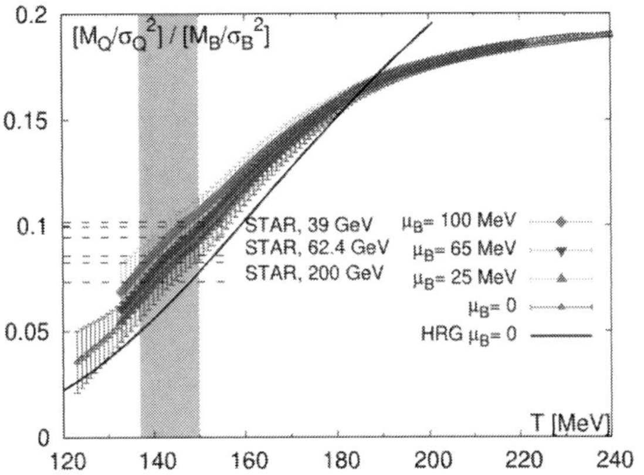

Figure 5. R_{12}^{Q}/R_{12}^{B}: the colored symbols correspond to continuum-extrapolated lattice QCD simulations at different values of μ_B. The dashed lines shows the recent experimental measurements by the STAR collaboration [4, 5] for a 0-10% centrality and different collision energies. The green-shaded area shows the valid temperature range, $T_f = (144 \pm 6)$ MeV.

The thermometer in Fig. 5 represents also an important consistency criterion: it agrees to the experimental data in the temperature range which needs to be assumed to have a common chemical potential between electric charge and baryon number. For high enough energies ($\sqrt{s} \geq$ 39 GeV) this consistency is granted if freeze-out occurs in the range $T_f = 144 \pm 6$ MeV. Notice that this temperature range lies just below the upper limit that we determined independently in Fig. 1.

3. Conclusions

In the present paper we have extracted the freeze-out conditions of heavy-ion collisions (temperature and chemical potential) from first principles, by comparing our continuum extrapolated lattice QCD results to the experimental moments of net-charge and net-proton multiplicity distribution by the STAR collaboration. Due to the present uncertainty in the lattice data, as well as the difficulty of simulating at finite chemical potential, we were able to obtain an upper limit for the chemical freeze-out temperature, and freeze-out chemical potentials for the four highest collision energies at RHIC. These new, efficiency corrected, experimental data point at a lower freeze-out temperature compared to previous estimates. This is compatible with the expectation that the freeze-out should occur just below the transition. The freeze-out chemical potentials obtained from an independent analysis of electric charge and baryon number fluctuations show a remarkable consistency with each other, as well as with previous results from the Statistical Hadronization Model. This comparison is possible for the first time, and the consistency of the results is of fundamental importance to validate the hypothesis on which this method is based, namely that the experimentally created system is close to thermal equilibrium at the freeze-out and can be described by lattice QCD simulations, at least in the light quark sector.

Acknowledgements

C. Ratti acknowledges fruitful discussions with Francesco Becattini, Rene Bellwied and Bill Llope. This project was funded by the DFG grant SFB/TR55. The work of C. Ratti is supported by funds provided by the Italian Ministry of Education, Universities and Research under the Firb Research Grant RBFR0814TT. S. D. Katz is funded by the ERC grant ((FP7/2007-2013)/ERC No 208740) as well as the "Lendület" program of the Hungarian Academy of Sciences ((LP2012-44/2012). The numerical simulations were performed on the QPACE machine, the GPU cluster at the Wuppertal University and on JUQUEEN (the Blue Gene/Q system of the Forschungszentrum Juelich) and on MIRA at the Argonne Leadership Computing Facility through the Innovative and Novel Computational Impact on Theory and Experiment (INCITE) program of the U.S. Department of Energy (DOE).

References

[1] M. A. Stephanov, K. Rajagopal and E. V. Shuryak, Phys. Rev. D **60**, 114028 (1999)
[2] R. V. Gavai and S. Gupta, Phys. Rev. D **78**, 114503 (2008)
[3] M. Cheng, *et al.*, Phys. Rev. D **77**, 014511 (2008)
[4] L. Adamczyk *et al.* [STAR Collaboration], Phys. Rev. Lett. **112**, no. 3, 032302 (2014)
[5] L. Adamczyk *et al.* [STAR Collaboration], arXiv:1402.1558 [nucl-ex].
[6] F. Karsch, Central Eur. J. Phys. **10**, 1234 (2012)
[7] A. Bazavov, *et al.*, Phys. Rev. Lett. **109**, 192302 (2012)
[8] S. Borsanyi, Z. Fodor, S. D. Katz, S. Krieg, C. Ratti and K. K. Szabo, Phys. Rev. Lett. **111** (2013) 062005
[9] A. Andronic, P. Braun-Munzinger and J. Stachel, Phys. Lett. B **673** (2009) 142 [Erratum-ibid. B **678** (2009) 516]
[10] F. Becattini, J. Manninen and M. Gazdzicki, Phys. Rev. C **73**, 044905 (2006)
[11] J. Cleymans, H. Oeschler, K. Redlich and S. Wheaton, Phys. Rev. C **73** (2006) 034905
[12] J. Manninen and F. Becattini, Phys. Rev. C **78**, 054901 (2008)
[13] J. Noronha-Hostler and C. Greiner, arXiv:1405.7298 [nucl-th].
[14] A. Bazavov, H. -T. Ding, P. Hegde, O. Kaczmarek, F. Karsch, E. Laermann, Y. Maezawa and S. Mukherjee *et al.*, arXiv:1404.6511 [hep-lat].
[15] A. Bzdak and V. Koch, Phys. Rev. C **86** (2012) 044904
[16] P. Garg, D. K. Mishra, P. K. Netrakanti, B. Mohanty, A. K. Mohanty, B. K. Singh and N. Xu, Phys. Lett. B **726**, 691 (2013)
[17] P. Alba, W. Alberico, R. Bellwied, M. Bluhm, V. M. Sarti, M. Nahrgang and C. Ratti, arXiv:1403.4903 [hep-ph].
[18] J. Steinheimer, J. Aichelin and M. Bleicher, Phys. Rev. Lett. **110** (2013) 042501
[19] F. Becattini, M. Bleicher, T. Kollegger, T. Schuster, J. Steinheimer and R. Stock, Phys. Rev. Lett. **111** (2013) 082302
[20] R. Bellwied, S. Borsanyi, Z. Fodor, S. DKatz and C. Ratti, Phys. Rev. Lett. **111** (2013) 202302
[21] S. Borsanyi, Z. Fodor, S. D. Katz, S. Krieg, C. Ratti and K. K. Szabo, arXiv:1403.4576 [hep-lat].
[22] M. Kitazawa and M. Asakawa, Phys. Rev. C **85** (2012) 021901
[23] M. Kitazawa and M. Asakawa, Phys. Rev. C **86** (2012) 024904 [Erratum-ibid. C **86** (2012) 069902]
[24] M. Nahrgang, M. Bluhm, P. Alba, R. Bellwied and C. Ratti, arXiv:1402.1238 [hep-ph].
[25] G. Endrodi, Z. Fodor, S. D. Katz and K. K. Szabo, JHEP **1104**, 001 (2011)
[26] S. Mukherjee and M. Wagner, PoS CPOD **2013** (2013) 039
[27] Y. Aoki, S. Borsanyi, S. Durr, Z. Fodor, S. D. Katz, S. Krieg and K. K. Szabo, JHEP **0906** (2009) 088
[28] S. Borsanyi *et al.* [Wuppertal-Budapest Collaboration], JHEP **1009** (2010) 073
[29] S. Borsanyi, G. Endrodi, Z. Fodor, A. Jakovac, S. D. Katz, S. Krieg, C. Ratti and K. K. Szabo, JHEP **1011** (2010) 077
[30] The efficiency-corrected data for the lowest cumulant ratio (c_2/c_1) for net-protons can be found on the public STAR webpage.
[31] A. Andronic, P. Braun-Munzinger and J. Stachel, Nucl. Phys. A **772** (2006) 167
[32] J. Adams *et al.* [STAR Collaboration], Nucl. Phys. A **757** (2005) 102
[33] J. Rafelski, J. Letessier and G. Torrieri, Phys. Rev. C **72** (2005) 024905
[34] P. Braun-Munzinger, J. Stachel and C. Wetterich, Phys. Lett. B **596** (2004) 61
[35] M. Floris, Proceedings of Quark Matter 2014, to appear in Nucl. Phys. A.

Journal of Physics: Conference Series **535** (2014) 012031 doi:10.1088/1742-6596/535/1/012031

In-medium meson properties and screening correlators

A. Bazavov[a], **F. Karsch**[a,b], **Y. Maezawa**[b], **Swagato Mukherjee**[a] **and P. Petreczky**[a]

[a] Physics Department, Brookhaven National Laboratory, Upton, NY 11973, USA,
[b] Fakultät für Physik, Universität Bielefeld, D-33615 Bielefeld, Germany

Abstract. We study spatial meson correlation functions consisting of strange quarks, strange and charm quarks and charm quarks in $(2+1)$-flavor QCD using the highly improved staggered quark action. We find that the in-medium modification of the meson correlators decreases with increasing charm quark content and decreasing size. In particular, we find strong in-medium modification of ϕ and D_s meson correlators around the chiral transition temperature T_c, while J/ψ and η_c correlators show strong in-medium modification only at temperatures of $1.4T_c$.

1. Introduction

At high temperatures matter controlled by the strong force undergoes a deconfinement transition at which the relevant degrees of freedom change from hadrons to quarks and gluons (see e.g. Refs. [1, 2] for recent reviews). Hadronic correlation functions have been advocated since long as a tool to learn about the properties of this strong interaction matter [3, 4]. They encode the in-medium properties of hadrons and their dissolution. Moreover, through the comparison of lattice results with weak coupling calculations at high temperature [5, 6] they also provide information on the change from strongly to weakly interacting matter. In particular, the modifications of meson correlators in the light and strange quark sector reflect the in-medium change of meson properties and the partial restoration of chiral symmetry. The in-medium modification and dissolution of heavy quarkonium was suggested as the signal for creating a deconfined medium in heavy ion collisions by Matsui and Satz [7]. The existence of heavy light-mesons above the chiral transition temperature has also been proposed to explain the large energy loss and flow of heavy quarks observed in heavy ion collisions [8].

Spectral functions, defined in terms of the Fourier transform of the real time meson correlation functions, provide the most straightforward way to describe in-medium meson properties and their dissolution. Meson states appear as peaks in the corresponding spectral functions with the peak position equal to the meson mass. The width of the peak corresponds to the in-medium width of the meson. However, lattice QCD is formulated in Euclidean space time. Temporal meson correlation functions calculated on the lattice,

$$G(\tau,\vec{p},T) = \int d^3x e^{i\vec{p}.\vec{x}} \langle J_H(\tau,\vec{x}) J_H(0,\vec{0}) \rangle \ , \qquad (1)$$

have a simple relation to the spectral function, $\sigma(\omega, \vec{p}, T)$:

$$G(\tau, \vec{p}, T) = \int_0^\infty d\omega \sigma(\omega, \vec{p}, T) K(\omega, \tau), K(\omega, \tau) = \frac{\cosh(\omega(\tau - 1/2T))}{\sinh(\omega/2T)}. \tag{2}$$

Here J_H is a meson operator, typically of the form $J_H = \bar{q}\Gamma_H q$, with Γ_H being some combination of the Dirac matrices that specifies the quantum numbers of the meson. One way to obtain the spectral function from the above relation is to use the maximum entropy method (MEM) [9, 10, 11, 12, 13, 14, 15, 16]. The analysis of temporal correlation functions is difficult due to the limited extent, $1/T$, in Euclidean time direction. In the case of heavy quarkonium correlators, for instance, it turned out that the melting of bound states does not lead to large changes in the correlation functions [17, 16]. In order to become sensitive to the corresponding disappearance of a resonance peak in the spectral function high statistical accuracy and the analysis of the correlation function at a large number of Euclidean time separations is needed. At fixed temperature $T = 1/N_\tau a$, this requires large lattices with temporal extent N_τ and sufficiently small lattice spacing, a.

On the lattice one can also study spatial meson correlation functions

$$G(z, T) = \int_0^{1/T} d\tau \int dx dy \langle J_H(\tau, x, y, z) J_H(0, 0, 0, 0) \rangle. \tag{3}$$

These are related to the spectral functions in a more complicated way that also involves integration over the spatial momenta,

$$G(z, T) = \int_0^\infty \frac{2d\omega}{\omega} \int_{-\infty}^\infty dp_z e^{ip_z z} \sigma(\omega, p_z, T). \tag{4}$$

On the other hand the spatial separation is not limited by the inverse temperature and the spatial correlation function can be studied at separations larger than $1/T$. Therefore the spatial correlation functions can be more sensitive to in-medium modifications and/or the dissolution of mesons. While the relation between spectral functions and spatial meson correlators is more involved in general, in some limiting cases it becomes simple. At large distances the spatial correlation functions drop exponentially, $G(z, T) \sim \exp(-Mz/T)$, where M is known as the *screening mass*. At small enough temperatures when there exists a well-defined mesonic bound state, the spectral function has the form $\sigma(\omega, 0, 0, p_z, T) \sim \delta(\omega^2 - p_z^2 - m_0^2)$, and M becomes equal to the (pole) mass m_0 of the meson. On the other hand, at high enough temperatures, when the mesonic excitations are completely melted, the spatial meson correlation functions describe the propagation of a free quark-antiquark pair. The screening masses are then given by

$$M_{\text{free}} = \sqrt{m_{q1}^2 + (\pi T)^2} + \sqrt{m_{\bar{q}2}^2 + (\pi T)^2}\,, \tag{5}$$

where m_{q1} and $m_{\bar{q}2}$ are the masses of the quark and anti-quark that form the meson. This form of the screening mass in the non-interacting limit is a direct consequence of the anti-periodic boundary conditions in Euclidean time that are needed for the representation of fermions at non-zero temperature. This leads to the appearance of a smallest non-zero Matsubara frequency, πT, in the quark and anti-quark propagators. As the bosonic meson state is dissolved in the non-interacting limit the screening mass results as the contribution of two independently propagating fermionic degrees of freedom. Thus the transition between these two limiting values of the screening mass can be used as an indicator for the thermal modification and eventual dissolution of mesonic excitations.

In this contribution we study spatial strange-strange, charm-strange and charm-charm meson correlators and screening masses using the Highly Improved Staggered Quark (HISQ) action [24] with a strange quark mass tuned to its physical value and almost physical, degenerate up and down quark masses. The HISQ action is known to have discretization effect that are smaller than those observed with all other staggered-type actions currently used in studies of lattice QCD thermodynamics [25]. Moreover, the HISQ action is well suited to study heavy quarks on the lattice [24] and turned out to be successful in quantitative studies of charmonium [26] and D meson properties [27].

2. Lattice setup

We calculate meson correlation functions on gauge configurations generated in $(2+1)$-flavor QCD using the HISQ action [25]. The strange quark mass m_s is adjusted to its physical value and the light quark masses are fixed at $m_l = m_s/20$, corresponding to $m_\pi \simeq 160$ MeV and $m_K \simeq 504$ MeV at zero temperature in the continuum limit [25]. Charm quarks are introduced as valance quarks and we used the HISQ action with the so-called ϵ-term for the charm quark [24] which makes our calculations in the heavy quark sector free of tree-level $\mathcal{O}((am_c)^2)$ discretization errors. Our calculations have been performed on lattices of size $N_\sigma^3 \cdot N_\tau = 48^3 \cdot 12$. We consider lattice couplings in the range $\beta = 6.664$–7.280 which corresponds to temperatures $T = 138$–245 MeV. Discretization effects on these lattices and for the range of gauge couplings used by us are found to be quite small. To study the spatial correlators at higher temperatures we adopt the fixed-scale approach and perform calculations at $\beta = 7.280$ for $N_\tau = 10, 8, 6, 4$ which corresponds to the temperature range $T = 297$–743 MeV. In all our calculations the spatial extent of lattice was four times the temporal extent: $N_\sigma = 4N_\tau$. The lattice spacing and the resulting temperature values, $T = 1/N_\tau a$, have been determined using results for the kaon decay constant [25].

In the staggered formulation quarks come in four valence tastes and meson operators are defined as $J_H = \bar{q}(\Gamma^D \times \Gamma^F)q$, Γ^D and Γ^F being products of Dirac Gamma matrices which generate spin and taste structures, respectively [28]. In this study we focus only on local meson operators with $\Gamma^D = \Gamma^F = \Gamma$. By using staggered quark fields $\chi(\boldsymbol{x})$ at $\boldsymbol{x} = (x, y, z, \tau)$ the local meson operators can be written in a simple form $J_H(\boldsymbol{x}) = \tilde{\phi}(\boldsymbol{x})\bar{\chi}(\boldsymbol{x})\chi(\boldsymbol{x})$, where $\tilde{\phi}(\boldsymbol{x})$ is a phase factor depending on the choice of Γ. We calculate only the quark-line connected part of the meson correlators since the contribution of the disconnected part either vanishes or is expected to be small in most cases considered in this study (see discussions below). Since staggered meson correlator couples to two different meson excitations with opposite parity, the large distance behavior of the lattice correlator can be described by

$$G(\tau) = A_{\rm NO}^2 \left(e^{-M_-\tau} + e^{-M_-(N_\tau-\tau)}\right) - (-1)^\tau A_{\rm O}^2 \left(e^{-M_+\tau} + e^{-M_+(N_\tau-\tau)}\right), \qquad (6)$$

where the first (second) term on the right-hand-side characterizes a non-oscillating (oscillating) contribution governed by a negative (positive) parity state. Taking the square of the amplitudes ensures their positivity [29]. In Table 1 we summarize the different choices of the phase factor $\tilde{\phi}$ and the meson states they correspond to. We have considered four channels which we denote as scalar (S), pseudo-scalar (PS), axial vector (AV) and vector (V). The negative parity states in these channels correspond to different tastes of the same physical meson and will thus have nearly degenerate masses if lattice spacing is sufficiently small. For instance, in the $c\bar{c}$ sector the negative parity states in S and PS channels both correspond to the same η_c state. We will comment on this in more detail later.

In Eq. (6) as well as in Table 1 we assumed that the direction of propagation is the imaginary time τ. When discussing spatial correlation functions we assume that the direction of propagation is z. In this case τ in Table 1 and in Eq. (6) should be replaced by z. Furthermore, N_τ should then also be replaced by N_σ in Eq. (6). We calculate meson propagators using point sources as well as corner-wall sources, where on a given times slice the source is set to one at the origin of each 2^3 cube and zero elsewhere. The use of corner-wall sources reduces the contribution of higher excited states and thus allows for a more accurate determination of the screening masses, especially for the positive parity states.

As stated above, in this study we do not include the contribution from the disconnected diagrams. In the case of charmonium the contribution of disconnected diagrams is expected to be small, see e.g. Ref. [30]. For $s\bar{s}$ mesons disconnected diagrams will cause mixing with the light quark sector in the isospin zero channel. For vector mesons this mixing is know to be very small and the ϕ meson is almost a pure $s\bar{s}$ state. Thus, neglecting the disconnected diagrams seems to be justified also in this case. Mixing is however large for iso-singlet pseudo-scalar mesons and it is certainly important to include the contribution from the disconnected diagrams when comparing the lattice calculations with the experiment. Alternatively, one can estimate the mass of the un-mixed pseudo-scalar $s\bar{s}$ meson using leading order chiral perturbation theory $m^2_{\eta_{s\bar{s}}} = 2m_K^2 - m_\pi^2 = (686MeV)^2$, and use this for comparison with the experiment in the PS channel. Not much is known about the mixing between the light and strange sectors for iso-scalar mesons in scalar and axial-vector channels. It is expected that there is strong mixing in the scalar meson sector as well. The mass of the lowest lying $s\bar{s}$ scalar meson considered in our calculation is about 1.12 GeV as shown in Fig. 2 (explained below). It thus is considerably heavier than the lightest iso-scalar scalar meson $f_0(980)$ but lighter than the next-to-lightest iso-scalar scalar state $f_0(1370)$. However, for the axial-vector meson mass we find good agreement between our calculations and the mass of the $f_1(1420)$ meson, suggesting that the mixing between the light and strange quark sector is likely to be small in this case. At sufficiently high temperatures we expect that the contribution of disconnected diagrams will become small because of screening effects of the deconfined phase.

The strange quark mass as function of the gauge coupling β was determined in Ref. [25]. We also need to determine the charm quark mass m_c. For the determination of the charm quark mass we calculate the masses of J/ψ and η_c mesons for gauge couplings $\beta = 10/g^2$ in the interval $[6.39, 7.28]$. We calculate correlation functions at several trial values of the bare charm quark mass in the range $10 \leq m_c/m_s \leq 14$ using point sources. We then perform linear interpolations in the charm quark mass of the spin-averaged charmonium mass, $\frac{1}{4}(m_{\eta_c} + 3m_{J/\psi})$ and match them to the physical value. This determines the bare charm quark mass am_c and the quark mass ratio m_c/m_s for each value of β. Finally we fit the β dependence of am_c to a renormalization

Table 1. List of meson operators and corresponding physical states in the strange ($s\bar{s}$), strange-charm ($s\bar{c}$) and charm ($c\bar{c}$) sectors. The lightest $s\bar{s}$ pseudo-scalar state is defined as $M_{\eta_{s\bar{s}}} = \sqrt{2M_K^2 - M_\pi^2} \sim 686$ MeV which is used to determine the strange quark mass on the lattice.

	$-\phi(x)$	Γ	J^{PC}	$s\bar{s}$	$s\bar{c}$	$c\bar{c}$
M_-^{S}	1	$\gamma_4\gamma_5$	0^{-+}			
M_+^{S}		1	0^{++}		D^*_{s0}	χ_{c0}
M_-^{PS}	$(-1)^{x+y+z}$	γ_5	0^{-+}	$\eta_{s\bar{s}}$	D_s	η_c
M_+^{PS}		γ_4	0^{+-}	—		—
M_-^{AV}	$(-1)^x$, $(-1)^y$	$\gamma_i\gamma_4$	1^{--}			
M_+^{AV}		$\gamma_i\gamma_5$	1^{++}		D_{s1}	χ_{c1}
M_-^{V}	$(-1)^{x+z}$, $(-1)^{y+z}$	γ_i	1^{--}	ϕ	D_s^*	J/ψ
M_+^{V}		$\gamma_j\gamma_k$	1^{+-}			h_c

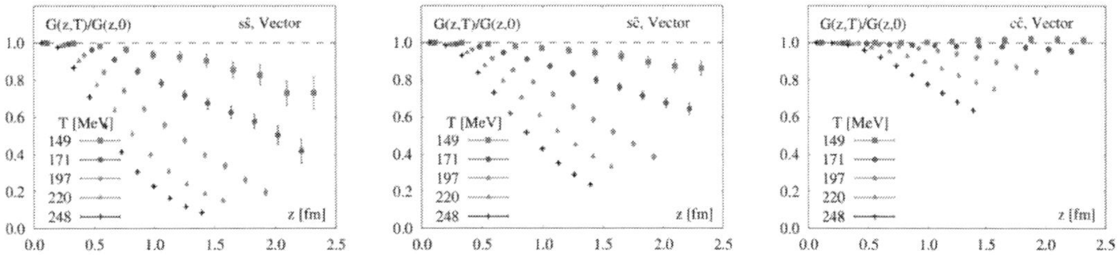

Figure 1. The ratio of the non-oscillating (negative parity) part of vector correlator in $s\bar{s}$ (left), $s\bar{c}$ (middle) and $c\bar{c}$ (right) sectors at different temperatures to the corresponding zero temperature results.

group inspired ansatz,

$$am_c^{\rm LCP} = \frac{c_0 f(\beta) + c_2(10/\beta) f^3(\beta)}{1 + d_2(10/\beta) f^2(\beta)}, \quad f(\beta) = \left(\frac{10 b_0}{\beta}\right)^{-b_1/(2b_0^2)} \exp(-\frac{\beta}{20 b_0}), \tag{7}$$

where b_0 and b_1 are the coefficients of the QCD beta function. The above formula defines the line of constant physics for the charm quark mass. From our fit we find $c_0 = 61.0(1.7)$, $c_2 = 2.76(26) \times 10^5$, and $d_2 = 3.3(3.7) \times 10^2$.

We performed calculations of meson correlation functions containing a charm quark in four different channels corresponding to local meson operators (see above) for $\beta = 6.74, 6.88, 7.03, 7.15$, and 7.28 using point and corner-wall sources. We extracted the zero temperature masses using the ansatz given in Eq. (6) and compared the corresponding results with the experiment. We find that within estimated errors the meson masses agree with the experiment. We also find that the difference between meson masses belonging to different taste multiplets is negligibly small.

3. Temperature dependence of spatial meson correlators

Having determined the charm quark masses we can perform calculations of the finite temperature $s\bar{s}$, $s\bar{c}$ and $c\bar{c}$ correlation functions in the four different quantum number channels that we have analyzed also at zero temperature.

We start the discussion of our results at non-zero temperature with the temperature dependence of meson correlators. As it was pointed out in Ref. [23], contrary to the temporal correlation functions, spatial correlation functions show visible changes already in the vicinity of the transition temperature even in the case of charmonium. Since staggered meson correlators in each channel contain both negative parity (non-oscillating) and positive parity (oscillating) states, it is important to separate these contributions before studying the temperature dependence of the correlators. One can then form ratios of these contributions at different temperatures and the corresponding zero temperature result. If there is no change in the meson spectral functions, these ratios will be equal to one. As a corollary, deviations of these ratios from unity indicate some medium modification of the meson spectral functions at non-zero temperature.

In Fig. 1, we show the ratio of the negative parity part of the vector correlator in $s\bar{s}$, $s\bar{c}$ and $c\bar{c}$ channels and the corresponding zero temperature result. At zero temperature, these non-oscillating parts of vector correlators are dominated by ϕ, D_s^* and J/ψ states, respectively. In the $s\bar{s}$ sector, We observe a strong decrease of this ratio starting at $T = 149$ MeV that becomes larger in magnitude with increasing temperature. A somewhat smaller but still significant decrease of

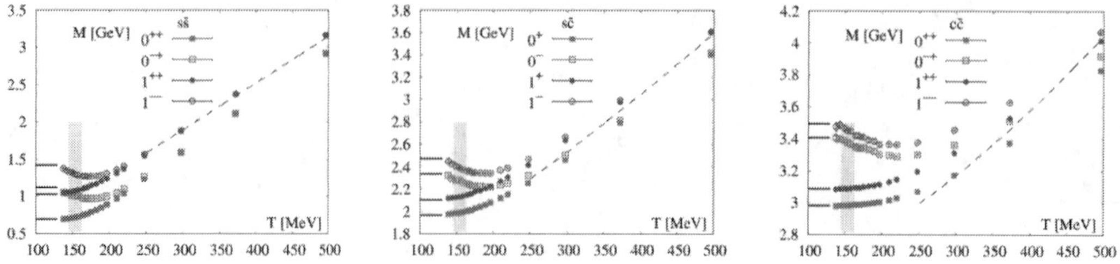

Figure 2. Screening masses for different channels in $s\bar{s}$ (left), $s\bar{c}$ (middle) and $c\bar{c}$ (right) sectors as function of the temperature. The solid horizontal lines on the left correspond to the zero temperature meson masses. The dashed line is the free field theory result (see text).

this ratio is also seen in the $s\bar{c}$ sector. For ratios of charmonium correlators no changes are visible at the lowest temperature (149 MeV), and even at $T = 171$ MeV the deviations of this ratio from unity are quite small. They become significant only at $T = 197$ MeV. The results shown in Fig. 1 therefore suggest that there is a sequential pattern of the medium modifications of the meson spectral functions which follows the heavy quark content and the corresponding mass of meson states. For $s\bar{c}$ and $c\bar{c}$ mesons we also studied ratios of finite and zero temperature correlators in the pseudo-scalar channel and found results that are very similar to those discussed above for vector mesons. Furthermore, we study the positive parity contributions to the correlators in the scalar channel and the axial-vector channels for $c\bar{c}$ mesons. At sufficiently low temperatures these ratios encode the in-medium properties of χ_{c0} and χ_{c1} states. For spatial separations $z > 0.75$ fm these ratios show a temperature dependence that is similar to that of the corresponding ratio in the vector channel which we have discussed above. The temperature dependence of the ratios at shorter distances, $z < 0.75$ fm is significantly stronger than that in the vector and pseudo-scalar channels, providing some hints for the expected sequential suppression pattern of different charmonium states. Clearly a more detailed analysis of the spatial correlators is needed to clarify this issue, which will be discussed in the next section.

4. Large distance behavior of spatial meson correlators and screening masses

We fit the large distance behavior of the spatial correlators using Eq. (6) and extract screening masses in various channels for $s\bar{s}$, $s\bar{c}$ and $c\bar{c}$ mesons. In our study of PS and V screening masses, we use point and corner-wall sources. For $c\bar{c}$ sector the difference between the point and corner-wall sources is quite small. It typically is around 0.4%, except for the three highest temperature, where it reaches 3%. Similarly in the $s\bar{c}$ sector, the difference between point and wall sources is typically about 1% for all temperatures except the three highest ones, where it is 3%. Larger differences between point and wall sources are seen in the $s\bar{s}$ sector, where they reach 3% at the highest five temperatures, and are about 1% at other temperatures. Therefore, we will use the results from corner-wall sources when presenting our results on the screening masses in the PS and V channels. In the S and AV channels screening masses can be reliably extracted only when using the corner-wall sources. The effects of taste symmetry breaking are only visible for the negative parity states in PS and S channels at low temperatures, where they are about 1.5%. For temperature above 200 MeV we do not find any statistically significant effect of taste splitting.

All our results on screening masses in the $s\bar{s}$, $s\bar{c}$ and $c\bar{c}$ are shown in Fig. 2. The error bars in the figure indicate the statistical and systematic errors added in quadrature. We expect that at very high temperature the screening masses are given by Eq. (5). We show the free theory (leading order perturbative) result as dashed lines in Fig. 2. For this we need to specify the quark

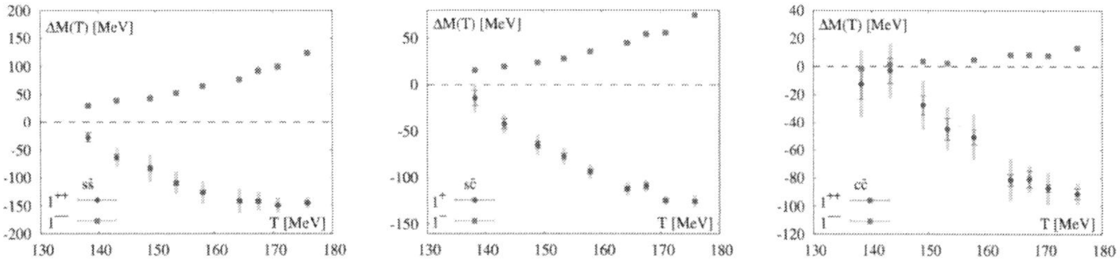

Figure 3. The difference $\Delta M(T) = M(T) - m_0$ for vector and axial-vector channels in $s\bar{s}$ (left), $s\bar{c}$ (middle) and $c\bar{c}$ (right) sectors as function of the temperature.

masses. The quark masses depend on the renormalization scale which is not fixed at leading order. A natural choice of the renormalization scale would be to identify it with the lowest scale in the problem provided that this lowest scale is still in the perturbative region. In our case there are two relevant energy scales, the charm quark mass m_c, and the thermal scale which here is taken to be $2\pi T$. For temperatures $T > 200$ MeV both scales are comparable. Therefore, we could take the charm quark mass as the renormalization scale and the corresponding value $m_c(\bar{\mu} = m_c) = 1.275$ GeV from PDG [31]. Using the ratio of charm to strange quark mass $m_c/m_s = 11.85$ [32] and the above value, we can determine the value of the strange quark mass at the same renormalization scale to be $m_s = 0.108$ GeV. This completely specifies our free theory prediction.

As one can see from Fig. 2, there are three distinct regions: the low temperature region, where the screening masses are close to the corresponding vacuum masses (solid lines), the intermediate temperature region, where we see significant changes in the value of the screening masses with respect to the corresponding vacuum masses, and finally, the high temperature region, where the screening masses are close to the free theory result (dashed lines). In the high temperature region, there clearly are no meson bound states anymore. The onset of the high temperature behavior is different in different sectors. In the $s\bar{s}$ sector it starts at around $T = 210$ MeV. In the $s\bar{c}$ sector it starts at $T = 250$ MeV, while in $c\bar{c}$ sector it starts at $T > 300$ MeV. As the temperature increases, we see that the screening masses corresponding to negative parity states increase monotonically, while the screening masses in the positive parity states first decrease before starting to rise towards the asymptotic high temperature values. In the intermediate temperature region the screening masses of opposite parity partners start to approach each other and we observe a significant rearrangement of the ordering of screening masses in different channels. At sufficiently high temperatures the PS and S as well as V and AV screening masses become degenerate. In the $s\bar{s}$ sector this is evident for $T > 220$ MeV, while for the two other sectors it happens at higher temperatures because in these cases the effect of the explicit breaking of parity by the quark masses is much larger. In the high temperature region the screening masses in the PS channel are smaller than the screening masses in the V channel. This behavior has been observed previously in lattice calculations [22] and in calculations using Dyson-Schwinger equations [33]. It persists to temperatures as high as 800 MeV, see e.g. Ref. [22].

To see in detail the modification of screening masses in the low and intermediate temperature regions it is convenient to consider the difference between the screening mass and the corresponding vacuum masses m_0 calculated at $T = 0$

$$\Delta M(T) = M(T) - m_0 . \tag{8}$$

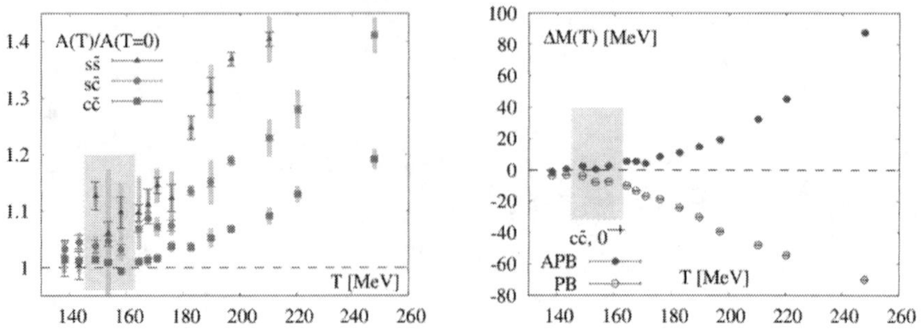

Figure 4. The right panel shows the ratio of the amplitudes of vector spatial meson correlators calcuated at finite temperature and zero temperature. The left panel shows the difference between the pseudo-scalar screening mass and the corresponding zero temperature mass $\Delta M(T) = M(T) - m_0$ for anti-periodic (filled symbols) and periodic (open symbols) boundary conditions

It is tempting to interpret this difference as the change in the binding energy of meson states, however, the relation between the screening mass and the pole mass only holds as long as there is a well defined bound state. Nonetheless, ΔM could provide some constraints on the change of the binding energy in the low and intermediate temperature regions. We show our results for $\Delta M(T)$ for vector and axial-vector $s\bar{s}$, $s\bar{c}$ and $c\bar{c}$ mesons in Fig. 3. The error bars and gray bands indicate the statistical and systematic errors, respectively. In all cases ΔM increases for negative parity states and decreases for positive parity states. The absolute value of ΔM is the largest for $s\bar{s}$ meson and is the smallest for $c\bar{c}$ mesons, i.e. it follows the expected sequential pattern with respect of the heavy quark content. Note, that the absolute size of ΔM is about the same for negative and positive parity states in the $s\bar{s}$ sector. In the charm-strange sector $|\Delta M|$ is slightly larger for positive parity state. For charmonium the picture is different, ΔM remains quite small for $T < 200$ MeV in the vector channel, while we see large decrease in ΔM for the axial-vector channel starting around $T = 160$ MeV. We also inspected the behavior of ΔM in the scalar and pseudo-scalar channels and found a behavior that is almost identical to that in the axial-vector and vector channels. Thus, in the charmonium case the temperature dependence of ΔM provides some hints for sequential suppression: it shows large decrease at $T > 160$ MeV for scalar and axial vector channels corresponding to 1P charmonium states (χ_{c0} and χ_{c1}) and very little change in the pseudo-scalar and vector channels corresponding to 1S charmonium states (η_c and J/ψ).

The amplitudes A_{NO} and A_O appearing in Eq. (6) are related to the wave function of meson states in the zero temperature limit. In the case of point sources and mesons consisting of heavy quarks they are proportional to the wave function at the origin and the derivative of the wave function at origin. Therefore, if we are interested in melting of meson states at high temperature it is worth to study the temperature dependence of this amplitudes. In Fig. 4 we show the ratios of the amplitude A_{NO} for the spatial meson correlators in the vector channel to the corresponding zero temperature result. If meson states exist in the medium with little or no modifications this ratio should be close to one. For the strange and strange-charm mesons we see that there are deviations of this ratio from one already at relatively low temperatures and these deviations are increasing with increasing temperature. For charmonium the ratio of the amplitude to the zero temperature amplitude is very close to one up to temperature of about 170 MeV, and slowly increases above that temperature. Only at temperatures above 200 MeV the deviations of this ratio from unity are similar to the ones observed in strange and strange-charm sectors in the transition region. This suggests that melting of the J/ψ happens for $T > 200$ MeV.

Finally, let us discuss the dependence of charmonium screening masses on the temporal boundary conditions. So far we used anti-periodic boundary conditions for valence charm quarks, i.e. we considered them to be in thermal equilibrium. In heavy ion experiments charm quarks are most likely not in thermal equilibrium due to their large mass and large relaxation times. The existence of charmonium bound state in the medium with a certain temperature does not depend on whether the charm quarks are thermalized or not. If the temperature is not too high one can use NRQCD to study heavy quark bound state. In this approach no boundary conditions are imposed on the heavy quark [34] but the sequential suppression pattern has been observed. Thus, if charmonium state exists in the medium the spatial correlation function and screening masses should be insensitive to the temporal boundary conditions. It was pointed out in Ref. [23] if bound states are melted meson screening masses will be very sensitive to the temporal boundary conditions. Asymptotically the quadratic difference of the screening masses calculated with anti-periodic and periodic boundary conditions will approach $(2\pi T)^2$ [23]. In Fig. 4 we show the charmonium screening masses in the pseudo-scalar channel calculated using anti-periodic and periodic boundary conditions. There is very little sensitivity to the boundary conditions for $T < 170$ MeV. Above that temperature we see clear sensitivity of the pseudo-scalar screening mass to the boundary conditions, which becomes quite large for $T > 200$ MeV possibly indicating dissolution of the η_c state at these temperatures. The results in the vector channel are very similar. We also see sensitivity to the boundary conditions in the scalar and axial-vector charmonium screening masses. But due to large errors in the corresponding screening masses it is more difficult to quantify this sensitivity.

5. Conclusions

We studied spatial meson correlation functions at non-zero temperature for $s\bar{s}$, $s\bar{c}$ and $c\bar{c}$ mesons with the aim to find out at what temperatures these mesons dissolve. We studied the spatial correlation functions in different ways: comparing them directly to the zero temperature results, studying their large distance behavior and extracting the corresponding screening masses and amplitudes, as well as studying the sensitivity to the temporal boundary conditions. We see that the size of medium effects in the spatial correlators and screening masses in different sectors follows a sequential pattern that corresponds to the heavy quark content. We find, however, that both $s\bar{s}$ and charm strange mesons are effected by the medium already at relatively low temperatures and dissolve at temperature close to the transition temperature. Contrary, the 1S charmonium states (J/ψ and η_c) do not seem to be affected by the medium up to temperatures of about 170 MeV and most likely dissolve at temperatures somewhat higher than 200 MeV. The correlators and the screening masses corresponding to 1P charmonium (χ_c) states show larger temperature dependence than the ones corresponding to 1S states. In particular, the screening masses corresponding to 1P charmonium states show large temperature dependence just above the transition temperature. Thus, we see some hints of the expected charmonium suppression from our study of spatial correlation functions.

Acknowledgments

Numerical calculations were carried out on the USQCD Clusters at the Jefferson Laboratory, USA, the NYBlue at the Brookhaven National Laboratory, USA and in NERSC at Lawrence Berkeley National Laboratory, USA. This work was supported by through the Contract No. DE-AC02-98CH10886 with the U.S. Department of Energy and the Bundesministerium für Bildung und Forschung under grant 05P12PBCTA, and the EU Integrated Infrastructure Initiative Hadron-Physics3. Partial support for this work was also provided through Scientific Discovery through Advanced Computing (SciDAC) program funded by U.S. Department of Energy, Office of Science, Advanced Scientific Computing Research (and Basic Energy Sciences/Biological and Environmental Research/High Energy Physics/Fusion Energy Sciences/Nuclear Physics).

[1] Petreczky P 2012 *J.Phys.* **G39** 093002 (*Preprint* 1203.5320)
[2] Philipsen O 2013 *Prog.Part.Nucl.Phys.* **70** 55–107 (*Preprint* 1207.5999)
[3] Detar C E and Kogut J B 1987 *Phys.Rev.Lett.* **59** 399
[4] Detar C E and Kogut J B 1987 *Phys.Rev.* **D36** 2828
[5] Laine M and Vepsalainen M 2004 *JHEP* **0402** 004 (*Preprint* hep-ph/0311268)
[6] Brandt B, Francis A, Laine M and Meyer H 2014 *JHEP* **1405** 117 (*Preprint* 1404.2404)
[7] Matsui T and Satz H 1986 *Phys.Lett.* **B178** 416
[8] Sharma R, Vitev I and Zhang B W 2009 *Phys.Rev.* **C80** 054902 (*Preprint* 0904.0032)
[9] Asakawa M, Hatsuda T and Nakahara Y 2001 *Prog. Part. Nucl. Phys.* **46** 459–508 (*Preprint* hep-lat/0011040)
[10] Wetzorke I, Karsch F, Laermann E, Petreczky P and Stickan S 2002 *Nucl.Phys.Proc.Suppl.* **106** 510–512 (*Preprint* hep-lat/0110132)
[11] Asakawa M, Hatsuda T and Nakahara Y 2003 *Nucl. Phys.* **A715** 863–866 (*Preprint* hep-lat/0208059)
[12] Asakawa M and Hatsuda T 2004 *Phys. Rev. Lett.* **92** 012001 (*Preprint* hep-lat/0308034)
[13] Datta S, Karsch F, Petreczky P and Wetzorke I 2004 *Phys. Rev.* **D69** 094507 (*Preprint* hep-lat/0312037)
[14] Jakovac A, Petreczky P, Petrov K and Velytsky A 2007 *Phys. Rev.* **D75** 014506 (*Preprint* hep-lat/0611017)
[15] Aarts G, Allton C, Oktay M B, Peardon M and Skullerud J I 2007 *Phys. Rev.* **D76** 094513 (*Preprint* 0705.2198)
[16] Ding H, Francis A, Kaczmarek O, Karsch F, Satz H *et al.* 2012 *Phys.Rev.* **D86** 014509 (*Preprint* 1204.4945)
[17] Petreczky P, Miao C and Mocsy A 2011 *Nucl.Phys.* **A855** 125–132 (*Preprint* 1012.4433)
[18] de Forcrand P *et al.* (QCD-TARO Collaboration) 2001 *Phys.Rev.* **D63** 054501 (*Preprint* hep-lat/0008005)
[19] Boyd G, Gupta S, Karsch F and Laermann E 1994 *Z.Phys.* **C64** 331–338 (*Preprint* hep-lat/9405006)
[20] Pushkina I *et al.* (QCD-TARO Collaboration) 2005 *Phys.Lett.* **B609** 265–270 (*Preprint* hep-lat/0410017)
[21] Iida H, Maezawa Y and Yazaki K 2010 *PoS* **LATTICE2010** 189 (*Preprint* 1012.2044)
[22] Cheng M, Datta S, Francis A, van der Heide J, Jung C *et al.* 2011 *Eur.Phys.J.* **C71** 1564 (*Preprint* 1010.1216)
[23] Karsch F, Laermann E, Mukherjee S and Petreczky P 2012 *Phys.Rev.* **D85** 114501 (*Preprint* 1203.3770)
[24] Follana E *et al.* (HPQCD Collaboration, UKQCD Collaboration) 2007 *Phys.Rev.* **D75** 054502 (*Preprint* hep-lat/0610092)
[25] Bazavov A, Bhattacharya T, Cheng M, DeTar C, Ding H *et al.* 2012 *Phys.Rev.* **D85** 054503 (*Preprint* 1111.1710)
[26] Davies C, Donald G, Dowdall R, Koponen J, Follana E *et al.* 2012 *PoS* **ConfinementX** 288 (*Preprint* 1301.7203)
[27] Bazavov A *et al.* (Fermilab Lattice and MILC Collaborations) 2013 (*Preprint* 1312.0149)
[28] Altmeyer R *et al.* (MT(c) collaboration) 1993 *Nucl.Phys.* **B389** 445–512
[29] Lepage G, Clark B, Davies C, Hornbostel K, Mackenzie P *et al.* 2002 *Nucl.Phys.Proc.Suppl.* **106** 12–20 (*Preprint* hep-lat/0110175)
[30] Levkova L and DeTar C 2011 *Phys.Rev.* **D83** 074504 (*Preprint* 1012.1837)
[31] Beringer J *et al.* (Particle Data Group) 2012 *Phys.Rev.* **D86** 010001
[32] Davies C, McNeile C, Wong K, Follana E, Horgan R *et al.* 2010 *Phys.Rev.Lett.* **104** 132003 (*Preprint* 0910.3102)
[33] Wang K l, Liu Y x, Chang L, Roberts C D and Schmidt S M 2013 *Phys.Rev.* **D87** 074038 (*Preprint* 1301.6762)
[34] Aarts G, Kim S, Lombardo M, Oktay M, Ryan S *et al.* 2011 *Phys.Rev.Lett.* **106** 061602 (*Preprint* 1010.3725)

Journal of Physics: Conference Series **535** (2014) 012032 doi:10.1088/1742-6596/535/1/012032

Kinetic Simulations of Rayleigh-Taylor Instabilities

Irina Sagert[1], Wolfgang Bauer[2,3], Dirk Colbry[3], Jim Howell[3], Alec Staber[3], Terrance Strother[4]

[1] Center for Exploration of Energy and Matter, Indiana University, Bloomington, Indiana, 47308, USA

[2]Department of Physics and Astronomy, Michigan State University, East Lansing, Michigan, 48824, USA

[3]Institute for Cyber-Enabled Research, Michigan State University East Lansing, Michigan 48824, USA

[4]XTD-6, Los Alamos National Laboratory, Los Alamos, New Mexico 87545, USA

E-mail: `isagert@indiana.edu`

Abstract. We report on an ongoing project to develop a large scale Direct Simulation Monte Carlo code. The code is primarily aimed towards applications in astrophysics such as simulations of core-collapse supernovae. It has been tested on shock wave phenomena in the continuum limit and for matter out of equilibrium. In the current work we focus on the study of fluid instabilities. Like shock waves these are routinely used as test-cases for hydrodynamic codes and are discussed to play an important role in the explosion mechanism of core-collapse supernovae. As a first test we study the evolution of a single-mode Rayleigh-Taylor instability at the interface of a light and a heavy fluid in the presence of a gravitational acceleration. To suppress small-wavelength instabilities caused by the irregularity in the separation layer we use a large particle mean free path. The latter leads to the development of a diffusion layer as particles propagate from one fluid into the other. For small amplitudes, when the instability is in the linear regime, we compare its position and shape to the analytic prediction. Despite the broadening of the fluid interface we see a good agreement with the analytic solution. At later times we observe the development of a mushroom like shape caused by secondary Kelvin-Helmholtz instabilities as seen in hydrodynamic simulations and consistent with experimental observations.

1. Introduction

Kinetic transport simulations are applied to study systems that have large Knudsen numbers $K > 0.01$ and therefore cannot be described by fluid dynamics [1]. Instead, they require the explicit solution of the transport equations for the particles they are composed of. The most famous example of a transport equation is the Boltzmann equation:

$$\frac{\partial f_i(\vec{r},\vec{p},t)}{\partial t} + \frac{\vec{p}}{m_i} \cdot \nabla f_i(\vec{r},\vec{p},t) + \vec{F} \cdot \frac{\partial f_i(\vec{r},\vec{p},t)}{\partial \vec{p}} = I_{i,\mathrm{coll}}, \tag{1}$$

where $\vec{r}$, $\vec{p}$, and m_i are the position, momentum, and mass of particle i respectively, $\vec{F}$ is an external force field, and $I_{i,\mathrm{coll}}$ takes into account the changes in the phase space distribution function $f_i(\vec{r},\vec{p},t)$ caused two-body collisions.

Typical examples of non-equilibrium environments are nuclei in heavy-ion collisions [2, 3, 4, 5, 6, 7, 8], hypersonic flow in aerospace research [9], inertial confinement fusion (ICF) capsules

[10, 11, 12], and astrophysical systems. The latter range from flow in accretion discs [13], cluster formation in the crust of neutron stars [14], to cosmological simulations [15]. Depending on the system, different numerical approaches are chosen (with extensions to include quantum and relativistic effects), such as Molecular Dynamics [16], Particle-In-Cell [17], and Direct Simulation Monte Carlo [18]. The latter approximates $f_i(\vec{r}, \vec{p}, t)$ by so-called test-particles in form of delta-functions:

$$f(\vec{r}, \vec{p}, t) = \sum_{i=0}^{N} \delta^3(\vec{r} - \vec{r}_i(t))\delta^3(\vec{p} - \vec{p}_i(t)). \quad (2)$$

For small Knudsen numbers $K = l/L \ll 1$, when the particles' mean free path l becomes smaller than a characteristic length scale L of the system, kinetic simulations have been shown to reproduce hydrodynamic phenomena such as relativistic shock wave evolution [19, 7] and fluid instabilities [20, 21, 22]. Therefore, transport models seem to offer the intriguing possibility to study such phenomena in equilibrium and for systems where the Navier-Stokes equations are not applicable anymore.

Our main interest lies in the study of core-collapse supernovae and inertia confinement fusion capsules. Despite the differences in both systems - ranging from scale to micro-physical processes - similar mechanisms might be at play to drive the dynamical evolution. Core-collapse supernovae are explosions of massive stars caused by the gravitational collapse of the iron core at the end of their lives [23, 24]. Despite its rich research history [25], understanding the mechanism that drives the supernova explosion remains an outstanding problem and modern computer simulations point towards a crucial role of fluid instabilities coupled to neutrino-matter interactions [25, 26, 27, 28, 29]. However, the modeling of both, especially in 3D and in the general relativistic regime is a major computational challenge and requires supercomputer facilities.
Inertia confinement fusion is a candidate to create fusion energy in the laboratory with current experiments being run at the National Ignition Facility [30, 31, 32, 33]. The goal is to reach fusion via the compression and heating of fusion fuel in form of capsules containing deuterium and tritium with high-energy laser beams. Though significant progress has been made towards achieving conditions necessary for ignition, only recently a greater than unity fuel gain was reported, i.e. the energy generated through fusion reactions exceeded the amount of deposited energy into the fusion fuel [34]. Similar to core-collapse supernovae, the difficulties seem to lie in the control of fluid instabilities [35] and non-equilibrium particle transport [36] (and references therein).

2. Rayleigh-Taylor Instability

Motivated by the challenges of both, core-collapse supernovae and inertia confinement fusion capsules, we are developing a large scale Direct Simulation Monte Carlo code that is written to run in parallel and can model the evolution of $> 10^6$ test-particles [37, 38, 39]. The ability of the code to reproduce hydrodynamic shock wave phenomena has already been demonstrated for the continuum limit as well as for non-equilibrium matter [38, 39]. In this work we focus on the modeling of fluid instabilities. As a first test-case we choose the Rayleigh-Taylor instability [40, 41]. It is created when a dense fluid lies on top of a light one in the presence of a gravitational acceleration. Both fluids are initially at rest, however, perturbations of the interface become unstable, grow with time, and eventually become chaotic.
For early times the growth of an single mode perturbation $\eta_0(x)$ of wavelength λ and amplitude B can be predicted from linear theory [42, 43]:

$$\eta(x, t) = \frac{1}{2}\left(e^{\gamma t} + e^{-\gamma t}\right)\eta_0(x). \quad (3)$$

Here, $\gamma = \sqrt{Ag\alpha}$ is the growth rate and is given by the Atwood number $A = \frac{\rho_2 - \rho_1}{\rho_2 + \rho_1}$, the wave number $\alpha = 2\pi/\lambda$, and the gravitational acceleration g. It can be seen from eq.(3) and $\gamma \propto \lambda^{-0.5}$ that short wavelength perturbations will grow most rapidly. The linear theory solution is valid up to a point, when the amplitude of the perturbation is of the order of $\sim 0.5\,\lambda$. After that, the non-linear phase begins and the interface develops bubbles of light fluid entering the dense one. The latter, on the other hand, sinks in finger-shaped structures into the light fluid. This relative movement causes Kelvin-Helmholtz instabilities to develop, primarily in form of vortices at the tip of the fingers, leading to a typical mushroom shape. At later times, more secondary instabilities set in, eventually resulting in chaotic behavior. Hereby, for the same initial conditions, the evolution of secondary instabilities seem to depend on the resolution of the simulation and the applied finite difference scheme [40].

3. Simulation Setup

In this work we study a 2D single mode Rayleigh-Taylor fluid instability applying $4 \cdot 10^7$ test-particles. Details of our code, for example the collision algorithm, can be found in [38]. For the current simulation we use periodic boundary conditions and a value of $g = 0.1$ for the gravitational acceleration. The width and height of the simulations box are $0 \leq x \leq 0.25$ and $0 \leq y \leq 1.6$, respectively, whereas we divide the simulation space into 800×5120 bins for the calculation and 200×640 bins for the output. Hereby, the upper half of the box with $y \geq 0.8$ (2) is filled with the high density gas while the lower half (1) contains low density matter with $\rho_2 = 2\rho_1$. The volumes V_2 and V_1 are the same and, to save computational time, we fill them with the same number of test-particles $N_2 = N_1$. Consequently, the latter are given different masses $m_2 = 2\,m_1$. Furthermore, to better follow the evolution of the instability, particles in the upper half are assigned a type-value $\tau = 2$ while particles in the lower half have $\tau = 1$. To initialize the single mode instability we perturb the separation layer between fluid (2) and (1) according to:

$$\eta(x) = 0.8 + B\cos(2\pi x/\lambda), \tag{4}$$

where we set the perturbation amplitude to $B = 0.01$ and its wavelength to $\lambda = 0.5$. A particle absolute velocity v_i is then determined according to the Maxwell-Boltzmann distributions with a height dependent temperature that is given by the barometric formula and the particle's y-position y_i:

$$T(h_i) = (V_{1,2}/N_{1,2})\,(P_0 - \rho_{1,2}\,g\,h_i)\,, \quad h_i = y_i - 0.8, \tag{5}$$

whereas we choose $P_0 = 2.5$.

For the initial distribution of particle velocities we use a Monte Carlo algorithm that works in the following way: From the height h_i we determine the temperature and root-mean-square velocity of particle i:

$$v_{\mathrm{rms,i}} = \sqrt{2\,T(h_i)/m_i}. \tag{6}$$

We assign a velocity $v_i \in [0 : 6\,v_{\mathrm{rms}}]$ and determine its probability according to the normalized Maxwell-Boltzmann velocity distribution:

$$\mathcal{P}(v_i) = \sqrt{\frac{m_i\,e}{kT}}\,v_i\,\exp\left(-\frac{m_i}{2\,kT}v_i^2\right). \tag{7}$$

Then, we choose a random probability $\mathcal{P}_r \in [0 : 1]$. If $\mathcal{P}_r \leq \mathcal{P}(v_i)$, v_i is assigned to particle i, otherwise we repeat the procedure.

Journal of Physics: Conference Series **535** (2014) 012032 doi:10.1088/1742-6596/535/1/012032

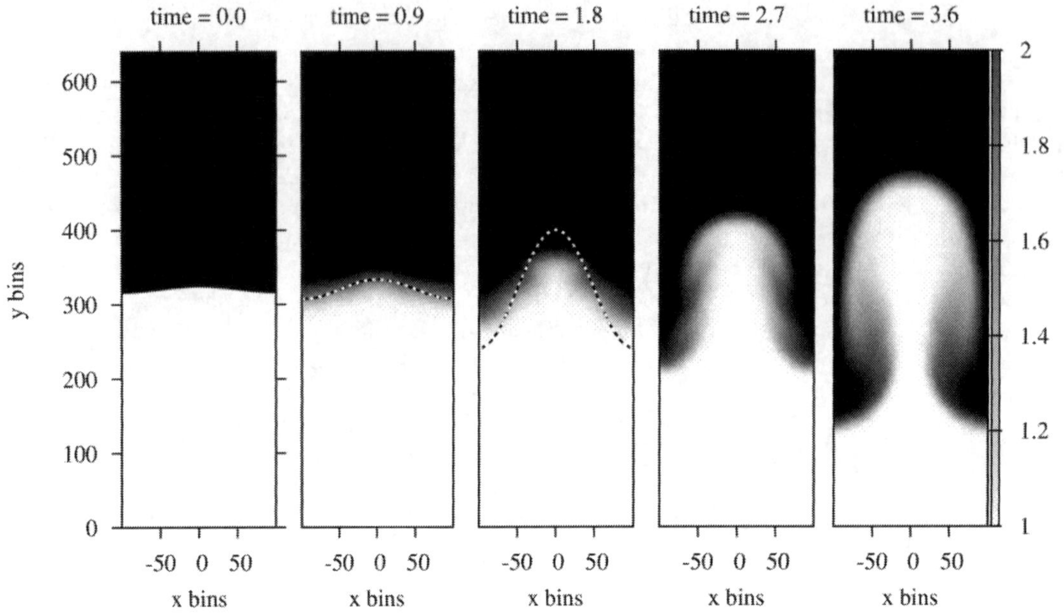

Figure 1. Average particle type τ in the Rayleigh-Taylor instability with $4 \cdot 10^7$ test particles and $\lambda = 2\,dx$ at different simulation times together with the analytic prediction for the fluid interface taken from eq.(3). The simulation space is mirrored at $x = 0$.

Unlike hydrodynamics codes, a particle-based approach always introduces some unevenness of the fluid interface due to the finite number of particles. This can serve as a potential seed for small-wavelength instabilities that, as mentioned previously, have the fastest growth rate and could thereby impact the evolution of the single mode perturbation. In order to prevent the development of these instabilities we set the particle mean free path to $l = 2.0\,dx$, where dx is the width of a bin. With that, we decrease the resolution of our simulation and blur possible irregularities of the interface. But, it should also be noted that a large mean free path increases the amount of particles diffusing from one fluid into the other. This leads to the development of a diffusion layer that could also influence the evolution of the single-mode instability.

4. Results

Figure 1 shows the average particle type τ per bin in our 2D Rayleigh-Taylor simulation at different times together with the analytic prediction for $\eta(x,t)$. The number of time steps to reach a simulation time of $t = 3.6$ is of the order of 10^5 whereas the size of the time steps is adaptive and determined from the maximal particle velocity [38]. For $t \lesssim 0.5$ we observe a broadening of the interface as particles diffuse from one fluid into the other. Despite the latter and the difficulty to locate its precise position, the layer seems to stay within the analytic prediction. For times $t \gtrsim 1.5$, the simulation enters the non-linear regime as the amplitude of the perturbation becomes comparable to $\sim 0.5\,\lambda$. Although we blur the fluid interface via the large particle mean free path, small structures are developing and are visible starting at $t \sim 1.1$. At later times, we observe the formation of the low density bubble and its rise in into the high-density matter whereas the latter sinks at the edges of the simulation box into the light fluid. Figure 2 shows the average x and y particle velocities per bin for the same simulation times as in Figure 1. Here, the y-velocity is plotted in the left half of the simulation boxes (i.e.

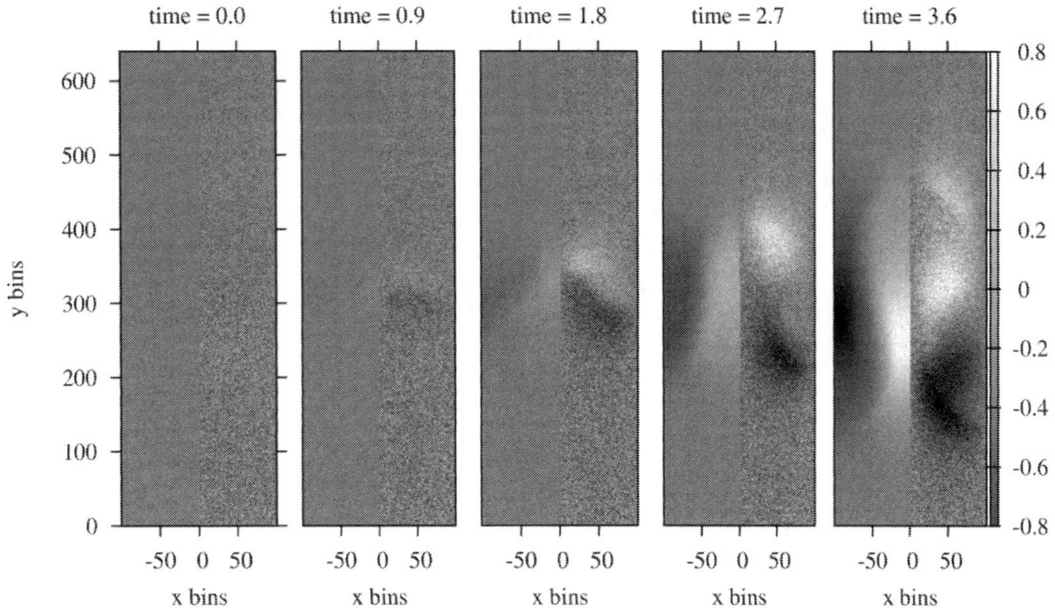

Figure 2. Average x and y particle velocities per bin at different simulation times. y-velocities are shown for $x \leq 0$, while x-velocities are shown for $x > 0$ and are scaled by a factor of 4 for better comparison.

for $x < 0$) while the x-velocity is plotted on the right hand side. For better comparison we scale the x-velocity by a factor of 4. It can be seen that a weak vorticity seems to develop leading to the expected mushroom shape of the instability. However, the dynamics is dominated by the motion into the positive and negative y-directions.

Overall, we observe the appearance of the typical Rayleigh-Taylor structure on expected time scales. Differences to hydrodynamic simulations lie in the growth rate which seems to overestimate the position of the instability in our simulation for $t \gtrsim 1.5$ as well as the strength of secondary Kelvin-Helmholtz instabilities. It is interesting to note that hydrodynamic simulations find a decrease in the growth rate of the single mode Rayleigh-Taylor instability and a delayed appearance of Kelvin-Helmholtz instabilities for low Reynolds numbers $\mathrm{Re}_p = (\lambda/\nu)\sqrt{Ag\lambda/(1+A)}$ where ν is the viscosity [44]. More kinetic studies should be conducted to examine any possible growth rate dependence on quantities such as test-partcle mean free path or the number of test-particles used to represent the fluids being modeled [45].

5. Summary and Outlook

In this work we present the first study of a single mode Rayleigh-Taylor instability performed with our large-scale Direct Simulation Monte Carlo code. The simulation uses $4 \cdot 10^7$ test-particles and follows the evolution of the instability from the linear to the non-linear regime for a simulation time of $t = 3.6$ in $\sim 10^5$ time steps. To prevent the early formation of small perturbations caused by irregularities of the interface between the light and heavy fluids we set the particle mean free path to a value of $l = 2\,dx$. With that, small scale irregularities are blurred, however, particles are also able to propagate from one fluid into the other leading to the formation of a diffusion layer. For simulation times $t \lesssim 1.5$ we find that the position of the interface generally agrees with the analytic prediction from linear theory. For later times a growing overestimate of the latter can be seen as the amplitude of the simulated instability lags behind the analytic solution. A possible explanation can be that at these times, linear theory is not valid anymore as the instability enters the non-linear regime. Weak secondary Kelvin-Helmholtz instabilities develop at later times and result in the formation of the typical mushroom shape. However, the effect is weak as the dynamics are dominated by the motion in the positive and negative y directions. We find that the observed evolution is similar to hydrodynamic simulations with low Reynolds numbers. Future studies will include a more detailed analysis of the Rayleigh-Taylor instability applying different values for the particle mean free path as well as the total number of test-particles.

6. Acknowledgements

This work used the Extreme Science and Engineering Discovery Environment (XSEDE), which is supported by National Science Foundation grant number OCI-1053575. Furthermore, I.S. acknowledges the support of the High Performance Computer Center and the Institute for Cyber-Enabled Research at Michigan State University.

References

[1] Agarwal R K, Yun K Y and Balakrishnan R 2001 *Physics of Fluids* **13** 3061–3085
[2] Bauer W, Bertsch G F, Cassing W and Mosel U 1986 *Physical Review C* **34** 2127–2133
[3] Bertsch G F, Kruse H and Gupta S D 1984 *Physical Review C* **29** 673–675
[4] Kruse H, Jacak B V and Stöcker H 1985 *Physical Review Letters* **54** 289–292
[5] Aichelin J and Bertsch G 1985 *Physical Review C* **31** 1730–1738
[6] Aichelin J and Stöcker H 1986 *Physics Letters B* **176** 14–19
[7] Bouras I, El A, Fochler O, Niemi H, Xu Z and Greiner C 2012 *Physics Letters B* **710** 641–646
[8] Kortemeyer G, Bauer W, Haglin K, Murray J and Pratt S 1995 *Physical Review C* **52** 2714–2724
[9] Li Z H and Zhang H X 2009 *Journal of Computational Physics* **228** 1116–1138
[10] Casanova M, Larroche O and Matte J P 1991 *Physical Review Letter* **67**(16) 2143–2146
[11] Vidal F, Matte J P, Casanova M and Larroche O 1993 *Physics of Fluids B: Plasma Physics* **5** 3182–3190
[12] Vidal F, Matte J P, Casanova M and Larroche O 1995 *Physical Review E* **52**(4) 4568–4571
[13] Matsuda T, Mizutani H and Boffin H M J 2002 *The Physics of Cataclysmic Variables and Related Objects* (*Astronomical Society of the Pacific Conference Series* vol 261) ed Gänsicke B T, Beuermann K and Reinsch K p 505
[14] Schneider A S, Horowitz C J, Hughto J and Berry D K 2013 *Physical Review C* **88** 065807
[15] Hernquist L and Ostriker J P 1992 *The Astrophysical Journal* **386** 375–397
[16] Ercolessi F 1997 A molecular dynamics primer Spring College in Computational Physics, ICTP, Trieste
[17] Tskhakaya D 2008 *Computational Many-Particle Physics* (*Lecture Notes in Physics* vol 739) ed Fehske H, Schneider R and Weie A (Springer Berlin Heidelberg) pp 161–189 ISBN 978-3-540-74685-0
[18] Bird G A 1994 *Molecular gas dynamics and the direct simulation of gas flows* (Clarendon Press)
[19] Bouras I, Molnár E, Niemi H, Xu Z, El A, Fochler O, Greiner C and Rischke D H 2009 *Physical Review Letters* **103** 032301
[20] Kadau K, Barber J L, Germann T C, Holian B L and Alder B J 2010 *Royal Society of London Philosophical Transactions Series A* **368** 1547–1560
[21] Gan Y, Xu A, Zhang G and Li Y 2011 *Physical Review E* **83** 056704
[22] Ashwin J and Ganesh R 2010 *Physical Review Letters* **104** 215003
[23] Bethe H A 1990 *Reviews of Modern Physics* **62** 801–866
[24] Janka H T, Langanke K, Marek A, Martínez-Pinedo G and Müller B 2007 *Physics Reports* **442** 38–74
[25] Janka H T 2012 *Annual Review of Nuclear and Particle Science* **62** 407–451
[26] Ott C D, Abdikamalov E, Mösta P, Haas R, Drasco S, O'Connor E P, Reisswig C, Meakin C A and Schnetter E 2013 *The Astrophysical Journal* **768** 115
[27] Burrows A 2013 *Reviews of Modern Physics* **85** 245–261
[28] Sumiyoshi K and Yamada S 2012 *The Astrophysical Journal Supplement* **199** 17
[29] Bruenn S W, Mezzacappa A, Hix W R, Lentz E J, Bronson Messer O E, Lingerfelt E J, Blondin J M, Endeve E, Marronetti P and Yakunin K N 2013 *The Astrophysical Journal Letters* **767** L6
[30] Lindl J 1995 *Physics of Plasmas* **2** 3933–4024
[31] Lindl J D, Amendt P, Berger R L, Glendinning S G, Glenzer S H, Haan S W, Kauffman R L, Landen O L and Suter L J 2004 *Physics of Plasmas* **11** 339–491

[32] Glenzer S H, Callahan D A, MacKinnon A J, Kline J L, Grim G, Alger E T, Berger R L, Bernstein L A, Betti R and Bleuel et al 2012 *Physics of Plasmas* **19** 056318
[33] Edwards M J, Patel P K, Lindl J D, Atherton L J, Glenzer S H, Haan S W, Kilkenny J D, Landen O L, Moses E I, Nikroo A and et al 2013 *Physics of Plasmas (1994-present)* **20** 070501
[34] Hurricane O A, Callahan D A, Casey D T, Celliers P M, Cerjan C, Dewald E L, Dittrich T R, Döppner T, Hinkel D E, Hopkins L F B and et al 2014 *Nature* **506** 343–348
[35] Bodner S E 1974 *Physical Review Letters* **33** 761–764
[36] Rosenberg M J, Rinderknecht H G, Hoffman N M, Amendt P A, Atzeni S, Zylstra A B, Li C K, Séguin F H, Sio H, Johnson M G and et al 2014 *Physical Review Letters* **112** 185001
[37] Howell J, Bauer W, Colbry D, Pickett R, Staber A, Sagert I and Strother T 2014 *Proceedings of Nuclear Physics: Presence and Future* (*FIAS Interdisciplinary Science Series, Springer Verlag* vol 2)
[38] Sagert I, Bauer W, Colbry D, Howell J, Pickett R, Staber A and Strother T 2014 *Journal of Computational Physics* **266** 191–213
[39] Sagert I, Bauer W, Colbry D, Pickett R and Strother T 2013 *Journal of Physics Conference Series* **458** 012031
[40] Liska R and Wendroff B 2003 *Hyperbolic Problems: Theory, Numerics, Applications* ed Hou T and Tadmor E (Springer Berlin Heidelberg) pp 831–840 ISBN 978-3-642-62929-7
[41] Jun B I, Norman M L and Stone J M 1995 *The Astrophysical Journal* **453** 332
[42] Chandrasekhar S 1961 *Hydrodynamic and Hydromagnetic Instabilities* (Oxford: Clarendon Press)
[43] Frieman E A 1954 *The Astrophysical Journal* **120** 18
[44] Wei T and Livescu D 2012 *Phys. Rev. E* **86**(4) 046405
[45] Sagert I and et al 2014 in preparation

Journal of Physics: Conference Series **535** (2014) 012033 doi:10.1088/1742-6596/535/1/012033

Understanding non-linear hydrodynamic response in heavy-ion collisions via event-plane correlations

Soumya Mohapatra
Department of Physics, Columbia University, New York, NY 10027
E-mail: soumya@cern.ch

Abstract. Recently the ATLAS collaboration measured correlations between pairs and triplets of event-plane angles in Pb+Pb collisions at $\sqrt{s_{\mathrm{NN}}}$=2.76 TeV and observed interesting centrality dependent patterns. In this work, the origin of these event-plane correlations and the quantative physical implications of their measured values are discussed. Such correlations can originate due to correlations between the eccentricities, ϵ_n, of different orders in the initial geometry. They can also be generated dynamically during the hydrodynamic expansion of the medium. Comparisons to a Glauber initial geometry model, and to AMPT transport model calculations show that the measured correlations are generated during hydrodynamic expansion of the medium. The measured correlations are shown to unambiguously establish that the higher order flow harmonics, v_4-v_6, are in fact largely driven by non-linear hydrodynamic response to lower order eccentricities ϵ_2-ϵ_3 of the initial geometry.

1. Introduction

The azimuthal distribution of particles in heavy-ion collisions are often expressed in a Fourier series as:

$$dN/d\phi \propto (1 + 2\Sigma_{n=1}^{\infty} v_n \cos(n(\phi - \Phi_n)) \quad (1)$$

where ϕ is the azimuthal angle of the particle momentum, and the harmonics v_n and phases Φ_n represent the magnitude and direction of the nth order anisotropy. The v_n are commonly termed flow harmonics due to their hydrodynamic origin, while the phases Φ_n are termed as event-plane angles The v_n contain information about both the initial geometry as well as the expansion mechanics of the medium produced in heavy-ion collisions. Thus they are important observables and have been studied in detail at RHIC and at the LHC [1, 2]. They have been instrumental, in our understanding of the expansion dynamics of the produced medium. Until recently however, very little attention was paid to studying the event-plane angles Φ_n. The ATLAS collaboration recently measured a large set of correlations involving two or three event-planes [3]. These event-plane correlations were measured in detail as a function of centrality and showed interesting centrality dependent patterns. The event-plane correlations are quite novel measurements as they contain information that can not be accessed using only the flow harmonics v_n. As will be shown later, these event-plane correlations give a clear understanding of non-linear correlations between the initial geometry and the final particle distributions.

Journal of Physics: Conference Series **535** (2014) 012033 doi:10.1088/1742-6596/535/1/012033

2. Origin of event-plane correlations

Hydrodynamic calculations show that the lower order harmonics v_2 and v_3 and their phases are directly correlated with the magnitude and phase of the second and third-order eccentricities of the initial geometry:

$$v_n e^{i\Phi_n} \propto \epsilon_n e^{i\Phi_n^*}, \qquad n = 2, 3 \tag{2}$$

where, the eccentricities of the initial geometry ϵ_n and their phases Φ_n^* (called participant-planes) are given in terms of the transverse positions r, ϕ of the participating nucleons relative to their common center of mass as:

$$\begin{aligned} \epsilon_n &= \frac{\sqrt{\langle r^n \cos(n\phi)\rangle^2 + \langle r^n \sin(n\phi)\rangle^2}}{\langle r^n \rangle} \\ n\Phi_n^* &= \arctan\left(\frac{\langle r^n \sin(n\phi)\rangle}{\langle r^n \cos(n\phi)\rangle}\right) + \pi \end{aligned} \tag{3}$$

where, the $\langle ... \rangle$ implies averaging over all participating nucleons. However, for higher order harmonics the response can be driven by lower order ϵ_n as well [4]. For example, v_4 not only receives a contribution from ϵ_4, but also a non-linear contribution from ϵ_2. Similarly, v_5 is not only driven by ϵ_5, but also by the product $\epsilon_2\epsilon_3$. Taking into consideration both the linear and leading non-linear components, the total v_4 and v_5 can be written as:

$$\begin{aligned} v_4 e^{i4\Phi_4} &= \alpha_4 \epsilon_4 e^{i4\Phi_4^*} + \alpha_{2,4}\epsilon_2^2 e^{i4\Phi_2^*} + ... \\ v_5 e^{i5\Phi_5} &= \alpha_5 \epsilon_5 e^{i5\Phi_5^*} + \alpha_{2,3,5}\epsilon_2 e^{i2\Phi_2^*}\epsilon_3 e^{i3\Phi_3*} + ... \end{aligned} \tag{4}$$

where, the coupling constants α are functions of centrality and the +.... at the end indicates that there can be additional non-linear terms. Using Eq. 2, Eq. 4 can be recast as:

$$\begin{aligned} v_4 e^{i4\Phi_4} &= \alpha_4 \epsilon_4 e^{i4\Phi_4^*} + \beta_{2,4} v_2^2 e^{i4\Phi_2} + ... \\ v_5 e^{i5\Phi_5} &= \alpha_5 \epsilon_5 e^{i5\Phi_5^*} + \beta_{2,3,5} v_2 e^{i2\Phi_2} v_3 e^{i3\Phi_3} + ... \end{aligned} \tag{5}$$

Equation 4 and its modified form, Eq. 5, explain the origin of event-plane correlations. For example, since ϵ_2 drives both v_2 and v_4, correlation between the pair of phases Φ_2 and Φ_4 should be expected. Similarly, since ϵ_2 and ϵ_3 drive v_5, correlations are expected between the triplet Φ_2-Φ_3-Φ_4. These strength of these correlations as a function of centrality can be quantified by the mean values of their cosines, $\langle\cos(4\Phi_2 - 4\Phi_4)\rangle$ and $\langle\cos(2\Phi_2 + 3\Phi_3 - 5\Phi_5)\rangle$ averaged over all events in that centrality.

These correlations can be quite large in certain circumstances. For example viscous effects are known to damp out the effects higher moments (or higher ϵ_n) in the initial geometry. In such a scenario, the linear terms in Eq. 4 would be much smaller than the non-linear terms leading to strong correlations between the event-plane angles.

It should be pointed out that non-linear hydrodynamic response in not the only mechanism by which the event-plane correlations can be generated. Such correlations can exist even in the linear response limit, i.e. when only the first terms on the right hand side in Eq. 4 are present. This can happen if there are correlations between the different Φ_n^* in the initial geometry. These initial-geometry correlations can then be simply transferred to correlations between the Φ_n. However, as it will be shown, the correlations between the participant-plane angles calculated in initial geometry models, are very different than the measured correlations between the event-planes. However, hydrodynamic and transport model calculations starting with the same initial geometry models reproduce the measured correlations quite well. Thus the measured correlations themselves can be used to determine their origin.

3. ATLAS measurements

Figure 1 shows the dependence of the eight two-plane correlators on the number of participating nucleons, N_{part}, as measured by ATLAS. The correlators are calculated using the traditional event-plane method (shown as open symbols) as well as the scalar-product (SP) method (shown as solid symbols). The SP method is similar to the EP method but explicitly takes into account effects of event-by-event flow fluctuations, by applying weights to each event that depend on the magnitude of the observed v_n in that event [5]. On the other hand, in the EP method, each event is given equal weight. For this reason the EP and SP methods are also referred to as unweighted and weighted methods respectively.

Also shown for comparison in Fig. 1 are the corresponding correlations between the participant-plane angles Φ_n^* calculated in a Glauber geometry model. These model calculations are also done using procedures that mimic the EP and SP methods. Figure 2 shows comparison of the measured correlators to AMPT transport model calculations [5] for the six non-zero correlations.

The first three panels of Fig. 1 show the first three moments of the Φ_2-Φ_4 correlation: $\langle\cos(4j(\Phi_2 - \Phi_4))\rangle$ for j=1-3. It is seen that the correlations are quite small in more central events (large N_{part}), but increase to fairly large values in mid-central and peripheral collisions. That all three moments are large, implies that the Φ_4 plane is strongly aligned with the Φ_2 plane. Comparing with the correlations between the participant planes Φ_n^*, it is seen that the measured correlations are very different than the correlations in the initial geometry. However the AMPT model calculations shown on Fig. 2 that start from a Glauber-like initial geometry model, but include final-state interactions through parton and hadron transport models reproduce the measured correlations quite well. This leads to the conclusion that the $\Phi_2 - \Phi_4$ correlations are generated during collective expansion due to non-linear hydrodynamic response. Further, the large magnitude of the measured correlator in mid-central and peripheral events indicates that a larger fraction of the v_4 is in fact generated by the second order moment ϵ_2, rather than the fourth order moment ϵ_4, of the initial geometry.

The fourth panel of Fig. 1 shows the $\langle\cos(6(\Phi_2\text{-}\Phi_3))\rangle$ correlator. The value of this correlator is much smaller than the Φ_2-Φ_4 correlators but still is significant. Due to the small magnitude of this correlator, it is not possible to draw strong physics conclusion from it. But it is interesting to note that even here, the AMPT results are in much better agreement with the measured correlator than the Glauber model calculation of the corresponding participant-plane correlations.

The fifth and sixth panels of Fig. 1 show the $\langle\cos(6(\Phi_2\text{-}\Phi_6))\rangle$ and $\langle\cos(6(\Phi_3\text{-}\Phi_6))\rangle$ correlators respectively. It is interesting to note that while the Φ_2 and Φ_3 are weakly correlated with one another, they are both strongly correlated with Φ_6. In both cases the measured correlators are very different than the Glauber model calculations, but are recovered in AMPT calculations (Fig. 2), clearly showing that these are also generated during the medium expansion. The structure of the correlators can be understood by writing v_6 as :

$$\begin{aligned} v_6 e^{i6\Phi_6} &= \alpha_6\epsilon_6 e^{i6\Phi_6^*} + \alpha_{2,6}\epsilon_2^3 e^{i6\Phi_2^*} + \alpha_{3,6}\epsilon_3^2 e^{i6\Phi_3^*} + ... \\ &= \alpha_6\epsilon_6 e^{i6\Phi_6^*} + \beta_{2,6} v_2^3 e^{i6\Phi_2} + \beta_{3,6} v_3^2 e^{i6\Phi_3} + ... \end{aligned} \quad (6)$$

The $\langle\cos(6(\Phi_2\text{-}\Phi_6))\rangle$ correlator is zero in most central events and increases in magnitude for mid-central and peripheral events. This behavior is similar to the $\langle\cos(4(\Phi_2 - \Phi_4))\rangle$ correlator and can be understood as as the contribution from the second term on the right hand side of Eq. 6. As the v_2 increases from central to mid-central events, the contribution of this term increases and leads to stronger $\Phi_2 - \Phi_6$ correlations. On the other hand, the $\langle\cos(6(\Phi_3 - \Phi_6))\rangle$ correlator is non-zero in most central events and decreases in magnitude for mid-central and peripheral events. In most central events, as the v_n are small, the v_3^2 term is larger than the v_2^3

term, leading to strong $\Phi_3 - \Phi_6$ correlation. With decreasing $N_{\rm part}$, the v_2 rapidly increases, but v_3 changes only weakly. This leads to a relative decrease in the contribution from the v_3 term leading to a decrease in the $\Phi_3 - \Phi_6$ correlation.

The last two measured correlators shown in Fig. 1 are $\langle\cos(12(\Phi_3 - \Phi_4))\rangle$ and $\langle\cos(10(\Phi_2 - \Phi_5))\rangle$. The measured correlators are consistent with zero, and also consistent with the Glauber model calculations over the measured $N_{\rm part}$ range. These correlations are not expected to have any non-linear components, as it is not possible to write equations of the form given in Eq. 5 that only couple v_3-v_4 or v_5-v_2.

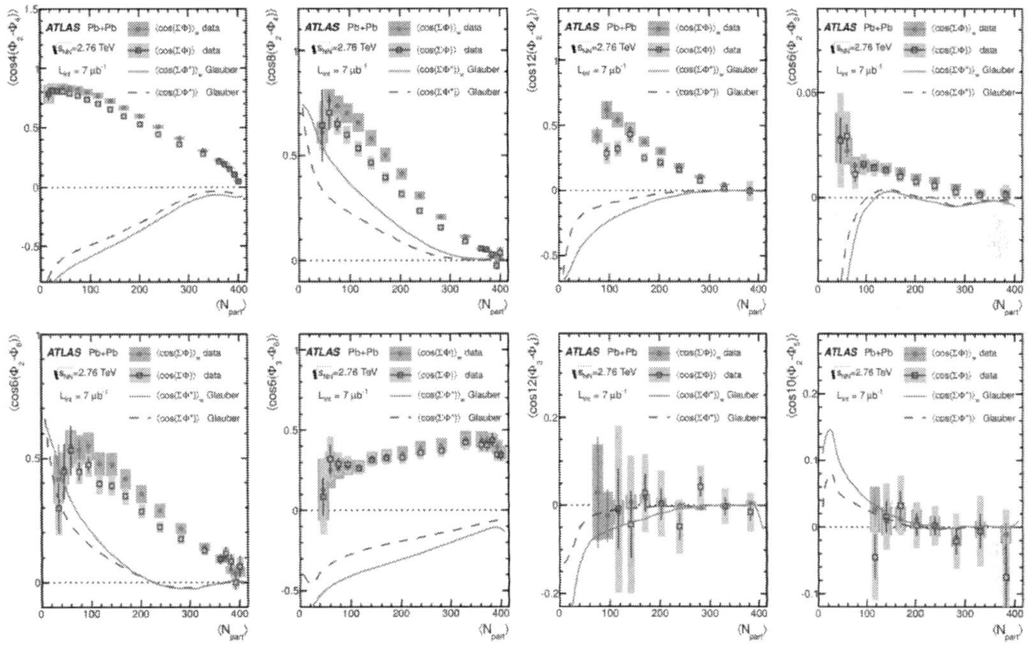

Figure 1. The set of two-plane correlators measured by ATLAS. The data points correspond to measurements via the scalar-product method (solid symbols) and the event-plane method (open symbols). The error bars and bands represent the statistical and systematic uncertainties respectively. The correlations between the participant-plane angles Φ_n^* calculated in a Glauber model are shown by the continuous curves for the weighted case, and by the dashed curves for the unweighted case. Figure taken from [3].

Figure 3 shows the three-plane correlators measured by ATLAS. Also shown for comparison are the corresponding participant-plane correlators calculated in a Glauber model. It is seen that the four correlators that are significant in strength are very different than the Glauber model calculations. Figure 4 compares the four non-zero three-plane correlators to the AMPT calculations, which reproduce the correlators remarkably well. These results clearly show that like the two-plane correlations, these are also generated during hydrodynamic expansion. In particular the large value of the $\langle\cos(2\Phi_2 + 3\Phi_3 - 5\Phi_5)\rangle$ is quite informative. For $N_{\rm part} < 330$ this correlator is well above 0.5, indicating that in this range a larger part of v_5 in fact is generated from non linear response to $\epsilon_2\epsilon_3$ rather than from ϵ_5.

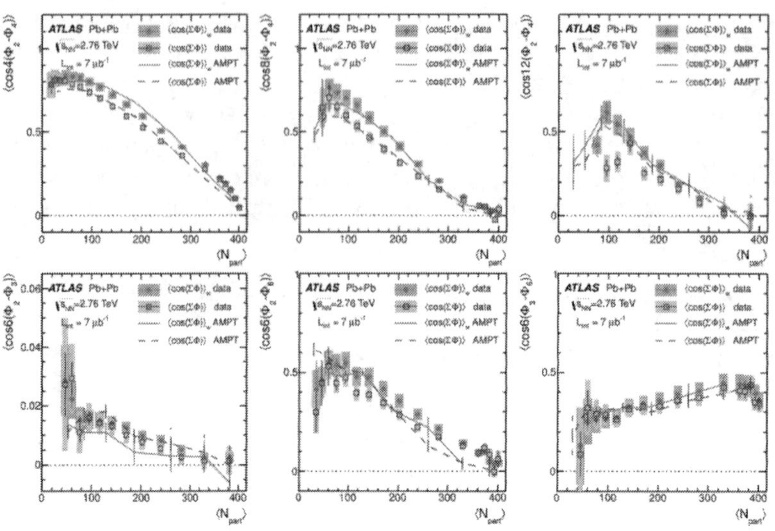

Figure 2. Comparison of six of the two-plane correlators measured by ATLAS to AMPT calculations. Figure taken from [3].

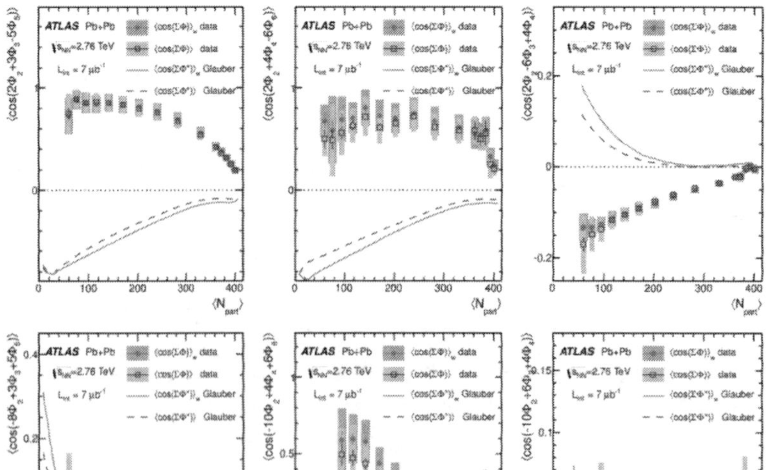

Figure 3. The set of three-plane correlators measured by ATLAS. Comparisons are shown to the corresponding correlations between the participant-plane angles calculated in a Glauber model. Figure taken from [3].

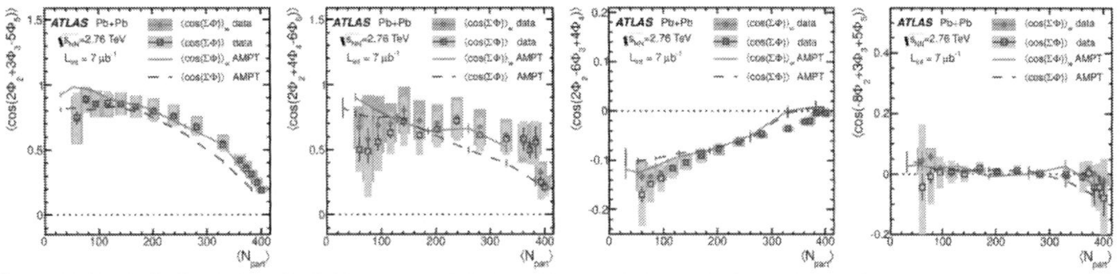

Figure 4. Comparison of four of the two-plane correlators measured by ATLAS to AMPT calculations. Figure taken from [3].

4. Summary

The results presented here clearly demonstrate the ability of the event-plane correlations to elucidate the effects of non-linear hydrodynamic response in heavy ion collisions. Most of the measured correlators are very different than the corresponding correlators between participant-planes of the initial geometry evaluated in Glauber model calculations. However they are quite well recovered in AMPT calculations, indicating that they are the results of collective dynamics, and generated during the expansion of the medium. These correlations show that significant fraction of v_4, v_5 and v_6 are in fact generated by the lower order moments ϵ_2 and ϵ_3 of the initial geometry. It should be pointed out that these correlations are a completely new observable and upon further study may reveal additional interesting insights into the collective dynamics in heavy ion collisions.

The event-plane correlations can also be used to constrain other medium properties such as the viscosity to entropy density ratio η/s. This is because the viscous effects affect the linear and non-linear terms differently. Similarly these correlations can be used to constrain initial geometry models as well, as they are sensitive to the eccentricities in the initial geometry. This has been demonstrated by Qui and Heinz [6].

References

[1] Aad G *et al.* (ATLAS Collaboration) 2012 *Phys.Rev.* **C86** 014907 (*Preprint* 1203.3087)
[2] Aad G *et al.* (ATLAS Collaboration) 2013 *JHEP* **1311** 183 (*Preprint* 1305.2942)
[3] Aad G *et al.* (ATLAS Collaboration) 2014 (*Preprint* 1403.0489)
[4] Gardim F G, Grassi F, Luzum M and Ollitrault J Y 2012 *Phys.Rev.* **C85** 024908 (*Preprint* 1111.6538)
[5] Bhalerao R S, Ollitrault J Y and Pal S 2013 *Phys.Rev.* **C88** 024909 (*Preprint* 1307.0980)
[6] Qiu Z and Heinz U 2012 *Phys.Lett.* **B717** 261–265 (*Preprint* 1208.1200)

30th Winter Workshop on Nuclear Dynamics (WWND2014) IOP Publishing
Journal of Physics: Conference Series **535** (2014) 012034 doi:10.1088/1742-6596/535/1/012034

Numerical Simulation of Thermal Noise in Heavy Ion Collisions

C. Young
School of Physics & Astronomy, 116 Church Street S.E., Minneapolis, MN 55455 USA
E-mail: young@physics.umn.edu

J. I. Kapusta
School of Physics & Astronomy, University of Minnesota, Minneapolis, MN 55455, U.S.A.
E-mail: kapusta@physics.umn.edu

C. Gale
Department of Physics, McGill University, 3600 rue University Montréal QC H3A 2T8 Canada
E-mail: gale@physics.mcgill.ca

S. Jeon
Department of Physics, McGill University, 3600 rue University Montréal QC H3A 2T8 Canada
E-mail: jeon@physics.mcgill.ca

B. Schenke
Physics Dept, Bldg. 510A, Brookhaven National Laboratory, Upton, NY 11973
E-mail: bschenke@quark.phy.bnl.gov

Abstract.
Thermal noise is present in any viscous fluid, making the simulation of relativistic noise in heavy ion collisions a necessity. It is likely possible to use it to make an independent measurement of viscosity in heavy ion collisions. The size, energy densities, and time scales of the collisions determine the relative importance of thermal noise. This causes a non-trivial contribution to two-particle correlations as well as event-by-event fluctuations in observables.

1. Fluctuations, dissipation, and noise in heavy ion collisions

The production of mesons and baryons in heavy ion collisions is consistent with the creation of a nearly perfect, relativistic fluid [1, 2]. Like any viscous fluid at finite temperature, the fluid created in heavy ion collisions has thermal fluctuations: stochastic contributions to the energy-momentum tensor $T^{\mu\nu}$. These fluctuations are related to the statistical fluctuations that must exist after coarse-graining: for example, while it is possible to approximate the number of particles in a box N with an average $\bar{N}$, there are fluctuations away from this average with variance $\sim \sqrt{N}$. The energy-momentum tensor has a fluctuating part, but the total is conserved:

$$T^{\mu\nu} = T^{\mu\nu}_{ave} + \delta T^{\mu\nu}, \quad \partial_\mu (T^{\mu\nu}_{ave} + \delta T^{\mu\nu}) = 0; \tag{1}$$

in the later sections the fluctuating part is separated further into ideal, viscous, and noise terms $\delta T^{\mu\nu} = \delta T^{\mu\nu}_{id} + \delta W^{\mu\nu} + \Xi^{\mu\nu}$.

Thermal fluctuations can be understood from the very general methods used to derive the fluctuation-dissipation relation. As an example familiar from quantum mechanics, consider a bosonic operator $\hat{\phi}$. The damping of the retarded Green function G_R becomes the imaginary part of $G_R(\omega)$ in Fourier space, and as discussed in [3]:

$$\begin{aligned}
\mathrm{Im}G_R(\omega) &= -\frac{i}{2}\Big[-i\int dt\,\theta(t)e^{i\omega t}\left\langle[\hat{\phi}(t),\hat{\phi}(0)]\right\rangle - i\int dt\,\theta(t)e^{-i\omega t}\left\langle[\hat{\phi}(t),\hat{\phi}(0)]\right\rangle^*\Big] \\
&= -\frac{1}{2}\Big[\int dt\,\theta(t)e^{i\omega t}\left\langle[\hat{\phi}(t),\hat{\phi}(0)]\right\rangle + \int dt\,\theta(t)e^{-i\omega t}\left\langle[\hat{\phi}(0),\hat{\phi}(t)]\right\rangle\Big] \\
&= -\frac{1}{2}\int dt\,e^{i\omega t}\left\langle[\hat{\phi}(t),\hat{\phi}(0)]\right\rangle = -\frac{1}{2}(1-e^{-\omega/T})\int dt\,e^{i\omega t}\left\langle\hat{\phi}(t)\hat{\phi}(0)\right\rangle \\
&= -\frac{i}{2}(1-e^{-\omega/T})G_>(\omega).
\end{aligned} \tag{2}$$

Functions that describe different physical phenomena (the retarded Green function G_R and the expectation value of a transition $G_>$) are related near thermal equilibrium. Similarly, the autocorrelation function $G_S \equiv \frac{1}{2}\langle\{\hat{\phi}(t),\hat{\phi}(t')\}\rangle$, which describes random fluctuations in a system, is related to the response function after a few more steps:

$$G_S(\omega) = -(1+2n_B(\omega))\mathrm{Im}G_R(\omega) \approx \frac{-2T}{\omega}\mathrm{Im}G_R(\omega),$$

where the final approximate equality comes from assuming the system is nearly classical. In [4], the equal-time autocorrelation of variances in velocity δv^i and in mass density ρ were found in first-order non-relativistic hydrodynamics:

$$\begin{aligned}
\left\langle\delta v^i(\mathbf{x})\delta v^j(\mathbf{x}')\right\rangle &= (T/\rho)\delta^{ij}\delta^3(\mathbf{x}-\mathbf{x}'), \\
\langle\delta\rho(\mathbf{x})\delta\rho(\mathbf{x}')\rangle &= \rho T(\tfrac{\partial\rho}{\partial p})_T\delta^3(\mathbf{x}-\mathbf{x}'),
\end{aligned} \tag{3}$$

showing how the strengths of these variances are related to temperature and other properties of the fluid.

Thermal hydrodynamical fluctuations are often neglected in physical applications; for example, for the average variance of δv^i over a volume of size ΔV

$$\left\langle\frac{\int d^3x\delta v^i(\mathbf{x})}{\Delta V}\frac{\int d^3x'\delta v^j(\mathbf{x}')}{\Delta V}\right\rangle = \frac{T}{\rho\Delta V}\delta^{ij} \tag{4}$$

is very small whenever $T \ll \rho\Delta V$. Thermal fluctuations are usually unimportant at everyday length scales, temperatures, and densities. However, in heavy ion collisions, the equation of state is ultrarelativistic while the fluid created is small, typically about 10 fm across. Therefore, a heavy ion collision should be relatively noisy. Thanks to the fluctuation-dissipation theorem, this noise might be used to provide an independent measurement of transport coefficients.

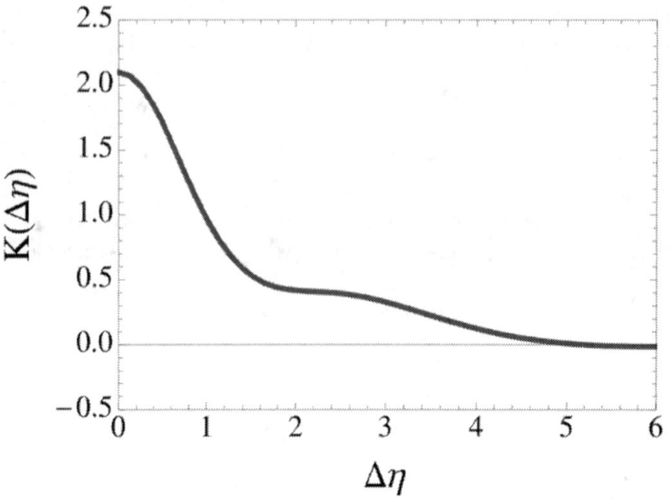

Figure 1. (Color online) $K(\Delta\eta)$, proportional to the two-particle correlation as a function of rapidity difference, calculated for a viscous Bjorken expansion typical for heavy-ion collisions, from [5].

Heavy ion collisions produce a fluid expanding at relativistic speeds, and thermal fluctuations need to be determined in this setting. Recently, Kapusta, Müller, and Stephanov determined the correlation functions for thermal noise in first-order relativistic hydrodynamics [5]. The thermal noise was calculated for both the Landau-Lifshitz frame, where the velocity u^μ represents the flow of energy density in the fluid, and the Eckart frame, where u^μ is the flow of a particle number density n. In first-order hydrodynamics and assuming a Bjorken expansion, analytic solutions are possible, and in [5] the authors found a contribution of thermal noise to two-particle correlation functions at rapidity gaps as large as 3, shown in Fig. 1.

2. Thermal fluctuations in Israel-Stewart hydrodynamics

In first order viscous hydrodynamics, shear modes propagate faster than light. This fact is problematic in heavy-ion collisions where the flow velocities approach the speed of light. The Israel-Stewart equations describe second order hydrodynamics with a system of equations:

$$\begin{aligned} \partial_\mu (T^{\mu\nu}_{\rm id} + W^{\mu\nu}) &= 0, \\ \Delta^\mu_\alpha \Delta^\nu_\beta (u\cdot\partial) W^{\alpha\beta} &= -\frac{1}{\tau_\pi}(W^{\mu\nu} - S^{\mu\nu}) - \frac{4}{3}(\partial\cdot u) W^{\mu\nu}, \end{aligned}$$

where $S^{\mu\nu} = \eta(\Delta^\mu u^\nu + \Delta^\nu u^\mu) + (\zeta - 2\eta/3)\Delta^{\mu\nu}(\partial\cdot u)$ is the first order viscous correction to the energy-momentum tensor, $\Delta^{\mu\nu} = g^{\mu\nu} - u^\mu u^\nu$, and $\Delta^\mu = \Delta^{\mu\nu}\partial_\nu$. The equations are closed by an equation of state relating energy density and pressure $e(p)$ and the requirement that the viscous correction $W^{\mu\nu}$ is transverse: $u_\mu W^{\mu\nu} = 0$. With an appropriate choice of relaxation time τ_π, this system does not have modes propagating faster than light; one can imagine the second equation as slowing the medium's response to a shear stress. Heavy-ion collisions are often described with second order hydrodynamics, and thermal fluctuations need to be determined again for this system of equations.

Up to linear order in response, the equations for thermal fluctuations in a system described

Journal of Physics: Conference Series **535** (2014) 012034 doi:10.1088/1742-6596/535/1/012034

with the Israel-Stewart equations were determined in [8]:

$$\partial_\mu(\delta T^{\mu\nu}_{id} + \delta W'^{\mu\nu}) = 0,$$

$$\begin{aligned}
(u\cdot\partial)\delta W'^{\mu\nu} &= -\frac{1}{\tau_\pi}(\delta W'^{\mu\nu} - \delta S^{\mu\nu} - \xi^{\mu\nu}) - \frac{4}{3}(\partial\cdot\delta u)W^{\mu\nu} - \frac{4}{3}(\partial\cdot u)\delta W'^{\mu\nu} - (\delta u\cdot\partial)W^{\mu\nu} \\
&\quad -\delta u^\mu((u\cdot\partial)u_\alpha)W^{\alpha\nu} - u^\mu((\delta u\cdot\partial)u_\alpha)W^{\alpha\nu} + (u\cdot\partial)\delta u_\alpha)W^{\alpha\nu} + (u\cdot\partial)u_\alpha)\delta W'^{\alpha\nu}) \\
&\quad -\delta u^\nu((u\cdot\partial)u_\alpha)W^{\alpha\mu} - u^\nu((\delta u\cdot\partial)u_\alpha)W^{\alpha\mu} + (u\cdot\partial)\delta u_\alpha)W^{\alpha\mu} + (u\cdot\partial)u_\alpha)\delta W'^{\alpha\mu}),
\end{aligned}$$

$$\begin{aligned}
\left\langle \xi^{\mu\nu}(x)\xi^{\alpha\beta}(x')\right\rangle &= \Big[2\eta T(\Delta^{\mu\alpha}\Delta^{\nu\beta} + \Delta^{\mu\beta}\Delta^{\nu\alpha}) \\
&\quad +2(\zeta - 2\eta/3)T\Delta^{\mu\nu}\Delta^{\alpha\beta}\Big]\delta^4(x-x'). \qquad (5)
\end{aligned}$$

Here, $\delta T^{\mu\nu}_{id} = -\delta p g^{\mu\nu} + \delta p\left(1 + (\partial e/\partial p)_{n/s}\right)u^\mu_0 u^\nu_0 + (e_0 + p_0)(u^\mu_0\delta u^\nu + u^\nu_0\delta u^\mu)$ is the fluctuation in the ideal part of the energy-momentum tensor, $\delta S^{\mu\nu}$ represents the change in $S^{\mu\nu}$ up to linear order in fluctuations, $\delta W'^{\mu\nu} = \delta W^{\mu\nu} + \Xi^{\mu\nu}$ is the sum of the thermal fluctuation of the viscous energy-momentum tensor and the noise $\Xi^{\mu\nu}$, and $\xi^{\mu\nu}(x)$ is a stochastic source term driving these fluctuations to thermal expectation values. The noisy part of the energy-momentum tensor $\Xi^{\mu\nu}$ now has a non-trivial autocorrelation in time:

$$\left\langle \Xi^{\mu\nu}(x)\Xi^{\alpha\beta}(x')\right\rangle \propto \exp(-|t-t'|/\tau_\pi).$$

Eqs. 5 are stochastic, due to the presence of the random variables $\xi^{\mu\nu}$. Even for a single initial condition, the solution to these equations is not a single event but more properly an ensemble of events, which in practice is approximated with a large set sampled with Monte Carlo event generation. In the following section, each sampling corresponds to a single heavy ion collision while a set of samplings is similar to a centrality class.

3. Results at the Large Hadron Collider

Where can effects from thermal fluctuations be expected in heavy ion collisions? One can determine this by considering the thermal expectation values of fluctuations. First of all, $\langle\delta T^{\mu\nu}\rangle = 0$: while each event has thermal fluctuations, the average of the fluctuations across many events is zero. As a result, thermal fluctuations should have a vanishingly small effect on one-particle observables averaged over an event class, such as average yields of pions as a function of p_T and η. However, $\left\langle\delta T^{\mu\nu}\delta T^{\alpha\beta}\right\rangle \neq 0$, and thermal fluctuations contribute to two-particle correlations, as discussed in [5]. Finally, in a given event, $\delta T^{\mu\nu} \neq 0$: thermal fluctuations are a source for event-by-event variations in *all* observables, even if the average effect for a given observable should be zero.

The solution of the thermally fluctuating Israel-Stewart equations describing heavy-ion collisions is discussed in [8]: the hydrodynamical code MUSIC has been extended to include linearized fluctuations. MUSIC [9, 10] solves relativistic hydrodynamics in τ-η coordinates, using initial conditions determined with the Glauber model for heavy-ion collisions and an equation of state determined with lattice QCD calculations [12]. Equations [5] are solved using the MacCormack method [11]. For the remainder of this work, we will discuss simulations using this extension of MUSIC meant to describe lead-lead collisions at $\sqrt{s}/A = 2.76$ TeV created and

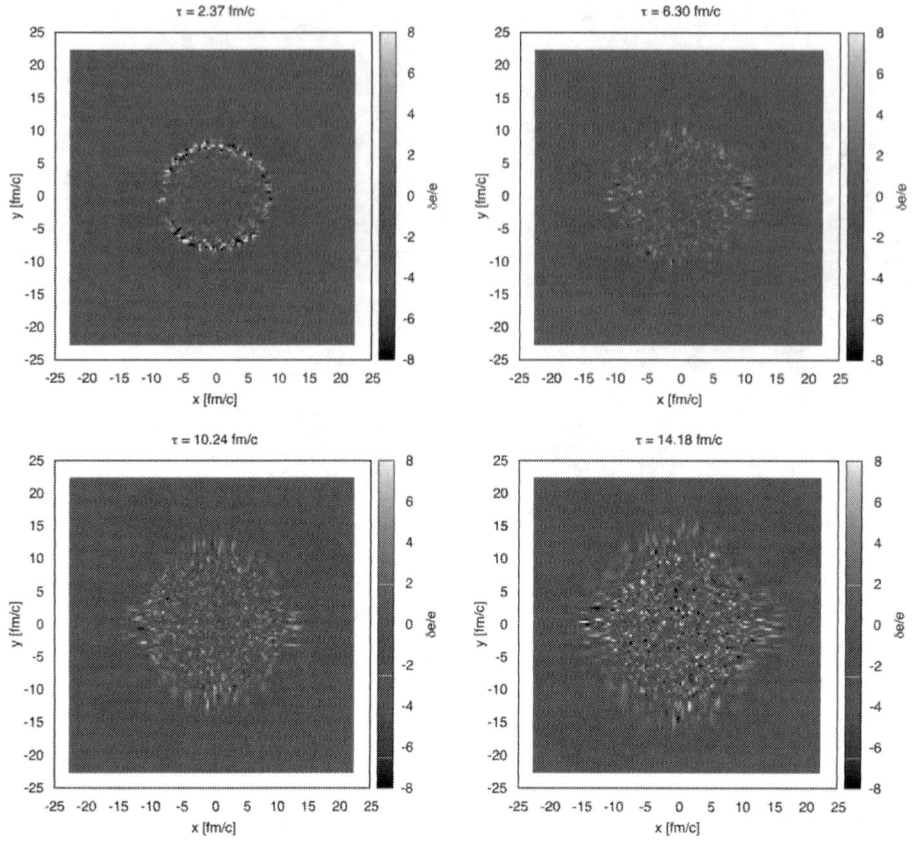

Figure 2. (Color online) Fluctuations in energy density in the transverse plane as a function of time in a typical Pb+Pb collision for $b = 6$ fm at LHC energies, from [8].

measured at the Large Hadron Collider (LHC), in both the 10-20% centrality class and for the "ultra-central" event class examined by CMS [13].

Figure 2 shows thermal fluctuations $\delta e/e$ at different proper times of the evolution of a lead-lead collision typical at the LHC. The magnitude of $\delta e/e$ increases with dropping temperature. At this point, these simulations should be tested in meaningful ways. One important test is to make sure that the average of a large set of thermally fluctuating events does indeed give the same results for particle spectra as does a single noiseless hydrodynamical simulation, as is discussed in the first paragraph of this section. Figure 3 demonstrates this: the distributions for dN/dp_T of π^+ mesons are compared for both a single simulation without noise and 200 simulations with thermal noise, and agreement between the two is found, as one would expect from the relation $\langle \delta T^{\mu\nu} \rangle = 0$. This is non-trivial: individual noisy events do not yield dN/dp_T distributions identical to the results of the noiseless simulation.

We also examine the momentum eccentricity ϵ_p, which is the addition in quadrature of $\langle T^{xx} - T^{yy} \rangle / \langle T^{xx} + T^{yy} \rangle$ and $\langle 2T^{xy} \rangle / \langle T^{xx} + T^{yy} \rangle$. These two quantities are, respectively, the expectation values $\langle \cos(2\phi) \rangle$ and $\langle \sin(2\phi) \rangle$ for the direction of the total transverse momentum in cells, weighted by the total energy of each cell. In [14], ϵ_p was used as a proxy for v_2 of hadrons after freeze-out. Figure 4 shows the distributions of ϵ_p for two different ensembles of events, one representing the 10-20% centrality class at the LHC and the other the ultra-central event class

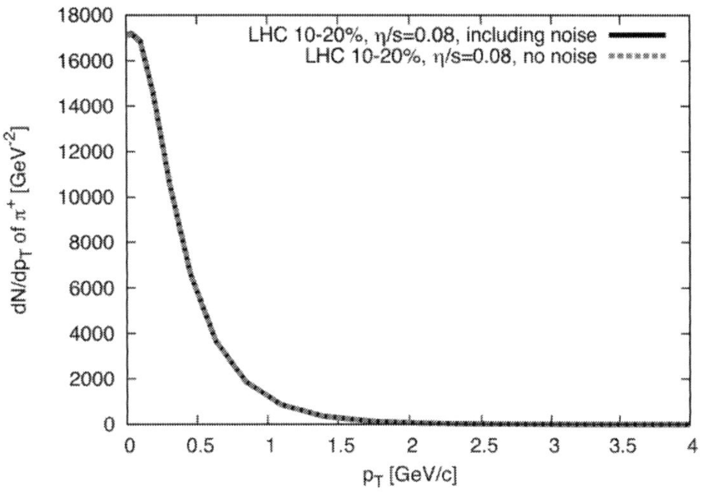

Figure 3. (Color online) Transverse momentum distribution for pions showing that the single particle distribution is unaffected by noise, from [8].

(in practice, the latter set of simulations just has the impact parameter set to zero).

Thermal noise affects both the average values of ϵ_p as well as the width of the distributions $dN/d\epsilon_p$ in a given centrality class. To see this, consider an event with zero impact parameter: in a simulation without thermal fluctuations, all azimuthal coefficients $v_n = 0$. However for a given fluctuating event, $v_n \neq 0$, and after defining the reaction plane, all v_n are positive. This makes the expectation values $\langle v_n \rangle$ also non-zero and positive. Thermal noise tends to increase average values of ϵ_p while contributing to the variance of this observable across events.

For the 10-20% centrality class, the variance in ϵ_p is subleading compared with the variance caused by fluctuating initial conditions [15], and the increase in the average of ϵ_p is modest. However the ensemble of events with zero impact parameter has an average value of ϵ_p comparable to the v_2 in the 0-0.2% centrality class examined by the CMS collaboration [13]. This suggests that a significant contribution to the azimuthal anisotropy in this event class may come from thermal fluctuations.

4. Conclusions

Thermal fluctuations are demonstrably present in viscous fluids, and the autocorrelation of the thermal noise is proportional to the shear viscosity. This suggests a complementary way of measuring the transport coefficients of the fluid created in heavy ion collisions. Previous work has examined the contribution of thermal noise to the two particle correlation function in a viscous Bjorken expansion. We have determined the equations for thermal fluctuations in second order viscous hydrodynamics and implemented a simulation of this noise that works well for heavy ion collisions at the Relativistic Heavy Ion Collider and the Large Hadron Collider.

5. Acknowledgments

CY thanks the organizers of the Winter Workshop for Nuclear Dynamics for kindly giving him time to present these results. JK and CY are supported by the U.S. DOE Grant No. DE-FG02-87ER40328. CG and SJ are supported by funding from the Natural Sciences and Engineering Research Council of Canada. BPS is supported under DOE Contract. No. DE-

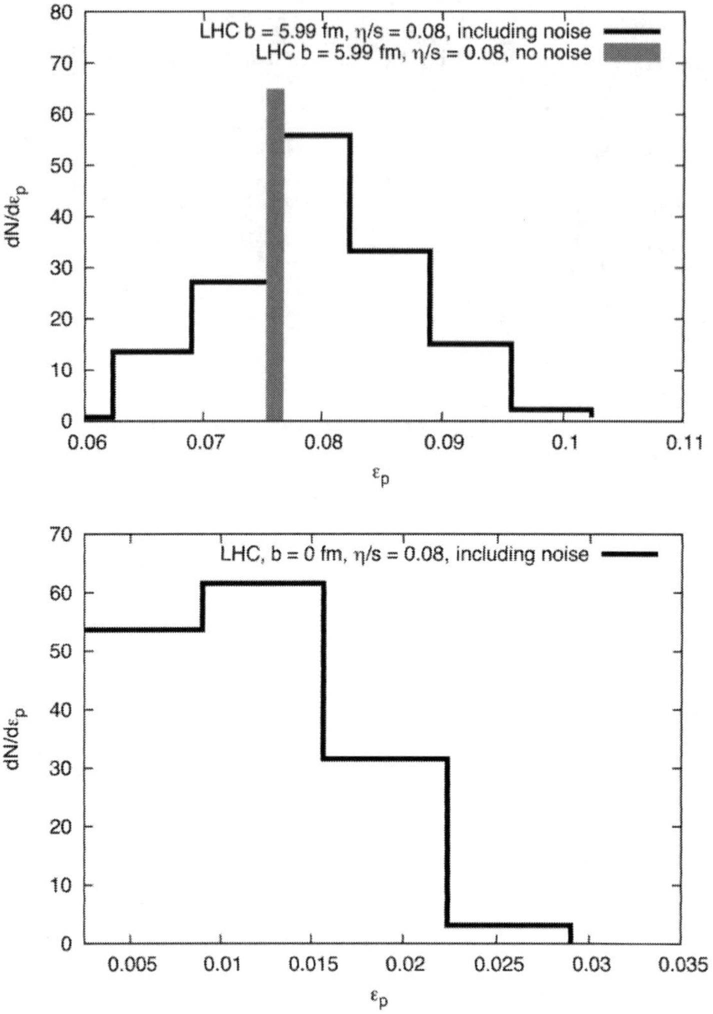

Figure 4. (Color online)Top panel ($b = 6$ fm): Inclusion of noise increases the average ϵ_p and broadens its distribution. Bottom panel ($b = 0$ fm): Even for exactly central collisions there is a nonzero average ϵ_p due to noise. From [8].

AC02-98CH10886. CY thanks E. Shuryak for very helpful suggestions. We are grateful for resources from the University of Minnesota Supercomputing Institute.

References

[1] K. Adcox *et al.* [PHENIX Collaboration], Nucl. Phys. A **757**, 184 (2005).
[2] J. Adams *et al.* [STAR Collaboration], Nucl. Phys. A **757**, 102 (2005).
[3] D. T. Son and D. Teaney, JHEP **0907**, 021 (2009) [arXiv:0901.2338 [hep-th]].
[4] L. D. Landau, E. M. Lifshitz, and L. P. Pitaevskii, *Statistical Physics Part 2, Landau and Lifshitz Course of Theoretical Physics Volume 9*, 1980 Pergamon Press plc.
[5] J. I. Kapusta, B. Müller and M. Stephanov, Phys. Rev. C **85**, 054906 (2012).
[6] C. Young, Phys. Rev. C **89**, 024913 (2014).
[7] W. Israel, Ann. Phys. (NY) **100**, 310 (1976); W. Israel and J. M. Stewart, Ann. Phys. (NY) **118**, 341 (1979).

[8] C. Young, J. I. Kapusta, C. Gale, S. Jeon and B. Schenke, arXiv:1407.1077 [nucl-th].
[9] B. Schenke, S. Jeon and C. Gale, Phys. Rev. C **82**, 014903 (2010) [arXiv:1004.1408 [hep-ph]].
[10] B. Schenke, S. Jeon and C. Gale, Phys. Rev. Lett. **106**, 042301 (2011).
[11] *Computational fluid dynamics: an introduction*, edited by John F. Wendt, 2009 Springer.
[12] P. Huovinen and P. Petreczky, Nucl. Phys. **A837**, 26 (2010).
[13] CMS Collaboration [CMS Collaboration], CMS-PAS-HIN-12-011.
[14] J.-Y. Ollitrault, Phys. Rev. D **48**, 1132 (1993).
[15] B. Schenke, P. Tribedy and R. Venugopalan, preprint arXiv:1312.5588.